Bayesian Process Monitoring, Control and Optimization

Bayesian Process Monitoring, Control and Optimization

edited by

Bianca M. Colosimo
Enrique del Castillo

Chapman & Hall/CRC
Taylor & Francis Group
Boca Raton London New York

Chapman & Hall/CRC is an imprint of the
Taylor & Francis Group, an informa business

Chapman & Hall/CRC
Taylor & Francis Group
6000 Broken Sound Parkway NW, Suite 300
Boca Raton, FL 33487-2742

Chapman & Hall/CRC is an imprint of Taylor & Francis Group, an Informa business

Printed in the United States of America on acid-free paper
10 9 8 7 6 5 4 3 2 1

International Standard Book Number-10: 1-58488-544-0 (Hardcover)
International Standard Book Number-13: 978-1-58488-544-3 (Hardcover)

Visit the Taylor & Francis Web site at
http://www.taylorandfrancis.com

and the CRC Press Web site at
http://www.crcpress.com

Preface

During the last two decades, the use of Bayesian techniques has spread in different fields thanks to the advent of newly developed simulation-based approaches (e.g., Markov Chain Monte Carlo-MCMC) which have reduced computational barriers to the use of Bayesian inference in many applied fields. Problems that were considered intractable in the past are now routinely solved thanks to these approaches and to the ever increasing computer power. Furthermore, the long-lasting disjunctive about selecting "the best" approach, either Bayesian or frequentist, is now outdated in applied circles and substituted by a more pragmatic point of view which allows the analyst to select the most suitable approach depending on the problem at hand. During this time, the quality profession has also seen some applications and developments in Bayesian statistics, but this has occurred mainly in the more technical journals with little of this work reflected in actual practice.

While there are many excellent and recent books on applied Bayesian Statistics, most of them are related to biostatistics and Econometric applications. In engineering, the Bayesian approach has certainly been used by electrical and chemical engineers who have studied or applied Kalman filtering techniques (an area where there are several excellent textbooks too) for many years. However, the Bayesian paradigm is unfortunately not familiar to industrial engineers or to most applied or "industrial" statisticians, perhaps with the only exception of persons involved in reliability studies. Many industrial problems related to quality control and improvement require an in-depth use of statistical approaches, but in this era of "six-sigmas" and "black belts" the use of advanced statistical methods in practice is difficult to carry on. Despite the slight attention to Bayesian methodology in the industrial engineering and applied statistics technical journals, it can be observed, in general, a lack of attention to opportunities arising from the adoption of Bayesian approaches in actual industrial practice. We believe the gap between application and development can be bridged by having available more books on Bayesian statistics with a perspective on engineering applications.

It is with this goal that the project for the present book originated. The aim is to provide a state-of-the-art survey of applications of Bayesian statistics in three specific fields of industrial engineering and applied or industrial statistics, namely, process monitoring, process control (or adjustment), and process optimization. The book is intended as a reference for applied statisticians working in industry, process engineers and quality engineers working in manufacturing or persons in academia, mainly professors and graduate students in industrial and manufacturing engineering, applied statistics, and operations research departments. This is reflected in the diversity of the contributors to this book, who come from both academia and industry, and are located in the USA, Europe, and Asia.

The book is organized in four parts. Part I contains two introductory chapters. The first chapter provides an introduction to Bayesian statistics, emphasizing basic inferential problems, and outlines how these methods are applied in process monitoring, control, and adjustment. This chapter contains references and brief descriptions to all other chapters in the book, where appropriate. The second chapter presents a general overview of methods developed in the past few decades for computing Bayesian analysis via simulation (such as Markov Chain Monte Carlo and Monte Carlo simulation). The use of MCMC is illustrated with reference to a classic hierarchical model, the variance component model, using available software packages (WinBUGS and CODA, which runs under R). It is our hope that readers not familiar with Bayesian methodology will find in these two chapters a useful introduction and a guide to more advanced references.

Part II contains five chapters covering Bayesian approaches for process monitoring. The advantages of a Bayesian approach to process monitoring arise from the sequential nature of Bayes' theorem. As pointed out by some of the authors in this part of the book, a Bayesian approach allows a more flexible framework, in particular with respect to the usual assumption made in classical statistical process control (SPC) charts about known parameters. This part of the book deals with Bayesian methods for SPC and considers both univariate and multivariate process monitoring techniques. Application and development of full Bayesian approaches and empirical bayes methods are discussed.

The chapters in Part III present some Bayesian approaches which can be used for time series data analysis (for instance in case of missing data) and process control (also known as engineering process control). Here the use of the Kalman filter as an estimator of the state in a state-space formulation is exploited for prediction and control. This is perhaps the best known application of Bayesian techniques in engineering (outside of Industrial), although it is curious that Kalman himself did not develop his celebrated filter from a Bayesian point of view. Applications to radar detection and discrete part manufacturing are included.

Finally, Part IV focuses on Bayesian methods for process optimization. The three chapters included in this part of the book show how Bayesian methods can be usefully applied in experimental design and response surface methods (RSM). This section presents and illustrates the application of Bayesian regression to sequential optimization, the use of Bayesian techniques for the analysis of saturated designs, and the use of predictive distributions for optimization. The predictive approach to response surface optimization represents a major advance in RSM techniques, as it incorporates the uncertainty of the parameter estimates in the optimization process and has no frequentist counterpart.

We wish to thank all the contributing authors, with whom we share our interest in development of Bayesian statistics in industrial applications. A special thought goes to the late Carol Feltz (Northern Illinois University, USA) who showed praiseworthy spirit and strength in dedicating her last weeks to one chapter of this book.

Bianca M. Colosimo

Enrique del Castillo

Contributors

Frank B. Alt
University of Maryland
Robert H. Smith School of Business
College Park, MD 20742
USA
falt@rhsmith.umd.edu

Marta.Y. Baba
School of Mathematical Sciences
Queen Mary, University of London
London E1 4NS
UK
myb@maths.qmul.ac.uk

Bianca M. Colosimo
Dipartimento di Meccanica
Politecnico di Milano
Piazza Leonardo da Vinci 32, 20133
Milano
Italy
biancamaria.colosimo@polimi.it

Enrique del Castillo
Department of Industrial & Manufacturing Engineering
Penn State University
310 Leonhard Building
University Park, PA 16802
USA
exd13@psu.edu

Carol J. Feltz[1]
Division of Statistics
Northern Illinois University
DeKalb, IL 60115
USA

Steven G. Gilmour
School of Mathematical Sciences
Queen Mary, University of London
London E1 4NS
UK
s.g.gilmour@qmul.ac.uk

Spencer Graves
PDF Solutions, Inc.
333 West San Carlos, Suite 700
San José, CA 95126
USA
spencerg@pdf.com

Douglas M. Hawkins
Department of Statistics
University of Minnesota
313 Ford Hall
224 Church Street S.E.
Minneapolis, MN 55455
USA
doug@stat.umn.edu

Melinda Hock
Naval Research Laboratory
Washington, DC 20375
USA

[1]Sadly, Professor Carol Feltz passed away while completing chapter 4 of this book.

Carlos Moreno
Ultramax Corporation
110 Boggs Lane, Suite 325
Cincinnati, OH 45249
USA
carlos.moreno@ultramax.com

George Nenes
Aristotle University of Thessaloniki
Department of Mechanical Engineering
54124 Thessaloniki
Greece
gnenes@auth.gr

Rong Pan
Department of Industrial Engineering
Arizona State University
PO Box 875906
Tempe AZ 85287–5906
Rong.Pan@asu.edu

John J. Peterson
Statistical Sciences Department (UW281A)
GlaxoSmithKline Pharmaceuticals, R&D
709 Swedeland Road
King of Prussia, PA 19406–0939
USA
john.peterson@gsk.com

Jyh-Jen Horng Shiau
Institute of Statistics
National Chiao Tung University
Hsinchu
Taiwan
jyhjen@stat.nctu.edu.tw

Refik Soyer
School of Business
The George Washington University
Washington, DC 20052
USA
soyer@gwu.edu

George Tagaras
Aristotle University of Thessaloniki
Department of Mechanical Engineering
54124 Thessaloniki
Greece
tagaras@auth.gr

Panagiotis Tsiamyrtzis
School of Statistics
Athens University of Economics and Business
76 Patission Str, 10434
Athens
Greece
pt@aueb.gr

Contents

Part I

Introduction to Bayesian Inference

1

An Introduction to Bayesian Inference in Process Monitoring, Control and Optimization

Enrique del Castillo

Department of Industrial & Manufacturing Engineering, Pennsylvania State University

Bianca M. Colosimo

Dipartimento di Meccanica, Politecnico di Milano

CONTENTS

ABSTRACT We present in this first chapter a general overview of Bayesian inference. A brief account of the fundaments is given, after which we focus on the following problems: inference in normally distributed data (univariate and multivariate), Kalman filtering, and Bayesian linear regression. Applications are noted in process monitoring, control, and optimization. Whenever appropriate, we refer to all subsequent chapters in this book.

1.1 Introduction

This chapter presents a general description of Bayesian inference and assumes knowledge of undergraduate statistics. Appropriately, this description starts with telling who Bayes was and what his work was about. Unfortunately, not many biographical details are known about Bayes (for some of them, see Press [29]). Reverend Thomas Bayes, a Presbyterian minister, lived in England in the 18th century and wrote a manuscript on "inverse probability" related to making inferences about the proportion of a binomial distribution. This was published posthumously in 1763. In 1774, working independently, Laplace stated what is now known as Bayes' theorem in general form.

Bayesian inference combines prior beliefs about model parameters with evidence from data using Bayes' theorem. A subjective interpretation of probability exists in this approach, compared to the "frequentist" approach in which the probability of an event is the limit of a ratio of frequencies of events. The main criticisms of Bayesian analysis have been that it is not objective (a fact that has been debated for many years) and that the required computations are difficult. As it will be discussed in Chapter 2, the second criticism has been overcome to a large extent in the last 10 to 15 years due to advances in integration methods, particularly Markov Chain Monte Carlo (MCMC) methods. Interestingly, Fisher and other authors have speculated that Bayes' reluctance in publishing his manuscript was due to his own doubts about the principles behind it, in particular his interpretation of subjective probability and the specification of a prior distribution on the unknown parameter. However, Stliger [32] disagrees, indicating that Bayes was quite sure and direct about the use of subjective probability and the use of a prior distribution in his work, and that it was Bayes' difficulties with solving an integral (an incomplete beta function) that lead him not to publish. Thus, if Stigler is right, Bayes' himself would be very happy today about recent developments in the numerical solution of Bayesian inference problems.

In this chapter, we provide a succinct overview of Bayesian inference, mentioning applications in monitoring, control, and optimization of production processes. Whenever possible, we will make references to other contributed chapters in this volume.

1.2 Basics of Bayesian Inference

1.2.1 Notation

We first introduce some notation used in this chapter. Let θ denote unobserved quantities or population parameters of interest. This can denote a scalar or a vector of parameters as the context will make clear. In some sections, particularly on Kalman filtering, $\boldsymbol{\theta}$ will denote a vector of k parameters. Let y denote observable quantities (data), either a single observation or a vector of n observations $(y_1, \ldots, y_n)$ if the context does not require any other vector notation. If the context uses other vector notation (e.g., in multivariate data and regression analysis), then multivariate observations will be denoted by the column vector $\boldsymbol{y}$. If the data is time-ordered (as required in most process monitoring and control applications), we will use $Y^t = (y_t, y_{t-1}, y_{t-2}, \ldots)$ or $\mathbf{Y}^t = (\boldsymbol{y}_t, \boldsymbol{y}_{t-1}, \boldsymbol{y}_{t-2}, \ldots)$ to denote the data observed up to and including time instant t. The variable $\tilde{y}$ denotes a future observation of the same nature as y, and $\boldsymbol{X}$ denotes an $n \times k$ matrix of explanatory variables or covariates. $\boldsymbol{X}$ contains the "experimental design" in a regression analysis.

In this chapter, $p(\cdot)$ denotes a continuous density function and the notation $w|y$ denotes a conditional random variable w given y (the data). $P(\cdot)$ denotes the probability of some event defined over a sample space. Sometimes we will simply write "data" for all the data obtained from an experiment.

1.2.2 Goals of Bayesian Inference

The goal of Bayesian inference is to reach conclusions about — or perhaps, to make decisions based on — a parameter θ or future observation $\tilde{y}$ using probability statements *conditional* on the data y:

$$p(\theta|y) \rightarrow \text{posterior density of } \theta$$
$$p(\tilde{y}|y) \rightarrow \text{posterior predictive density of } y.$$

Bayesian inference considers all unknowns (parameters and future observations) as random variables. Classical (frequentist) statistical inference considers population parameters as fixed but data as random (due to sampling).

1.2.3 Bayes' Theorem for Events

Before looking at Bayes' theorem for densities, let us look first at the theorem for simple events. This should be familiar to any reader who has taken a basic probability course. For events A and B in some sample space $\mathcal{S}$, from the definition of conditional probability we have,

$$P(A|B) = \frac{P(A \cap B)}{P(B)}$$

and

$$P(B|A) = \frac{P(A \cap B)}{P(A)}.$$

These expressions are true if and only if

$$P(A \cap B) = P(A|B)P(B) = P(B|A)P(A)$$

from which

$$P(A|B) = \frac{P(A)P(B|A)}{P(B)}.$$

This formulation gives the essence of Bayes' theorem: if event B represents some additional information that becomes available, then $P(A|B)$ is the probability after this information becomes available, i.e., the *posterior* of A, and $P(A)$ is the probability before this information becomes available, i.e., the *prior* for A.

Suppose events A_i form a partition of $\mathcal{S}$, that is, $\bigcap_{\text{all}i} A_i = \mathcal{S}$; $A_i \cap A_j = \oslash$ for all $i, j, i \neq j$. Then the ("total") probability of B is given by

$$P(B) = \sum_{\text{all}\,j} P(B|A_j)P(A_j)$$

and therefore

$$P(A_i|B) = \frac{P(A_i)P(B|A_i)}{\sum_{\text{all}\,j} P(B|A_j)P(A_j)}. \tag{1.1}$$

Equation (1.1) is what Laplace [19] referred to as the problem of finding the "inverse probability" — given that the "effect" B is observed, find which of several potential "causes" was the true cause of the observed effect.

As clear from the above, nothing is incorrect in Bayes formula, as it is derived from the probability axioms. The debate concerns the interpretation of the probabilities involved. Classical statistics regards each probability in the formula (prior and posterior) as the limit of the ratio of frequencies; Bayesian statistics regards each probability as a subjective measure.

1.2.4 Bayes' Theorem for Densities

Bayes' theorem also holds for densities and its derivation parallels that of the previous section. If y denotes data and θ some parameter or vector of parameters, then from the definition of conditional density, we have that

$$p(y|\theta) = \frac{p(\theta, y)}{p(\theta)}$$

and

$$p(\theta|y) = \frac{p(\theta, y)}{p(y)}.$$

This implies the joint density is

$$p(\theta, y) = p(\theta)p(y|\theta) = p(y)p(\theta|y),$$

which implies

$$p(\theta|y) = \frac{p(\theta)p(y|\theta)}{p(y)} \tag{1.2}$$

where analogous to the total probability of an event

$$p(y) = \int_{\text{all } \theta} p(\theta)p(y|\theta)d\theta. \tag{1.3}$$

Equation (1.2) is Bayes' law for densities. The denominator is usually not computed because it is not a function of θ, which is integrated out in Equation (1.3) and only makes $p(\theta|y)$ integrate to one in Equation (1.2). Therefore, the Bayesian statistics literature uses the proportionality sign $\propto$, and Bayes' formula in its most common form is:

$$p(\theta|y) \propto p(\theta)p(y|\theta). \tag{1.4}$$

In words, *the posterior is proportional to the product of the prior times the likelihood of the data.* Note that the posterior probabilities are therefore proportional to the likelihood, and the likelihood is a central concept in classical statistics, e.g., in maximum likelihood estimation. What we do is to modify the likelihood according to our prior beliefs of the parameter. If the prior is very "flat," Bayesian inferences will be very close to likelihood inferences. In Bayesian statistics, likelihoods are considered as carrying the only useful information contained in the data; this is the *likelihood principle*. Bayesian statistics are then consistent with the likelihood principle; classical statistics are not. The typical example is the classic example of rejecting a hypothesis based on a p-value. Recall this is the probability of observing a sample as extreme or more extreme than the one we already have at hand. Thus Jeffreys, in a celebrated phrase, remarked that by using a p-value, "a hypothesis which may be true may be rejected because it has not predicted observable results which have not occurred" [14].

Other information of relevance to making inferences might be available, not included in the likelihood, which can be available *a priori*. Bayesian statistics assume this can be incorporated in the prior.

Bayes' formula provides a recursive mechanism for updating the posterior distribution that is very useful in applications where the observations are obtained sequentially. As each new observation is obtained, the posterior is updated, treating the previous posterior as the prior and forming a chain:

$$p(\theta) \Rightarrow p(\theta|y_1) \Rightarrow p(\theta|y_1, y_2) \Rightarrow p(\theta|y_1, y_2, y_3) \Rightarrow \cdots$$

Evidently, the theorem assures that if more than one observation is obtained at a time, it is possible to "jump" two or more steps in the chain above with identical results. For example,

$$p(\theta) \Rightarrow p(\theta|y_1, y_2)$$

will result in the same distribution that would be obtained if going from $p(\theta)$ to $p(\theta|y_1, y_2)$ via $p(\theta|y_1)$. The sequential application of Bayes' theorem is a central idea in some engineering applications — for example, in Kalman filtering [23], a theme we return to later on in this chapter. In this volume (see Chapter 7), Graves discusses the idea of information and how this relates to Bayesian sequential inference with application to process monitoring.

To make inferences on an unobservable θ, we can simply look at the posterior distribution $p(\theta|y)$. This provides a complete characterization of our state of knowledge about the parameter and we recommend reporting it, perhaps using a histogram from a simulation or graphs if a closed-form expression exists. Still, in some applications, a single value or guess is needed about the unknown parameter, or an interval of possible values is desired or needed. Such a single value is the analog of a classical "point estimate." Two usual Bayesian choices are the mode of the posterior distribution:

$$\hat{\theta} = \arg\max p(\theta|y),$$

which evidently implies a maximization problem, and the mean of the posterior distribution

$$\hat{\theta} = E[p(\theta|y)],$$

which implies an integration. Each choice has different properties. For example, the mean of $p(\theta|y)$ minimizes the expected square error. The mode maximizes the expected utility function when a unit benefit exists if $\hat{\theta} = \theta$ and zero benefit if $\hat{\theta} \neq \theta$.

A Bayesian interval estimator of a parameter θ or *Bayesian credibility interval* is given by the interval (a, b) such that $P(a < \theta < b|y) = 1 - \alpha$, where $1 - \alpha$ is the "credibility" level. This value, contrary to the confidence level of a classical interval, is a probability and gives an indication of how probable that the parameter is contained within the computed interval. This interpretation is easier to grasp — particularly for beginners — than the long-run coverage interpretation of a classical confidence interval. Sometimes, a "highest posterior density" (HPD) interval is desired, which is obtained from solving

$$\min\ b - a \quad s.t. \quad \int_a^b p(\theta|y)d\theta = 1 - \alpha.$$

For symmetric unimodal distributions, this is always a symmetric interval around the mode. For a multimodal distribution, HPD intervals are harder to obtain.

To make inferences about a future observation, $\tilde{y}$, we compute the *posterior predictive density* as follows:

$$\begin{aligned} p(\tilde{y}|y) &= \int_{\text{all } \theta} p(\tilde{y}, \theta|y)d\theta \\ &= \int_{\text{all } \theta} p(\tilde{y}|\theta, y)p(\theta|y)d\theta \\ &= \int_{\text{all } \theta} p(\tilde{y}|\theta)p(\theta|y)d\theta, \end{aligned} \tag{1.5}$$

where the last equality follows because $\tilde{y}$ and y are conditionally independent given θ, that is, the parameters, if known, summarize the data. Similarly for unobservable parameters, we can use summarizing measures to provide single estimates or intervals on $\tilde{y}$. Again, simply looking at a plot of the posterior predictive distribution is the most complete approach.

Contrasting the predictive density with the classical approach of making predictions on $\tilde{y}$ is useful. The classical approach uses $p(\tilde{y}|\hat{\theta})$, the data density evaluated at the maximum likelihood or least squares estimator, to make predictions. Unlike Equation (1.5), this distribution does not account for the uncertainty in estimating θ, a crucial issue in applications because, as mentioned before, different parameter estimates will result in different solutions of the problem at hand.

1.2.5 Predictivism

The posterior predictive density is the basis of Bayesian *predictivism*. It is a *weighted average* of the probability of a new observation $\tilde{y}$ given the parameter θ multiplied by the probability of θ given the data. This weighted average is useful in scientific inference (see Press [29]) and can be illustrated easily in the case with a discrete number of alternative "theories" we wish to test. Suppose $\theta = 1$ means "Theory A is true," and $\theta = 0$ means "Theory B is true," and only these two theories are entertained to explain a phenomena.

We collect measurements and after observing the data, we compute

$$p(\tilde{y}|data) = p(\tilde{y}|\theta = 1)p(\theta = 1|data) + p(\tilde{y}|\theta = 0)p(\theta = 0|data),$$

which is the predictive probability of a new observation $\tilde{y}$.

Predictivism is a school of thought in the philosophy of science that postulates that the value of a scientific theory is measured by its ability to predict some phenomena, regardless of its ability to explain it. Within Bayesian statistics, predictivism states that the important quantities are the *observable* ones, not the unobservable ones (parameters). The posterior predictive density is the means to make predictions about and test a hypothesis about an observable. In applied problems in science and engineering, "model diagnostics" are based on the posterior predictive densities. In this type of diagnostics, we compare simulated predicted $\tilde{y}$ values using the posterior predictive density versus the data and see how similar they look. If the data is very different than the simulated responses, this is an indication our model fails to represent reality well.

1.2.6 Simulation of Posterior Quantities

Given $p(\theta|y)$ and $p(\tilde{y}|y)$, we can obtain posterior probabilities for functions of θ or $\tilde{y}$ as complex as needed.

Example

To find the posterior distribution of the coefficient of variation CV$= \mu/\sigma$ of a N(μ, σ^2) distribution when both μ and σ are unknown, let $\boldsymbol{\theta} = (\mu, \sigma)'$. Given $p(\boldsymbol{\theta}|y)$, we can simulate instances of $\boldsymbol{\theta}$ as shown in Table 1.1. To perform the simulation of the posterior of the parameters (a widely used "trick" if the joint posterior is difficult to obtain analytically[1]) is to note that

$$p(\mu, \sigma^2|y) = p(\mu|\sigma^2, y)p(\sigma^2|y).$$

Thus, an algorithm for the simulation of the posterior of the CV will look like this[2] (see Table 1.1 and Figure 1.1):

1. Collect n observations and compute $\overline{y}$ and s^2.
2. Simulate $\sigma^2|y$.
3. Simulate $\mu|\sigma^2, y$ (pairs of simulated values above give $\mu, \sigma^2|y$).
4. Compute $\frac{\mu}{\sigma}$.
5. Go to step 2 until we iterate N times.

---

[1] The joint posterior is not difficult to obtain in this case, but we will use this simpler case for illustration of this approach.

[2] To be more precise, and as it will be seen later, the algorithm requires us to generate $\sigma^2|y \sim Inv - \chi^2(n-1, s^2)$ (a scaled inverse chi squared distribution) and $\mu|\sigma^2, y \sim N(\overline{y}, \sigma^2/n)$. This is based on noninformative priors.

TABLE 1.1

Simulation of the Coefficient of Variation

Draw Number	Parameters (θ)	CV
1	μ_1, σ_1	μ_1/σ_1
2	μ_2, σ_2	μ_2/σ_2
$\vdots$	$\vdots$	$\vdots$
m	μ_m, σ_m	μ_m/σ_m

In Figure 1.1, the values $\overline{y} = 100$ and $s^2 = 10$ were observed based on a sample of size $n = 5$.

1.2.7 How to Simulate the Posterior Predictive Density

Sometimes the integral required to compute the posterior predictive density is difficult to obtain. If $p(\tilde{y}|\theta)$ and $p(\theta|y)$ are available (if they are not, see Chapter 2 in this volume), simulation is an easy alternative. To simulate

$$p(\tilde{y}|y) = \int p(\tilde{y}|\theta)p(\theta|y)d\theta,$$

we do the following:

1. Simulate a θ from $p(\theta|y)$.
2. Simulate a $\tilde{y}$ from $p(\tilde{y}|\theta)$.
3. Go to step 1 unless N iterations are reached.

A histogram of the N $\tilde{y}$ values characterizes $\tilde{y}|y$.

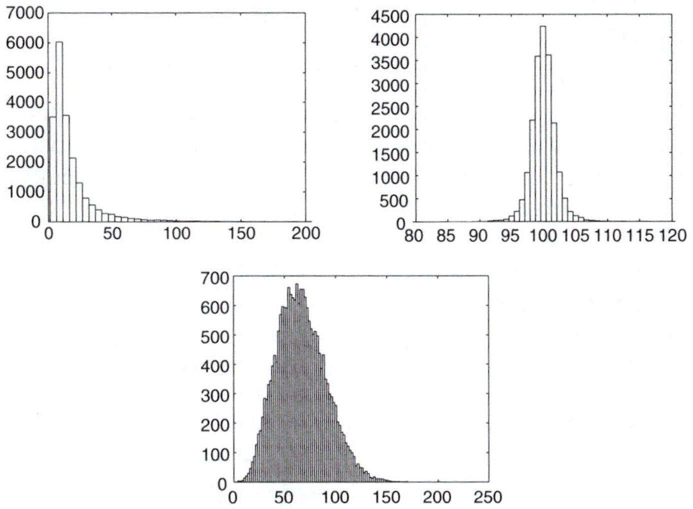

FIGURE 1.1
Simulation of the coefficient of variation of a Normal distribution, $N = 10,000$. First row: left $p(\sigma^2|y)$; right: $p(\mu|\sigma^2, y)$; Second row: $p(CV|y)$.

1.3 Choice of Prior Distribution

The question, "Where did the prior come from?" is the most common and valid criticism in Bayesian analysis, as any prior needs to be justified in practice. Three common choices of priors are:

1. Conjugate priors
2. Non-conjugate priors
3. Non-informative priors (these are non-conjugate as well)

Conjugate priors. If $\mathcal{F}$ is a class of sampling distributions $p(y|\theta)$ and $\mathcal{P}$ is a class of prior distributions for θ, $p(\theta)$, then $\mathcal{P}$ is said to be *conjugate* for $\mathcal{F}$ if

$$p(\theta|y) \in \mathcal{P} \;\; \forall \;\; p(y|\theta) \in \mathcal{F}, \, p(\theta) \in \mathcal{P}.$$

In words, this means that the prior and the posterior distributions of the parameter have the same form (with different parameters), so conjugacy is a closure property. The main merit of conjugate priors is that it simplifies computations, particularly in sequential applications of Bayes' theorem. With these distributions, the integral we need to compute to obtain the posterior has a familiar form, hence the computational advantage. However, in many applications "tuning" a conjugate prior to reflect the knowledge of the user is a difficult problem, or the conjugate priors might not be able to reflect this knowledge. Whereas some literature is found on "elicitation of priors" (see Kadane et al. [16] for the case of a regression model), elicitation has had little impact on statistical practice.

Non-conjugate priors do not have the closure property of conjugate priors; they result in posteriors that have a different parametric form than the prior. Until recently, they were not discussed frequently in the literature, given the hard integrals involved. With the advent of Markov Chain Monte Carlo (MCMC) methods in the last decade (see Chapter 2 in this volume), these priors have been used more often as they can now be chosen to better reflect prior knowledge of the parameters.

Non-informative priors are non-conjugate and try to reflect a situation with a complete lack of knowledge about a parameter. Therefore, they are called "objective priors" by some authors, and the resulting analysis is called objective Bayesian analysis. In applied problems, complete *a priori* ignorance hardly exists in an experiment, so a non-informative prior should, in practice, be regarded as an approximation to a situation where little is known *a priori* [3].

We will consider determining non-informative priors for *location* and *scale* parameters of a distribution, so we need the following definitions. Let $p(y)$

be any pdf. A *location-scale* family of density functions has the form

$$\frac{1}{\sigma} p \left(\frac{y - \mu}{\sigma} \right),$$

where $-\infty < \mu < \infty$ is a location parameter and $\sigma > 0$ is a scale parameter. That is, μ shifts the location of the distribution on the y axis and σ stretches (contracts) the graph of $p(y)$ if $\sigma > 1$ ($\sigma < 1$). In either case, changing these parameters does not change the shape of the distribution. Examples of location-scale distributions are the normal distribution and the double exponential. If $\sigma = 1$, then a density of the form $p(y - \mu)$ is called a *location density*, and if $\mu = 0$, a density of the form $p(y/\sigma)/\sigma$ is called a *scale density*.

1.3.1 Non-Informative Priors

First, let us note that a uniform distribution over a finite range is informative in the sense that values of the parameter are excluded (if the prior is zero over some range, the posterior will be zero over that range). Such a prior is non-informative only if the parameter has a range that coincides with that of the uniform distribution.

Laplace introduced the *principle of insufficient reason* for which he implied that in the absence of any information about a parameter, all values should be equally likely. Jeffreys generalized this reasoning into what is called the *invariance principle*. We will focus on invariance as a principle of finding non-informative priors, but readers should be aware that many other approaches have been put forward to define a non-informative prior (see Kass and Wasserman [18] for an excellent review). In particular, a lot of debate is found about what "non-informative" means, and this has resulted in the agreement that the non-informative priors used are more for convenience than as a true description of lack of information. They should be used as "reference priors" in the sense of being a default choice that makes sense in the situation when one knows little about a parameter. Unfortunately, well-known Bayesian software (e.g., Win BUGS) does not allow non-informative priors (see Chapter 2).

A summary of non-informative priors is as follows:

1. For parameters θ defined over a finite range of possible values $R \subset \mathbb{R}$, define the prior of θ to be uniform in R. An example of this was proposed by Bayes himself, who used a uniform (0,1) on the binomial proportion parameter p.
2. For parameters θ defined over all real numbers $\mathbb{R}$, use a uniform $(-\infty, \infty)$ distribution as prior.
3. For parameters θ defined over the positive real line $\mathbb{R}^+ \subset \mathbb{R}$, define a prior for $\log \theta$ to be uniform in $(-\infty, \infty)$.

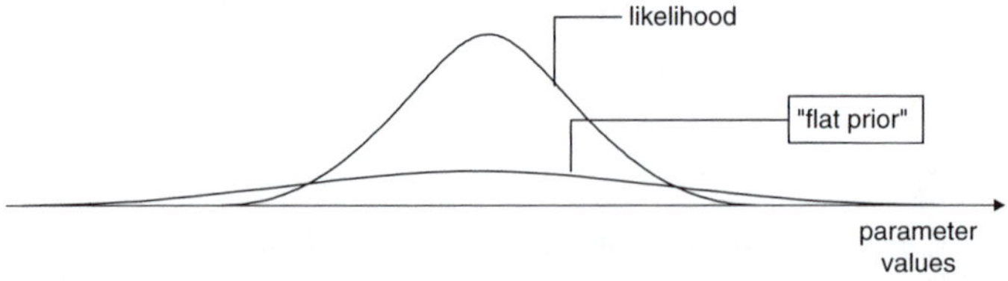

FIGURE 1.2

Hypothetical likelihood for a parameter θ defined over all real numbers and a flat but proper prior density.

4. When trying to set up a non-informative prior in multiple parameters $\boldsymbol{\theta} = (\theta_1, \theta_2, \ldots, \theta_k)'$, apply the criteria above to each parameter individually. This means that the parameters are independent *a priori*.

Evidently, only the first case above gives a proper prior distribution. Case 2 and case 3 lead to improper prior distributions, that is, density functions that do not have a finite integral. These improper priors can lead to proper posteriors in important cases, as we will show shortly. However, proper posteriors will not always result, so one should always check that the posterior is proper when using improper priors.

The important thing in practice, however, is that a non-informative prior should be flat where the likelihood is non-negligible. See Figure 1.2, where a "flat" but proper prior is set on a parameter where the likelihood of the data given the parameter is non-negligible. Because we cannot know the location of the likelihood before the data is collected, we will not be able to use such a flat, proper prior as we will not know where to locate it *a priori*. Instead, we use the improper priors in Case 1 and Case 2, which evidently are "everywhere uniform" over the possible values of either θ (Case 1) or $\log \theta$ (Case 2). The improper prior should be seen as a useful device to approximate the ideal case of Figure 1.2. For a justification of the choice of improper priors (Case 2 and Case 3 above) see Berger [1, pp. 82–90].

1.3.2 Jeffreys' Non-Informative Priors

The idea of using a formal rule to define a non-informative prior is due to Jeffreys. He used the concept of invariance as a formal rule. Thus far, we have not used a well-defined notion of invariance. We now define it more formally.

Invariance principle. If some rule led to $p(\theta)$ as a noninformative prior for θ, the same rule should lead to

$$p(\phi) = p(\theta)\left|\frac{d\theta}{d\phi}\right| \tag{1.6}$$

as a noninformative prior for ϕ, where $\phi = h(\theta)$ is a one-to-one transformation. If this is true, posterior inferences based on $p(\phi)$ will be the same as those made on $p(\theta)$.

Thus, what we request is that the conclusions we reach in our analysis do not change if we transform the parameter. This occurs only if the prior for a transformed quantity is consistent with the transformation of random variables formula (see Appendix). The principle particularly implies that regardless of the origin and scale of measurements of quantities of interest, which are always arbitrary, the conclusions that we reach or inferences we make will be invariant.

Jeffreys then showed that a prior that meets the invariance principle is

$$p(\theta) \propto I(\theta)^{1/2}, \tag{1.7}$$

where $I(\theta)$ is Fisher's information for the parameter θ, defined by

$$I(\theta) = -E\left[\frac{d^2 \log p(y|\theta)}{d\theta^2}\bigg|\,\theta\right].$$

This criterion, when applied individually to location and scale parameters as in the previous section, results in $p(\theta) \propto$ constant and $p(\sigma) \propto 1/\sigma$, respectively.[3]

That Jeffreys criterion of Equation (1.7) satisfies the invariance principle of Equation (1.6) is easy to see. The proof is as follows (this proof is presented in Zellner [35], who attributed it to Stone). Suppose $\phi = h(\theta)$, and write $p = p(y|\theta)$. Then, from the chain rule for differentiation,

$$\frac{d\ \log p}{d\theta} = \frac{d\ \log p}{d\phi}\frac{d\phi}{d\theta}.$$

Therefore,

$$I(\theta) = -\frac{d\phi}{d\theta}E\left(\frac{d\ \log p}{d\phi}\ \frac{d\log p}{d\phi}\right)\frac{d\phi}{d\theta} = I(\phi)\left(\frac{d\phi}{d\theta}\right)^2.$$

Thus,

$$I(\theta)^{1/2} = I(\phi)^{1/2}\left|\frac{d\phi}{d\theta}\right|.$$

This implies that if we set $p(\theta) \propto I(\theta)^{1/2}$ and use the same rule for setting a prior on ϕ, then

$$p(\phi) \propto I(\phi)^{1/2} = I(\theta)^{1/2}\left|\frac{d\theta}{d\phi}\right| \propto p(\theta)\left|\frac{d\theta}{d\phi}\right|$$

and this satisfies the invariance principle.

For multiple parameters $\boldsymbol{\theta} = (\theta_1, \theta_2, \ldots, \theta_k)'$, Jeffreys' prior is

$$p(\boldsymbol{\theta}) \propto |I(\boldsymbol{\theta})|^{1/2} \tag{1.8}$$

[3] Note that for the binomial parameter p (proportion), the prior we recommended earlier (Bayes' prior) is U(0,1), — equivalent to a Beta(1,1) — but Jeffreys' criterion of Equation (1.7) yields a Beta(1/2,1/2).

(the square root of the determinant of Fisher's information matrix), where

$$I_{ij} = -E\left[\frac{\partial \log p(y|\boldsymbol{\theta})}{\partial\theta_i\partial\theta_j}\right], \quad i, j = 1, 2, \ldots, k.$$

However, when $I(\boldsymbol{\theta})$ is not diagonal, this would imply that *a priori*, the parameters are dependent, and this idea is counter to our intuition of a non-informative prior on the parameters. Therefore, Jeffreys suggests that instead of using criterion of Equation (1.8), one should use his scalar criterion of Equation (1.7) one parameter at a time. This implies we assume the multiple parameters are *a priori* independent and individually noninformative but do not follow jointly the concept of what a "non-informative prior" should be according to Jeffreys himself.

1.3.3 Empirical Bayes Methods

A prior distribution for an unknown parameter θ, $p(\theta)$, has itself parameters, usually referred to as *hyperparameters*. The hyperparameters can be assumed known, as in a classical conjugate analysis. Alternatively, they can be assumed random variables that themselves have a prior distribution in an additional level of a hierarchy of parameters and prior distributions that can be repeated at several levels up to the highest level in the hierarchy in which all parameters are assumed known. The goal is to make inferences about all unknown parameters at all levels. These are the so-called hierarchical models, which we describe in Chapter 2 in this volume.

A third alternative for dealing with hyperparameters is to use the data to estimate them, for example, using maximum likelihood or method of moments. Such use of the data to estimate the prior receives the name *Empirical Bayes methods*, a term first coined by Robbins [30].

Using the data in the prior violates Bayes' theorem, which requires that the prior probabilities do not depend on the data. However, Empirical Bayes methods can be viewed as (well-behaved) approximations to full Bayes' methods. They are particularly useful in complex inferential problems, or in problems where assessing the hyperparameters from subjective user opinion is difficult or impossible.

In Chapter 4, Shiau and Feltz provide a detailed account of Empirical Bayes methods and their use in process monitoring problems. They consider monitoring univariate and multivariate continuous data, binary data, and polytomous data (data that has more than two categories, as in binary data).

1.4 Inferences on Normally-Distributed Data

We present next the application of Bayesian ideas to inference problems when the data is normally distributed. We consider both the univariate and the multivariate normal cases.

1.4.1 Inferences on Normally Distributed Data (Known Variance)

Consider data that is normal: $y \sim N(\theta, \sigma^2)$ with σ^2 known and θ unknown. This situation is probably not very realistic in practice but it is a useful first simple model to illustrate the ideas. Suppose, also for simplicity, that we observe one data point. The likelihood for one observation $y = y_1$ is

$$p(y|\theta) = \frac{1}{\sqrt{2\pi}\,\sigma}\, e^{-\frac{1}{2\sigma^2}(y-\theta)^2}.$$

1.4.1.1 Analysis with a Conjugate Prior

The *conjugate prior* for normal data is

$$\theta \sim \mathrm{N}\left(\mu_0, \tau_0^2\right),$$

where we assume that the "hyperparameters" μ_0 and τ_0 are known. That this is a conjugate prior for the normal can be seen because the *posterior density* is an exponential in a quadratic form in θ, which turns out to be a normal,

$$p(\theta|y) \propto p(\theta)p(y|\theta) = e^{-\frac{1}{2}\left[\left(\frac{y-\theta}{\sigma}\right)^2 + \left(\frac{\theta-\mu_0}{\tau_0}\right)^2\right]},$$

which can be simplified by completing the square[4] in θ to

$$p(\theta|y) \propto e^{-\frac{1}{2\tau_1^2}(\theta-\mu_1)^2}, \qquad \text{thus } \theta|y \sim N\left(\mu_1, \tau_1^2\right),$$

where the posterior mean is

$$\mu_1 = \frac{\frac{1}{\tau_0^2}\mu_0 + \frac{1}{\sigma^2}y}{\frac{1}{\tau_0^2} + \frac{1}{\sigma^2}}$$

and the posterior variance follows the relation

$$\frac{1}{\tau_1^2} = \frac{1}{\tau_0^2} + \frac{1}{\sigma^2}.$$

Thus, the posterior mean is a weighted average of the prior mean and data with weights equal to the precisions (i.e., the inverse of the variances). The posterior precision is the sum of the prior precision and data precision. The posterior mean can also be interpreted as

$$\mu_1 = \mu_0 + (y - \mu_0)\frac{\tau_0^2}{\sigma^2 + \tau_0^2},$$

[4] This is a very common step in Bayesian statistics. Recall that "completing the square" means we get a perfect binomial square of the form $(\theta - c)^2$. Once this is done, we treat anything not a function of the random variable as a constant, which is not shown due to the use of the proportionality sign.

so we can say that the prior mean is adjusted towards the observed y. Similarly,

$$\mu_1 = y - (y - \mu_0)\frac{\sigma^2}{\sigma^2 + \tau_0^2},$$

so some authors (e.g., [11]) say that the data "shrinks" towards the prior mean. Some interesting cases are:

- If $\tau_0^2 = 0$, then $\mu_1 = \mu_0$, i.e., the prior mean is "infinitely precise" and dominates.
- if $\sigma^2 = 0$, then $\mu_1 = y$, i.e., the data is "infinitely precise" and dominates.
- if $y = \mu_0$, then $\mu_1 = \mu_0 = y$, i.e., the data and prior means agree and so does the posterior mean.
- if $\tau_0^2 \to \infty$, then $\mu_1 \to y$, i.e., we approach a "non-informative" prior on the mean parameter.

The *posterior predictive density* is obtained from

$$p(\tilde{y}|y) = \int p(\tilde{y}|\theta)p(\theta|y)d\theta \propto \int e^{-\frac{1}{2\sigma^2}(\tilde{y}-\theta)^2} e^{-\frac{1}{2\tau_1^2}(\theta-\mu_1)^2} d\theta$$

$$= \int e^{-\frac{1}{2}\theta^2\left(\frac{1}{\sigma^2}+\frac{1}{\tau_1^2}\right)+\theta\left(\frac{\tilde{y}}{\sigma^2}+\frac{\mu_1}{\tau_1^2}\right)-\frac{1}{2}\left(\frac{\tilde{y}^2}{\sigma^2}+\frac{\mu_1^2}{\tau_1^2}\right)} d\theta.$$

Integrating with respect to θ yields

$$p(\tilde{y}|y) \propto e^{-\frac{1}{2}\left(\frac{\tilde{y}^2}{\sigma^2}+\frac{\mu_1^2}{\tau_1^2}\right)} e^{\frac{1}{2}\left(\frac{\tilde{y}}{\sigma^2}+\frac{\mu_1}{\tau_1^2}\right)^2\left(\frac{1}{\frac{1}{\sigma^2}+\frac{1}{\tau_1^2}}\right)}.$$

Completing the square in $\tilde{y}$ in the exponent gives

$$p(\tilde{y}|y) \propto e^{-\frac{1}{2}\frac{(\tilde{y}-\mu_1)^2}{\sigma^2+\tau_1^2}}.$$

Therefore, the posterior predictive density is given by

$$\tilde{y}|y \sim \mathrm{N}\left(\mu_1, \sigma^2 + \tau_1^2\right).$$

Case of several observations. If n normally-distributed data points $y = (y_1, \ldots, y_n)$ are observed, the *posterior* of the mean is

$$p(\theta|y) \propto p(\theta)p(y|\theta) = p(\theta)\prod_{i=1}^{n} p(y_i|\theta)$$

$$\propto e^{-\frac{1}{2\tau_0^2}(\theta-\mu_0)^2} \prod_{i=1}^{n} e^{-\frac{1}{2\sigma^2}(y_i-\theta)^2}$$

$$\propto e^{-\frac{1}{2}\left(\frac{1}{\tau_0^2}(\theta-\mu_0)^2+\frac{1}{\sigma^2}\sum_{i=1}^{n}(y_i-\theta)^2\right)}.$$

Completing the square on θ (placing terms not a function of θ in the proportionality constant),

$$P(\theta|y) = p(\theta|\overline{Y}) = \mathrm{N}\left(\mu_n, \tau_n^2\right)$$

where the first equality follows because the sample mean $\overline{Y}$ is a sufficient statistic, i.e., the posterior is only a function of the data through the sample average. Here we have that

$$\mu_n = \frac{\frac{1}{\tau_0^2}\mu_0 + \frac{n}{\sigma^2}\overline{Y}}{\frac{1}{\tau_0^2} + \frac{n}{\sigma^2}} \tag{1.9}$$

and the posterior precision is

$$\frac{1}{\tau_n^2} = \frac{1}{\tau_0^2} + \frac{n}{\sigma^2}.$$

Note how as $n \to \infty$ or as $\tau_0 \to \infty$, $\mu_n \to \overline{Y}$ and $\tau_n^2 \to \sigma^2/n$. This coincides with the frequentist results.

Example

Inferences on normal distributed data, variance known. Suppose $\sigma^2 = 50$ is known from previous experience, and we take $n = 5$ observations, from which $\overline{Y} = 370$. Figure 1.3 shows the results for a conjugate prior $\theta \sim N(500, 2^2)$ distribution. The prior is very precise, and the posterior has a mean close to the prior mean ($\mu_n = 462.85$, $\tau_n = 1.69$). The posterior predictive density also has mean equal to $\mu_n = 462.85$, but its standard deviation is $\sqrt{\sigma^2 + \tau_1^2} = 7.27$. Therefore, the prior dominates the likelihood.

In contrast, Figure 1.4 shows the corresponding results for a conjugate prior $\theta \sim N(500, 100^2)$, a much flatter distribution. The prior is relatively imprecise, and the posterior has a mean close to the data mean but with larger variability ($\mu_n = 370.12$, $\tau_n = 3.16$). The posterior predictive density also has mean equal to $\mu_n = 370.12$ with a standard deviation equal to $\sqrt{\sigma^2 + \tau_1^2} = 7.74$. Here, the likelihood dominates the prior.

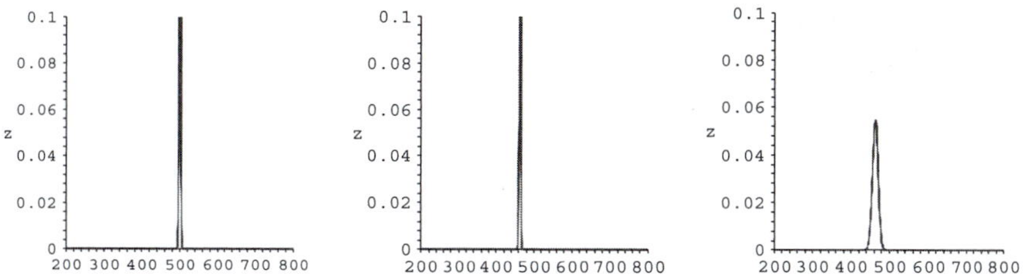

FIGURE 1.3
Inferences on normally distributed data, σ^2 known, "informative" conjugate prior, $\overline{Y} = 370$, $n = 5$. Left: $p(\theta)$; center: $p(\theta|y)$; right: $p(\tilde{y}|y)$.

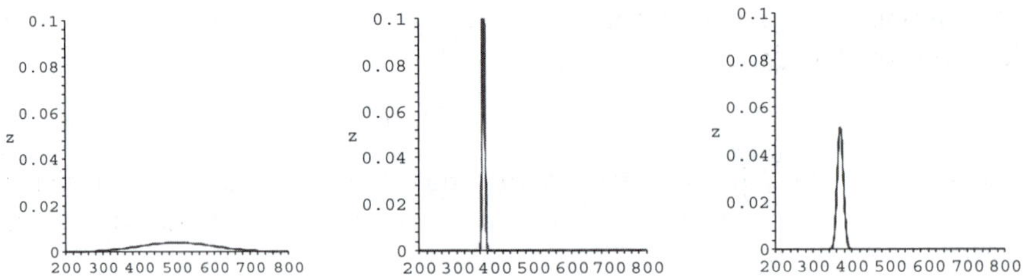

FIGURE 1.4
Inferences on normally distributed data, σ^2 known, "less informative" conjugate prior, $\overline{Y} = 370$, $n = 5$. Left: $p(\theta)$; center: $p(\theta|y)$; right: $p(\tilde{y}|y)$.

1.4.1.2 Analysis with a Non-Informative Prior

For the normal model with σ^2 known, consider now the non-informative prior $p(\theta) \propto$ constant. The posterior after n observations $y = (y_1, \ldots, y_n)$ is

$$\begin{aligned} p(\theta|y) &\propto p(y|\theta)p(\theta) \\ &\propto p(y|\theta) \\ &\propto e^{-\frac{1}{2\sigma^2}\sum_{i=1}^{n}(y_i-\theta)^2} \\ &= e^{-\frac{1}{2\sigma^2}[n\theta^2-2\theta\sum y_i+\sum y_i^2]} \\ &= e^{-\frac{n}{2\sigma^2}(\theta^2-2\theta\overline{Y}+\sum y_i^2/n)}. \end{aligned}$$

Completing the square on θ in the exponent (placing constant terms in the proportionality constant),

$$p(\theta|y) \propto e^{-\frac{n}{2\sigma^2}(\theta-\overline{Y})^2}.$$

Therefore,

$$\theta|y \sim \mathrm{N}(\overline{Y}, \sigma^2/n), \tag{1.10}$$

which evidently is a proper posterior. Note how we also obtain this distribution in the conjugate prior case when $\tau_0 \to \infty$.

1.4.2 Inferences on Normally Distributed Data, both Parameters Unknown

1.4.2.1 Analysis with a Conjugate Prior

The *conjugate prior* for this case is

$$\mu|\sigma^2 \sim \mathrm{N}(\mu_0, \sigma^2/\kappa_0)$$

and

$$\sigma^2 \sim \mathrm{Inv}\text{-}\chi^2\left(\nu_0, \sigma_0^2\right)$$

(a scaled inverse χ^2 distribution, see Appendix). Notice that μ depends *a priori* on σ^2, so no *a priori* independence assumption is made. The hyperparameter κ_0 can be thought of as the number of data points we believe our prior beliefs to be "worth." If $\kappa_0 > n$ (sample size), then the prior will have a strong influence on the posterior and vice versa. The joint prior density is

$$\begin{aligned} p(\mu, \sigma^2) &\propto \sigma^{-1} e^{-\frac{\kappa_0}{2\sigma^2}(\mu_0-\mu)^2} (\sigma^2)^{-(\nu_0/2+1)} e^{-\nu_0\sigma_0^2/(2\sigma^2)} \\ &= \sigma^{-1}(\sigma^2)^{-(\nu_0/2+1)} e^{-\frac{1}{2\sigma^2}(\nu_0\sigma_0^2+\kappa_0(\mu_0-\mu)^2)}, \end{aligned}$$

which Gelman et al. [11] refer to a Normal-inverse-χ^2 distribution (a four parameter distribution), that is,

$$(\mu, \sigma^2) \sim \text{N-Inv-}\chi^2\left(\mu_0, \sigma_0^2/\kappa_0; \nu_0, \sigma_0^2\right).$$

We need to show that the joint posterior distribution after observing $y = (y_1, \ldots, y_n)$ is also of the Normal-inv-χ^2 form. We have that:

$$\begin{aligned} p(\mu, \sigma^2 | y) &\propto \underbrace{\sigma^{-1}(\sigma^2)^{-(\nu_0/2+1)} e^{-\frac{1}{2\sigma^2}(\nu_0\sigma_0^2+\kappa_0(\mu_0-\mu)^2)}}_{p(\mu,\sigma^2)} \\ &\times \underbrace{(\sigma^2)^{-n/2} e^{-\frac{1}{2\sigma^2}\left(\sum_{i=1}^{n}(y_i-\overline{Y})^2+n(\overline{Y}-\mu)^2\right)}}_{p(y|\mu,\sigma^2)}. \end{aligned} \tag{1.11}$$

Thus, we need to show that $(\mu, \sigma^2 | \overline{Y}, s^2) \sim \text{N-Inv-}\chi^2(\mu_n, \sigma_n^2; \nu_n, \sigma_n^2)$. Note that the posterior is a function of the data only through $\overline{Y}$ and s^2 (i.e., these are sufficient statistics). The four parameters of the posterior are given by

$$\begin{aligned} \mu_n &= \frac{\kappa_0}{\kappa_0+n}\mu_0 + \frac{n}{\kappa_0+n}\overline{Y} \\ \kappa_n &= \kappa_0 + n \\ \nu_n &= \nu_0 + n \\ \nu_n\sigma_n^2 &= \nu_0\sigma_0^2 + (n-1)s^2 + \frac{\kappa_0 n}{\kappa_0+n}(\overline{Y}-\mu_0)^2. \end{aligned} \tag{1.12}$$

Equation (1.12) can be understood as the posterior sum of squares being equal to the prior sum of squares, plus the sample sum of squares, plus the sum of squares due to differences between sample data average and prior mean. The derivation to find this posterior requires taking the exponents of the exponential terms in Equation (1.11), completing the square in $(\mu_n - \mu)^2$, and treating all terms not a function of μ or σ constants that are gathered in the proportionality constant, which is not shown. With this, one obtains

$$p(\mu, \sigma^2 | \overline{Y}, s^2) \propto \sigma^{-1} e^{-\frac{\kappa_n}{2\sigma^2}(\mu_n-\mu)^2} (\sigma^2)^{-(\nu_n/2+1)} e^{-\frac{1}{2\sigma^2}\nu_n\sigma_n^2}. \tag{1.13}$$

Comparing Equation (1.11) and Equation (1.13), we confirm that $(\mu, \sigma^2 | \overline{Y}, s^2) \sim \text{N-Inv-}\chi^2(\mu_n, \sigma_n^2; \nu_n, \sigma_n^2)$.

From the above, the conditional posterior distribution of the mean is

$$\mu | \sigma^2, \overline{Y}, s^2 \sim \text{N}\left(\mu_n, \frac{\sigma^2}{\kappa_n}\right) = \text{N}\left(\frac{\kappa_0\mu_0 + n\overline{Y}}{\kappa_0+n}, \frac{\sigma^2}{\kappa_0+n}\right).$$

This can be written as

$$N\left(\frac{\frac{\kappa_0}{\sigma^2}\mu_0 + \frac{n}{\sigma^2}\overline{Y}}{\frac{\kappa_0}{\sigma^2} + \frac{n}{\sigma^2}}, \frac{1}{\frac{\kappa_0}{\sigma^2} + \frac{n}{\sigma^2}}\right),$$

which is in agreement with the posterior obtained for μ in the case that σ^2 is known — see Equation (1.9), using $\frac{\kappa_0}{\sigma^2} = \frac{1}{\tau_0^2}$ or $\tau_0^2 = \frac{\sigma^2}{\kappa_0}$.

The marginal posterior distributions of the parameters, obtained from integrating the joint posterior, are

$$\sigma^2|\overline{Y}, s^2 \sim \text{Inv-}\chi^2\left(\nu_n, \sigma_n^2\right) \tag{1.14}$$

and

$$\mu|\overline{Y}, s^2 \sim t_{\nu_n}\left(\mu_n, \sigma_n^2/\kappa_n\right). \tag{1.15}$$

Finally, the *predictive posterior distribution* of a new observation $\tilde{y}$ is obtained from

$$p(\tilde{y}|y) = \int\int \underbrace{p(\tilde{y}|\theta, \sigma^2, y)}_{N(\theta,\sigma^2)} p(\theta, \sigma^2|y)d\theta d\sigma^2,$$

which can be solved analytically, yielding

$$\tilde{y}|y = \tilde{y}|\overline{Y}, \quad s^2 \sim t_{\nu_n}\left(\mu_n, \frac{\sigma_n^2(\kappa_n + 1)}{\kappa_n}\right).$$

Reporting a graph of the posterior distributions is the most complete approach for making inferences. If needed, the posterior predictive density can be simulated either directly from the Student t distribution or approximated from the posterior of the parameters as follows:

1. Simulate σ^2 from an Inv-$\chi^2(\nu_n, \sigma_n^2)$. If no inverse χ^2 generator is at hand, one can do the following:
   (a) Simulate $X \sim \chi^2_{\nu_n}$.
   (b) Let $\sigma^2 = \nu_n\sigma_n^2/X$.
2. Simulate $\mu \sim N(\mu_n, \sigma^2/\kappa_n)$.
3. Simulate $\tilde{y} \sim N(\mu, \sigma^2)$.

Example

Inferences on normal data, both parameters unknown, conjugate priors. Suppose we think *a priori* that the joint distribution prior of the parameters has hyperparameters $\mu_0 = 150$, $\kappa_0 = 1$, $\nu_0 = 1$, and $\sigma_0 = 10$. We then collect 10 observations from which $\overline{Y} = 100$ and $s^2 = 20$. With this prior and data, the posterior parameters are $\mu_{10} = 104.54$, $\kappa_{10} = 11$, $\nu_{10} = 11$, and $\sigma_{10}^2 = 232.06$. We will simulate the predictive density of a new observation, $\tilde{y}|y$, using a) the distributions of $\sigma^2|y$ and $\theta|\sigma^2, y$ and b) using the t distribution directly. Figure 1.5 shows histograms obtained using 10,000 draws from the marginal

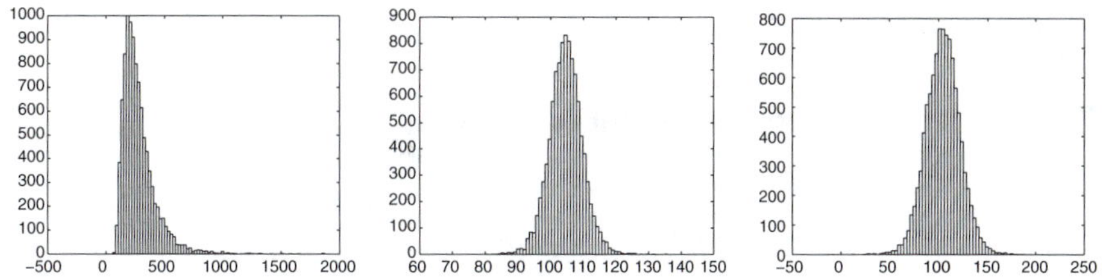

FIGURE 1.5
Inferences on normally distributed data, θ and σ^2 unknown, conjugate prior, $\overline{Y} = 100$, $s^2 = 20$, $n = 10$. Left: $p(\sigma^2|y)$; center: $p(\theta|\sigma^2, y)$; right: $p(\tilde{y}|y)$.

distributions; Figure 1.6 shows the corresponding posterior predictive density generated directly from the t distribution's closed form. The two simulated predictive densities are practically the same as expected. The distribution of $\sigma^2|y$ is a scaled inverse chi-square, with expected value $\nu_{10}/(\nu_{10}-2)\sigma_{10}^2 = 283.5$ and mode at $\nu_{10}/(\nu_{10}+2)\sigma_{10}^2 = 196.3$. Although in this case, using the closed form of the predictive density is easy (to report it using a graph, from example), this illustrates a useful approach to generate the distribution of $\tilde{y}|y$ when no closed-form expression exists.

1.4.2.2 Analysis with a Non-Informative Prior

We will only sketch the main results for a non-informative prior. For two unknown parameters, we apply Jeffreys' rule one parameter at a time, i.e.,

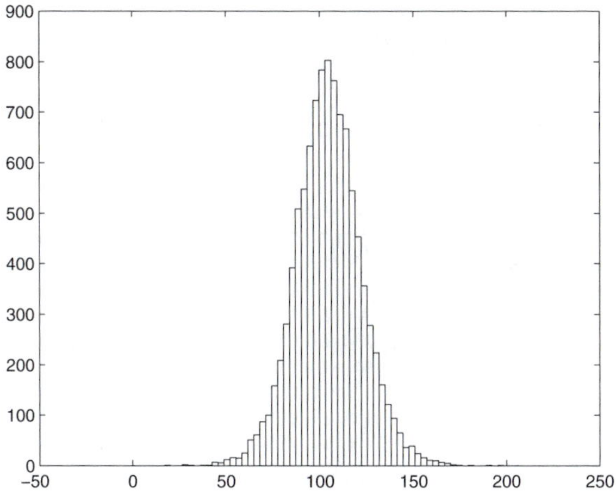

FIGURE 1.6
Predictive density $p(\tilde{y}|y)$ obtained by simulating directly from the t distribution, θ and σ^2 unknown, conjugate prior, $\overline{Y} = 100$, $s^2 = 20$, $n = 10$. Compare with the rightmost graph on Figure 1.5.

we assume *a priori* that the parameters are independent, in contrast to the conjugate case:

$$p(\theta) \propto \text{ constant } \quad \text{and} \quad p(\sigma^2) \propto 1/\sigma^2,$$

which yields

$$p(\theta, \sigma^2) = p(\theta)p(\sigma^2) \propto 1/\sigma^2.$$

The *joint posterior distribution* after observing $y = (y_1, \ldots, y_n)$ is obtained by making use of the expression

$$p(\theta, \sigma^2|y) = p(\theta|\sigma^2, y)p(\sigma^2|y)$$

where, if σ^2 is given, we have shown — see Equation (1.10) — that for a noninformative prior on θ,

$$\theta|\sigma^2, y \sim N(\overline{Y}, \sigma^2/n).$$

We can also show that

$$\sigma^2|y \sim Inv - \chi^2(n-1, s^2).$$

With these two distributions, the joint posterior can easily be simulated. Note how the distribution of $\sigma^2|y$ is analogous to the classical (frequentist) result $(n-1)s^2/\sigma^2 \sim \chi^2_{n-1}$ where s^2 is the random variable. In the Bayesian case, σ^2 is the random variable.

The marginal posterior for the mean is

$$\theta|y \sim t_{n-1}(\overline{Y}, s^2/n),$$

where the statistics $\overline{Y}$ and s^2 are sufficient. The distribution is a noncentral t distribution with location parameter $\overline{Y}$ and scale parameter s^2/n. Note this implies the Bayesian result

$$\frac{\theta - \overline{Y}}{s/\sqrt{n}} \sim t_{n-1},$$

a central t, where θ is the random variable. Compare this with the classical result:

$$\frac{\overline{Y} - \theta}{s/\sqrt{n}} \sim t_{n-1},$$

where $\overline{Y}$ is the random variable.

Similarly, as in the conjugate case, the *posterior predictive distribution* is given by

$$p(\tilde{y}|y) = \int\int \underbrace{p(\tilde{y}|\theta, \sigma^2, y)}_{N(\theta,\sigma^2)} p(\theta, \sigma^2|y) d\theta d\sigma^2.$$

The integral can be solved analytically, after some algebra, yielding

$$\tilde{y}|y \sim t_{n-1}\left(\overline{Y}, \left(1 + \frac{1}{n}\right)s^2\right).$$

This equation can be graphed and reported. If needed, the posterior predictive density can be simulated either by directly drawing from the t distribution or from the parameter posteriors as follows:

1. Draw (θ, σ^2) from $p(\theta, \sigma^2|y)$:
   (a) Draw σ^2 from $\sigma^2|y \sim \text{Inv-}\chi^2(n-1, s^2)$.
   (b) Draw θ from $\theta|\sigma^2, y \sim \text{N}(\overline{Y}, \sigma^2/n)$.
2. Draw $\tilde{y}$ from $\tilde{y} \sim \text{N}(\theta, \sigma^2)$.

1.4.3 Inferences on Multivariate Normally Distributed Data, Mean and Covariance Matrix Unknown, Conjugate Prior

The developments for the multivariate normal data parallel those of the univariate case. We will consider only the conjugate prior case. If $\boldsymbol{y}$ is a $p \times 1$ vector of observables (in applications, the p elements typically refer to p different responses observed simultaneously) with distribution

$$\boldsymbol{y}|\boldsymbol{\mu}, \quad \boldsymbol{\Sigma} \sim N(\boldsymbol{\mu}, \boldsymbol{\Sigma}),$$

where $\boldsymbol{\Sigma}$ is a $p \times p$ covariance matrix and $\boldsymbol{\mu}$ is a $p \times 1$ vector of means, the likelihood for one multivariate observation $\boldsymbol{y} = \boldsymbol{y}_1$ is

$$p(\boldsymbol{y}|\boldsymbol{\mu}, \boldsymbol{\Sigma}) \propto |\boldsymbol{\Sigma}|^{-1/2}\, e^{-\frac{1}{2}(\boldsymbol{y}-\boldsymbol{\mu})'\boldsymbol{\Sigma}^{-1}(\boldsymbol{y}-\boldsymbol{\mu})}.$$

For a sample of n points, the likelihood is

$$\begin{aligned} p(\boldsymbol{y}_1, \ldots, \boldsymbol{y}_n|\boldsymbol{\mu}, \boldsymbol{\Sigma}) &\propto |\boldsymbol{\Sigma}|^{-n/2}\, e^{-\frac{1}{2}\sum_{i=1}^{n}(\boldsymbol{y}_i-\boldsymbol{\mu})'\boldsymbol{\Sigma}^{-1}(\boldsymbol{y}_i-\boldsymbol{\mu})} \\ &= |\boldsymbol{\Sigma}|^{-n/2}\, e^{-\frac{1}{2}\, tr(\boldsymbol{\Sigma}^{-1}\mathbf{S}_0)}, \end{aligned}$$

where $\mathbf{S}_0 = \sum_{i=1}^{n}(\boldsymbol{y}_i - \boldsymbol{\mu})(\boldsymbol{y}_i - \boldsymbol{\mu})'$ (the last equality above can be shown going step by step over the products involved in the exponential).

The *conjugate prior* is

$$\begin{aligned} \boldsymbol{\mu}|\boldsymbol{\Sigma} &\sim N(\boldsymbol{\mu}_0, \boldsymbol{\Sigma}/\kappa_0) \\ \boldsymbol{\Sigma} &\sim Inv-Wishart\left(\nu_0, \boldsymbol{\Lambda}_0^{-1}\right), \end{aligned}$$

where the inverse Whishart distribution is the multivariate analogy of the inverse χ^2 distribution (see Appendix). Here $\boldsymbol{\Lambda}$ is a $p \times p$ matrix and ν_0 are the degrees of freedom. Just as in the univariate case, the prior of the mean vector depends on the covariance matrix, so no *a priori* assumption is made about the independence of these two parameters. Also, κ_0 can be understood as the number of observations we think our prior is "worth," with $\kappa_0 > n$ resulting in a prior that dominates the data and vice versa.

The *joint prior density* is

$$p(\boldsymbol{\mu}, \boldsymbol{\Sigma}) \propto |\boldsymbol{\Sigma}|^{-(\nu_0+p)/2+1}\, e^{-\frac{1}{2}\, tr\left(\boldsymbol{\Lambda}_0\boldsymbol{\Sigma}^{-1}\right)-\frac{\kappa_0}{2}(\boldsymbol{\mu}-\boldsymbol{\mu}_0)'\boldsymbol{\Sigma}^{-1}(\boldsymbol{\mu}-\boldsymbol{\mu}_0)},$$

which Gelman et al. [11] call a *Normal-Inverse Wishart* $(\boldsymbol{\mu}_0, \boldsymbol{\Lambda}_0/\kappa_0; \nu_0, \boldsymbol{\Lambda}_0)$. Following exactly the same steps as in the univariate case, the *joint posterior* density

$$p(\boldsymbol{\mu}, \boldsymbol{\Sigma}|y_1, \ldots, y_n) = p(\boldsymbol{\mu}, \boldsymbol{\Sigma}|\overline{\mathbf{Y}}, \mathbf{S})$$

is

$$\boldsymbol{\mu}, \boldsymbol{\Sigma}|\overline{\mathbf{Y}}, \quad \mathbf{S} \sim N - Inv - Wishart(\boldsymbol{\mu}_n, \boldsymbol{\Lambda}_n/\kappa_n; \nu_n, \boldsymbol{\Lambda}_n)$$

with

$$\begin{aligned}
\boldsymbol{\mu}_n &= \frac{\kappa_0}{\kappa_0 + n}\boldsymbol{\mu}_0 + \frac{n}{\kappa_0 + n}\overline{\mathbf{Y}} \\
\kappa_n &= \kappa_0 + n \\
\nu_n &= \nu_0 + n \\
\boldsymbol{\Lambda}_n &= \boldsymbol{\Lambda}_0 + \mathbf{S} + \frac{\kappa_0 n}{\kappa_0 + n}(\overline{\mathbf{Y}} - \boldsymbol{\mu}_0)(\overline{\mathbf{Y}} - \boldsymbol{\mu}_0)',
\end{aligned}$$

where $\boldsymbol{\Lambda}_n$, $\boldsymbol{\Lambda}_0$ and $\mathbf{S}$ are sum of squares for the posterior, the prior, and the data (just as in the univariate case), and $\mathbf{S} = \sum_{i=1}^{n}(y_i - \overline{\mathbf{Y}})(y_i - \overline{\mathbf{Y}})'$.

The *marginal posterior* of $\boldsymbol{\mu}$ is given by

$$\boldsymbol{\mu}|\overline{\mathbf{Y}}, \mathbf{S} \sim t_{\nu_n - p+1}\left(\boldsymbol{\mu}_n, \frac{\boldsymbol{\Lambda}_n}{\kappa_n(\nu_n - p + 1)}\right).$$

Finally, the *predictive posterior distribution* of a future observation $\tilde{\mathbf{y}}$ is

$$\tilde{\mathbf{y}}|\overline{\mathbf{Y}}, \mathbf{S} \sim t_{\nu_n - p+1}\left(\boldsymbol{\mu}_n, \frac{\boldsymbol{\Lambda}_n(\kappa_n + 1)}{\kappa_n(\nu_n - p + 1)}\right).$$

Example

Consider a bivariate case $(p = 2)$, where we believe *a priori* that $\boldsymbol{\mu}|\boldsymbol{\Sigma} \sim N([0, 0]', \boldsymbol{\Sigma}/10)$ and $\boldsymbol{\Sigma} \sim$ Inv-Wishart $(10, (10\mathbf{I}_2)^{-1})$. We collect 5 samples, with values y_i equal to $(6, 4)', (4, 7)', (3, 8)', 6, 3)'$, and $(5, 5)'$, thus $\overline{\mathbf{Y}} = (6.4, 7.2)'$. Therefore, we have that the posterior distribution has

$$\begin{aligned}
\boldsymbol{\mu}_5 &= \frac{10}{15}\begin{pmatrix} 0 \\ 0 \end{pmatrix} + \frac{5}{15}\begin{pmatrix} 4.8 \\ 5.4 \end{pmatrix} = \begin{pmatrix} 1.6 \\ 1.8 \end{pmatrix} \\
\kappa_5 &= 15 \\
\nu_5 &= 15 \\
\boldsymbol{\Lambda}_5 &= \boldsymbol{\Lambda}_0 + \mathbf{S} + \frac{50}{15}(\overline{\mathbf{Y}})(\overline{\mathbf{Y}})' = \begin{pmatrix} 93.6 & 75.8 \\ 75.8 & 124.4 \end{pmatrix}
\end{aligned}$$

where

$$\mathbf{S} = \begin{pmatrix} 6.8 & -10.6 \\ -10.6 & 17.2 \end{pmatrix}.$$

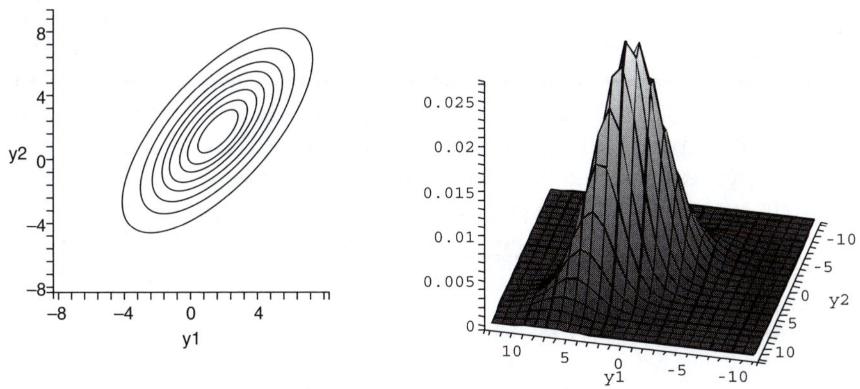

FIGURE 1.7
Posterior predictive density, $p(\tilde{\mathbf{y}}|\mathbf{y})$, bivariate Student t distribution, unknown parameters, conjugate prior. $\boldsymbol{\mu}_0 = (0, 0)'$, $\overline{\mathbf{Y}} = (6.4, 7.2)'$, $n = 5$, $\boldsymbol{\mu}_5 = (1.6, 1.8)'$. The prior dominates the data, so the posterior is located not too far from the former.

Figure 1.7 shows 2-D and 3-D plots of the corresponding posterior predictive distribution. Note that because $\kappa_0 > n$, the posterior mean is located closer to the prior mean (the origin) than to the data mean.

In this volume, the case when $\boldsymbol{\Sigma}$ is known is used by Alt (Chapter 5) to develop multivariate process monitoring schemes.

1.5 Applications in Process Monitoring

The practice of *statistical process control* (SPC) advises the use of one or more control charts to determine the stability of process parameters and to detect changes in them. Most of the control charting methods used in practice follow the work by Shewhart, which is frequentist. This is based on collecting data from the process while it was believed to be in a state of statistical control (by which is typically meant a state where the data distribution is stable over time) to estimate the model parameters. Tentative control limits are determined from these estimates, and if the initial data set is determined to be indeed in control, the limits are used to monitor future production data — otherwise, the tentative limits can be revised, and the procedure repeated.

Performance of a chart is based on the frequentist concept of *run length* (RL). This is the number of samples until detection, which we want to be large if the process is in-control (i.e., we want low false-alarm rate) and small if the process is really out of control (i.e., we want high power).

A problem that has been studied with considerable interest in the SPC literature in the last 15 years is the "Short Run SPC Problem." This problem points out the obvious difficulties one will encounter if the data set used to estimate the parameters (and control limits) is small, a problem that is encountered when manufacturing in small lots of parts. Rather than promising the impossible, namely, good RL performance (by good, we mean as good as if parameters were known) with small data sets, one could approach the short run SPC problem from a Bayesian perspective. Starting from a prior distribution on the process parameters, posterior distributions are obtained from which inferences can be made. This makes sense because it is precisely when little data is available when a prior belief could be incorporated in practice.

For example, considering monitoring the mean of a normally distributed process, when both parameters are unknown, using conjugate priors. In this case, we have seen from Equation (1.14) and Equation (1.15) that the marginal posterior distributions at time n are

$$\sigma^2|\overline{Y}, s^2 \sim Inv - \chi^2 \left(\nu_n, \sigma_n^2\right)$$

and

$$\mu|\overline{Y}, s^2 \sim t_{\nu_n} \left(\mu_n, \sigma_n^2/\kappa_n\right).$$

A graphical device proposed by Hoadley [13] (see also Crowder [6]) is to determine if the process is on target T by plotting box and whisker graphs with box limits at

$$\mu_n \pm L\sigma_n^2/\kappa_n$$

for some multiple L, which is a percentage point of the t distribution with ν_n degrees of freedom. If these boxes fail to cover the target value T, the process is deemed out of control (notice how "control with respect to a target" is a different notion than Shewhart's notion of a state of control, albeit one that is frequently used in industry). The value of L must be determined based on the desired RL performance. Crowder [6] provides RL analysis and tables for a more general model, which we describe in Section 1.7.

Despite how reasonable this approach may be in theory, for a real "short run" situation in practice, solely determining a state of control to target is probably not what a manufacturer wants. In most cases in industry, a major concern is also how to adjust the process to bring back to target a (short-run) process that might start off-target. This is related to the "Setup Adjustment Problem," which we will discuss in Section 1.8.

In Chapter 6 of this volume, Tagaras and Nenes give a review of Bayesian approaches to process monitoring. They present a model for the optimal design of $\overline{X}$ charts in which the time between samples, the sample size, and control limits can change adaptively, depending on the posterior probabilities of the process being in control or not. To obtain the solution to the adaptive $\overline{X}$ chart problem, they investigate numerical methods for the approximate solution of the underlying dynamic programming problem.

Parameter monitoring and adaptive chart designs are two instances where Bayesian methods have been used in the area of statistical process control. A third instance, of great importance in time series and econometrics applications, is changepoint detection. In this type of problem, the goal is to infer the point in time k at which a process changes one or more of its parameters. For example, Carlin et al. [4] study a changepoint detection problem in which data follows

$$y_t \sim N\left(\alpha_1 + \beta_1 t, \sigma_1^2\right), \quad t = 1, 2, \ldots, k$$

and

$$y_t \sim N\left(\alpha_2 + \beta_2 t, \sigma_2^2\right), \quad t = k + 1, k + 2, \ldots$$

A priori, the parameters (α_1, β_1) and (α_2, β_2) have independent normal distributions, whereas the variances have independent inverse χ^2 distributions. The key parameter of interest is k, the changepoint, which *a priori* follows a uniform distribution over all time indices. A Gibbs sampler (see Chapter 2) is utilized to find the posterior of the parameters, including k.

1.6 Kalman Filtering from a Bayesian Viewpoint

In this section, we look at a model that is central in process monitoring and control. The main result is the celebrated Kalman filter, which can be derived using a Bayesian approach, as we do here. We follow in this section the excellent account by Meinhold and Singpurwalla [23]. However, we point out that in the original derivation of his filter, Kalman [17] did not use a Bayesian approach.

Let $\mathbf{Y}^{t-1} = (y_{t-1}, y_{t-2}, y_{t-3}, \ldots)$ represent the collection of p-dimensional data vectors that have been observed from a process up to time instant $t-1$. (If the data is univariate, then $p = 1$ and the resulting y_i's are scalars. We look at the general vector case.) In contrast to previous sections, this is *time series data*, that is, to each data point we attach a subscript denoting a discrete point in time when the observation was made. Assume the data are generated according to the *observation equation:*

$$y_t = \mathbf{F}_t \boldsymbol{\theta}_t + \mathbf{v}_t, \quad \mathbf{v}_t \sim N(\mathbf{0}, \mathbf{V}_t) \tag{1.16}$$

where $\mathbf{F}_t$ is a $p \times q$ matrix of coefficients assumed known, $\mathbf{v}_t$ is a $p \times 1$ i.i.d. random vector with zero mean and covariance matrix $\mathbf{V}_t$, and the $q \times 1$ vector $\boldsymbol{\theta}_t$ varies according to the *state equation*:

$$\boldsymbol{\theta}_t = \mathbf{G}_t \boldsymbol{\theta}_{t-1} + \mathbf{w}_t, \quad \mathbf{w}_t \sim N(\mathbf{0}, \mathbf{W}_t) \tag{1.17}$$

where the $q \times q$ matrix $\mathbf{G}_t$ is also assumed known, and $\mathbf{w}_t$ is i.i.d and uncorrelated with $\mathbf{v}_t$. The two covariance matrices $\mathbf{V}_t$ and $\mathbf{W}_t$ are assumed known. We can only observe y_t, but we are interested in knowing the "state" of the process $\boldsymbol{\theta}_t$.

Although admittedly with many strong assumptions, the "state-space" model of Equation (1.16) and Equation (1.17) has found a very large number of applications in several areas. We detail one application to process quality control later in the next section. In the time series and econometrics literature, the state-space model is known as a *dynamic linear model* (DLM), see West and Harrison [33]. Later in this volume (Chapter 8), Hock and Soyer consider a DLM in conjunction with MCMC methods and apply it to a problem related to inferences in pulse train signals, such as radar signals, for application in defense systems.

In 1960, Kalman found a method for recursively estimating $\boldsymbol{\theta}_t$ based on the most current observation y_t and the past observations $\mathbf{Y}^{t-1}$. His method became known as the *Kalman filter*, which we now derive using Bayes' theorem.

From Bayes' theorem,

$$p(\boldsymbol{\theta}_t|y_t, \mathbf{Y}^{t-1}) \propto p(\boldsymbol{\theta}_t|\mathbf{Y}^{t-1})p(y_t|\boldsymbol{\theta}_t, \mathbf{Y}^{t-1}).$$

In this form, Bayes' theorem provides a recursive equation to update the posterior distribution of $\boldsymbol{\theta}_t$ given the data. We can use the mean of this posterior distribution as a "point estimate." In what follows, we will use the notation

$$\boldsymbol{\theta}_{t-1}|\mathbf{Y}^{t-1} \sim N(\widehat{\boldsymbol{\theta}}_{t-1}, \boldsymbol{\Sigma}_{t-1})$$

and, in general, $\widehat{\boldsymbol{\theta}}$ will refer to $E[\boldsymbol{\theta}]$ and $\boldsymbol{\Sigma}$ will denote $\mathrm{Var}(\boldsymbol{\theta})$.

At time t, the prior distribution of the state ($\boldsymbol{\theta}_t$) (prior to observing y_t) is given by

$$\boldsymbol{\theta}_t|\mathbf{Y}^{t-1} \sim N(\mathbf{G}_t\widehat{\boldsymbol{\theta}}_{t-1}, \mathbf{G}_t\boldsymbol{\Sigma}_{t-1}\mathbf{G}_t' + \mathbf{W}_t), \tag{1.18}$$

where we define henceforth the $q \times q$ matrix $\mathbf{R}_t = \mathbf{G}_t\boldsymbol{\Sigma}_{t-1}\mathbf{G}_t' + \mathbf{W}_t$, the variance of the prior of the state.

Suppose the state is given (fixed). From the observation equation, we then have that

$$y_t|\boldsymbol{\theta}_t, \mathbf{Y}^{t-1} \sim N(\mathbf{F}_t\boldsymbol{\theta}_t, \mathbf{V}_t),$$

noting that this gives the likelihood of the observation at time t. If $\boldsymbol{\theta}_t$ is not given but instead is unknown, then it is modeled as a random variable and we have that

$$y_t|\mathbf{Y}^{t-1} \sim N(\mathbf{F}_t\mathbf{G}_t\widehat{\boldsymbol{\theta}}_{t-1}, \mathbf{F}_t\mathbf{R}_t\mathbf{F}_t' + \mathbf{V}_t). \tag{1.19}$$

The derivation of the Kalman filter from a Bayesian point of view makes use of two key facts of conditional multivariate normal distributions. We indicate these two results in the Appendix to this chapter for completeness. The second result (which is actually the reverse of the first result) indicates the following. Let $\mathbf{X}_t$ be a $q \times 1$ vector and $\mathbf{X}_2$ be a $(p-q) \times 1$ vector. If the vector $\mathbf{X}_1$, given a value x_2 of the vector $\mathbf{X}_2$ (written $\mathbf{X}_1|\mathbf{X}_2 = x_2$) has (conditional) distribution

$$\mathbf{X}_1|\mathbf{X}_2 = x_2 \sim N_q\left(\boldsymbol{\mu}_1 + \boldsymbol{\Sigma}_{12}\boldsymbol{\Sigma}_{22}^{-1}(x_2 - \boldsymbol{\mu}_2), \quad \boldsymbol{\Sigma}_{11} - \boldsymbol{\Sigma}_{12}\boldsymbol{\Sigma}_{22}^{-1}\boldsymbol{\Sigma}_{21}\right) \tag{1.20}$$

and $\mathbf{X}_2 \sim N_{p-q}(\boldsymbol{\mu}_2, \boldsymbol{\Sigma}_{22})$, then we have that the joint distribution of $\mathbf{X}_1$ and $\mathbf{X}_2$ is

$$\begin{pmatrix} \mathbf{X}_1 \\ \mathbf{X}_2 \end{pmatrix} \sim \mathrm{N}_p\left(\begin{pmatrix} \boldsymbol{\mu}_1 \\ \boldsymbol{\mu}_2 \end{pmatrix}, \begin{pmatrix} \boldsymbol{\Sigma}_{11} & \boldsymbol{\Sigma}_{12} \\ \boldsymbol{\Sigma}_{21} & \boldsymbol{\Sigma}_{22} \end{pmatrix}\right).$$

Applying this result with $\mathbf{X}_1 \equiv y_t$ and $\mathbf{X}_2 \equiv \boldsymbol{\theta}_t$, we get that since we have the prior Equation (1.18) and Equation (1.19), then the joint prior distribution of the observation and the state is given by

$$\begin{pmatrix} y_t \\ \boldsymbol{\theta}_t \end{pmatrix} \sim \mathrm{N}\left(\begin{pmatrix} \mathbf{F}_t\mathbf{G}_t\widehat{\boldsymbol{\theta}}_{t-1} \\ \mathbf{G}_t\widehat{\boldsymbol{\theta}}_{t-1} \end{pmatrix}, \begin{pmatrix} \mathrm{Var}(y_t) = \mathbf{F}_t\mathbf{R}_t\mathbf{F}_t' + \mathbf{V}_t & \boldsymbol{\Sigma}_{12} = \mathrm{Cov}(y_t, \boldsymbol{\theta}_t) \\ \boldsymbol{\Sigma}_{21} = \mathrm{Cov}(\boldsymbol{\theta}_t, y_t) & \mathrm{Var}(\boldsymbol{\theta}_t) = \mathbf{R}_t \end{pmatrix}\right).$$

One remaining question to resolve is what is $\boldsymbol{\Sigma}_{12} = \mathrm{Cov}(y_t, \boldsymbol{\theta}_t) = \boldsymbol{\Sigma}_{21}'$. To find it, notice from Equation (1.16) and Equation (1.20) that

$$E[y_t|\boldsymbol{\theta}_t] = \underbrace{\mathbf{F}_t\mathbf{G}_t\widehat{\boldsymbol{\theta}}_{t-1}}_{\boldsymbol{\mu}_1} + \boldsymbol{\Sigma}_{12}\underbrace{\mathbf{R}_t^{-1}}_{\boldsymbol{\Sigma}_{22}^{-1}}(\underbrace{\boldsymbol{\theta}_t}_{x_2} - \underbrace{\mathbf{G}_t\widehat{\boldsymbol{\theta}}_{t-1}}_{\boldsymbol{\theta}_t}) = \mathbf{F}_t\boldsymbol{\theta}_t.$$

For this expression to hold, it is sufficient that

$$\boldsymbol{\Sigma}_{12}\mathbf{R}_t^{-1} = \mathbf{F}_t \quad \Rightarrow \quad \boldsymbol{\Sigma}_{12} = \mathbf{F}_t\mathbf{R}_t,$$

which also implies $\boldsymbol{\Sigma}_{21} = \mathbf{R}_t'\mathbf{F}_t'$. Therefore, the joint prior distribution of state and observation at time t is given by

$$\begin{pmatrix} y_t \\ \boldsymbol{\theta}_t \end{pmatrix} \sim \mathrm{N}\left(\begin{pmatrix} \mathbf{F}_t\mathbf{G}_t\widehat{\boldsymbol{\theta}}_{t-1} \\ \mathbf{G}_t\widehat{\boldsymbol{\theta}}_{t-1} \end{pmatrix}, \begin{pmatrix} \mathbf{F}_t\mathbf{R}_t\mathbf{F}_t' + \mathbf{V}_t & \mathbf{F}_t\mathbf{R}_t \\ \mathbf{R}_t'\mathbf{F}_t' & \mathbf{R}_t \end{pmatrix}\right).$$

However, what we need to make inferences about the state is $\boldsymbol{\theta}_t|y_t, \mathbf{Y}^{t-1}$, the *posterior distribution of the state* (posterior after observing y_t). To find this distribution, apply the first "key result" of multivariate conditional normals listed in the Appendix. The result says that if

$$\begin{pmatrix} \mathbf{X}_1 \\ \mathbf{X}_2 \end{pmatrix} \sim \mathrm{N}_p\left(\begin{pmatrix} \boldsymbol{\mu}_1 \\ \boldsymbol{\mu}_2 \end{pmatrix}, \begin{pmatrix} \boldsymbol{\Sigma}_{11} & \boldsymbol{\Sigma}_{12} \\ \boldsymbol{\Sigma}_{21} & \boldsymbol{\Sigma}_{22} \end{pmatrix}\right)$$

and $|\boldsymbol{\Sigma}_{22}| > 0$, then the marginal distribution of $\mathbf{X}_1$ given $\mathbf{X}_2 = x_2$ is given by

$$\mathbf{X}_1|\mathbf{X}_2 = x_2 \sim \mathrm{N}_q\left(\boldsymbol{\mu}_1 + \boldsymbol{\Sigma}_{12}\boldsymbol{\Sigma}_{22}^{-1}(x_2 - \boldsymbol{\mu}_2), \boldsymbol{\Sigma}_{11} - \boldsymbol{\Sigma}_{12}\boldsymbol{\Sigma}_{22}^{-1}\boldsymbol{\Sigma}_{21}\right).$$

Applying this result with $\mathbf{X}_1 = \boldsymbol{\theta}_t$ and $\mathbf{X}_2 = y_t$, we get the desired posterior distribution of the state:

$$\begin{aligned} \boldsymbol{\theta}_t|y_t, \mathbf{Y}^{t-1} \sim \mathrm{N}(&\underbrace{\mathbf{G}_t\widehat{\boldsymbol{\theta}}_{t-1}}_{\boldsymbol{\mu}_1} + \underbrace{\mathbf{R}_t'\mathbf{F}_t'}_{\boldsymbol{\Sigma}_{12}}\underbrace{(\mathbf{F}_t\mathbf{R}_t\mathbf{F}_t' + \mathbf{V}_t)^{-1}}_{\boldsymbol{\Sigma}_{22}^{-1}}(\underbrace{y_t}_{\mathbf{X}_2} - \underbrace{\mathbf{F}_t\mathbf{G}_t\widehat{\boldsymbol{\theta}}_{t-1}}_{\boldsymbol{\mu}_2}), \\ &\underbrace{\mathbf{R}_t}_{\boldsymbol{\Sigma}_{11}} - \underbrace{\mathbf{R}_t'\mathbf{F}_t'}_{\boldsymbol{\Sigma}_{12}}\underbrace{(\mathbf{F}_t\mathbf{R}_t\mathbf{F}_t' + \mathbf{V}_t)^{-1}}_{\boldsymbol{\Sigma}_{22}^{-1}}\underbrace{\mathbf{F}_t\mathbf{R}_t}_{\boldsymbol{\Sigma}_{21}}). \end{aligned}$$

The two moments of this distribution can be updated recursively as each new observation y_t is obtained, yielding the Kalman filter recursive equations. The mean of the posterior of the state is updated as follows:

$$\widehat{\boldsymbol{\theta}}_t = \mathbf{G}_t\widehat{\boldsymbol{\theta}_{t-1}} + \mathbf{R}_t'\mathbf{F}_t'(\mathbf{F}_t\mathbf{R}_t\mathbf{F}_t' + \mathbf{V}_t)^{-1}(y_t - \mathbf{F}_t\mathbf{G}_t\widehat{\boldsymbol{\theta}}_{t-1}) \tag{1.21}$$

or, more simply, as

$$\widehat{\boldsymbol{\theta}}_t = \mathbf{G}_t\widehat{\boldsymbol{\theta}_{t-1}} + \mathbf{K}_t\mathbf{e}_t \tag{1.22}$$

where

$$\mathbf{K}_t = \mathbf{R}_t'\mathbf{F}_t'(\mathbf{F}_t\mathbf{R}_t\mathbf{F}_t' + \mathbf{V}_t)^{-1} \tag{1.23}$$

is a $q \times p$ matrix called the *Kalman gain matrix* (note this matrix is not a function of the data nor a function of any unknown parameter), and $\mathbf{e}_t = y_t - \mathbf{F}_t\mathbf{G}_t\widehat{\boldsymbol{\theta}}_{t-1}$, a $p \times 1$ vector of one step ahead prediction errors. In the time series literature, this would be written $\mathbf{e}_t = y_t - \widehat{y}_{t|t-1}$ and called the *innovations*, because they depend only on the last observation. In other words, the innovations are so called because they are the newest "correction" introduced into the state estimate due to the latest observation.

The second recursive equation, which completes the description of the Kalman filter, is:

$$\boldsymbol{\Sigma}_t = \mathbf{R}_t - \mathbf{R}_t'\mathbf{F}_t'(\mathbf{F}_t\mathbf{R}_t\mathbf{F}_t' + \mathbf{V}_t)^{-1}\mathbf{F}_t\mathbf{R}_t = [\mathbf{I} - \mathbf{K}_t\mathbf{F}_t]\mathbf{R}_t, \tag{1.24}$$

which updates the variance-covariance matrix of the posterior distribution of the state.

The three prior equations evidently require starting values $\widehat{\boldsymbol{\theta}}_0$ and $\boldsymbol{\Sigma}_0$. These characterize the initial prior $\boldsymbol{\theta}_0 \sim \mathrm{N}(\widehat{\boldsymbol{\theta}}_0, \boldsymbol{\Sigma}_0)$. Note how at each point in time, only the latest observation is needed to update the expressions. This is a direct consequence of the recursive application of Bayes' theorem.

The optimality properties of Kalman filter estimates are well known. Duncan and Horn [8] show how under all the stated assumptions, the Kalman filter estimate minimizes the mean square error (i.e., it is MMSE optimal) among all possible estimators. If $\mathbf{v}_t$ and $\mathbf{w}_t$ are not normally distributed, the Kalman filter estimate is MMSE optimal only among all linear estimators. However, the filter is not robust against non-normality [24]. This is because the state estimate of Equation (1.22) is not a bounded function of the errors $\mathbf{e}_t$, and the variance matrix of Equation (1.24) is not a function of the data. Thus, an outlier can have a strong effect on the state estimate. Estimation of the state when errors are non-normal, when the state and or observation equations are not linear, or when there are unknown parameters needs to be accomplished with numerical methods. MCMC methods (see Chapter 2) have been used successfully for such purpose.

As mentioned earlier, Kalman filtering has had a tremendous range of applications in practice. Besides obvious applications in aircraft control, the Kalman filter is used, for example, to formulate the likelihood function of

ARIMA models in the time series literature, a formulation that allows to include missing data [2].

1.7 An Application of Kalman Filtering to Process Monitoring

To illustrate the use of some of the previous results on Kalman filtering to process monitoring, consider the following model, studied by Crowder [6]:

$$y_{it} = \theta_t + \varepsilon_{it}, \quad i = 1, 2, \ldots, n \tag{1.25}$$

$$\theta_t = \theta_{t-1} + v_t, \quad t = 1, 2, \ldots \tag{1.26}$$

where θ_t is the process mean and the $\varepsilon_{it}'s$ and v_ts are i.i.d random variables with zero means and variances σ_ε^2 and σ_v^2, respectively. This is a slight modification of the "steady model" studied by Smith [31]. It is a particular case of the state-space model shown in the previous section with $p = n, q = 1, \mathbf{F}_t = \mathbf{1}$ (an $n \times 1$ vector of ones), $\mathbf{V}_t = \sigma_\varepsilon^2 \mathbf{I}_n$, $G_t = 1$ and $W_t = \sigma_v^2$. Assuming all parameters known, the posterior distribution of the state is given by the Kalman recursive equations, which, after some algebra, reduce to

$$\widehat{\theta}_t = (1 - \lambda_t)\widehat{\theta}_{t-1} + \lambda_t \overline{Y}_t$$

$$\Sigma_t = \lambda_t \sigma_\varepsilon^2 / n$$

where

$$\lambda_t = \frac{\Sigma_{t-1} + \sigma_v^2}{\Sigma_{t-1} + \sigma_v^2 + \sigma_\varepsilon^2/n}.$$

Crowder [6] notes how the following limit is approached fast:

$$\lambda = \lim_{t \to \infty} \lambda_t = \frac{1}{\sqrt{\Delta + \frac{1}{4}} + \frac{1}{2}}$$

where $\Delta = \frac{\sigma_\varepsilon^2/n}{\sigma_v^2}$. Thus, in practice, one loses little if the state is estimated from the steady-state filter

$$\widehat{\theta}_t = (1 - \lambda)\widehat{\theta}_{t-1} + \lambda \overline{Y}_t,$$

which indicates that the MMSE optimal forecast of the process of Equation (1.25) and Equation (1.26) is an exponentially weighted moving average (EWMA) of the data with parameter λ.

The ratio Δ is a measure of how rapidly the process mean "moves." If $\Delta >> 1$, the process mean moves slowly, λ is near zero, and most weight is given to the older data. Conversely, if Δ is near zero, the process mean moves fast, λ is close to 1 and little weight is given to older data, with most weight given to the most recent sample mean. Crowder gives RL performance of EWMA control charts based on this model—although as mentioned recently by Crowder and Eshleman [7], if the process mean moves fast, the process monitoring question of whether a major shift has occurred is not relevant. The focus in such case is instead estimation of the process mean, providing vital information to process engineers about whether the process needs to be adjusted.

Crowder [6], (see also Crowder and Eshleman [7]) proposes "adaptive filtering" methods in case the process parameters are unknown. A different way to approach such case is to use MCMC or sequential Monte Carlo (SMC) methods of the kind described in Chapter 2. See Zilong and Del Castillo [36] for details.

In Chapter 3, Tsiamrytzis and Hawkins study a Bayesian monitoring problem where the objective is to monitor the state of the following model:

$$\begin{aligned} y_t &= \theta_{t-1} + \nu_t, \quad \nu_t \sim N(0, c\sigma^2) \\ \theta_t &= \theta_{t-1} + w_t, \end{aligned}$$

where

$$w_t \sim \begin{cases} N(0, \sigma^2) & \text{with probability } p \\ N(\delta, \sigma^2) & \text{with probability } 1-p \end{cases}$$

and δ is the expected magnitude of random "shocks" that occur with probability $1 - p$. Tsiamrytzis and Hawkins derive the posterior distribution of θ_t, σ^2, once the nuisance parameters are integrated out, and use it to determine if the process mean has crossed a given threshold or not.

1.8 An Application of Kalman Filtering to Process Control

We now describe an application of Kalman filtering to the area of engineering process control, or EPC. We consider the so-called *Setup Adjustment Problem*, first studied by Grubbs [12]. This relates to a machine that produces discrete parts in batches, with a setup operation taking place between batches. Let us initially assume we are interested in controlling a single property or quality characteristic in the product being produced by manipulating a single controllable factor. The setups might induce random errors in the quality characteristic. The goal is to adjust the process to minimize the effect of the setup errors using a mean square error objective. Let the level or "setpoint" of the controllable factor be U_t. Then $U_t - U_{t-1} \equiv \nabla U_t$ is the change or adjustment made at time t. The subscript t denotes the observation or part number.

If y_t denotes the deviation from target of the quality characteristic, in this type of problem the model describing the process is assumed

$$y_t = \mu_t + v_t, \quad v_t \sim N\left(0, \sigma_v^2\right), \quad i.i.d., \tag{1.27}$$

where

$$\mu_t = \mu_{t-1} + \nabla U_t. \tag{1.28}$$

Equation (1.28) indicates the assumption that adjustments modify the process mean μ_t, where the initial recursion will be assumed to be $\mu_1 = \mu_0 + U_0$, with U_0 the initial setpoint before production starts. Evidently the two previous equations correspond to an "observation" and "state" model, because we only observe y_t but not μ_t. The mean square error (MSE) objective, more appropriately called mean squared deviation objective in this control context, is with respect to the state of the process over some finite horizon n, so we wish to find the adjustments ∇U_t that minimize

$$E\left[\sum_{t=1}^{n}(\mu_t|Y^{t-1} - \text{target})^2\right] = E\left[\sum_{t=1}^{n}\mu_t^2|Y^{t-1}\right] \tag{1.29}$$

because the state μ has itself "target" zero as it measures deviation from the actual target. Each term in the sum is of the form

$$E\left[\mu_t^2|Y^{t-1}\right] = \text{Var}\left(\mu_t^2|Y^{t-1}\right) + \left(E\left[\mu_t|Y^{t-1}\right]\right)^2. \tag{1.30}$$

This is simply the usual formula which says that MSE equals variance plus squared bias applied to the prior distribution of the state at time t. Note how we wish to choose ∇U_t to control μ_t, so we cannot make use of the observation at time t, y_t, to do so, because y_t depends on μ_t. This is why the prior at time t appears in the above equations.

Comparing Equation (1.27) and Equation (1.28) with Equation (1.22) and Equation (1.24), we see that, using notation from the previous section, $p = q = 1$, $\widehat{\theta} = E[\mu]$, $\Sigma = \text{Var}(\mu)$, $G_t = F_t = 1$, $V_t = \sigma_v^2$, and $W_t = 0$. The state equation has the extra term ∇U_{t-1}, which we simply append to the Kalman filter expressions.

From Equation (1.18), the distribution of the state prior to observing y_t, needed in the two previous equations is:

$$\mu_t|Y^{t-1} \sim N(\widehat{\mu}_{t-1} + \nabla U_{t-1}, \Sigma_{t-1}),$$

thus

$$E\left[\mu_t^2|Y^{t-1}\right] = \Sigma_{t-1} + [\widehat{\mu}_{t-1} + \nabla U_{t-1}]^2.$$

To minimize this expression we set

$$\nabla_{t-1} = -\widehat{\mu}_{t-1}.$$

This also minimizes the sum, Equation (1.29) [9].

To see what effect this adjustment has, and to better understand how to compute the right-hand side of the expression above, we need to look at the distribution of the state after we observe y_t, i.e., $\mu_t|y_t, Y^{t-1}$, the posterior of μ. Its distribution is normal with moments given by the Kalman filter equations.

Using these equations with the equivalences $\widehat{\theta}_t \equiv \widehat{\mu}_t = E[\mu_t|y_t, Y^{t-1}]$ and $\Sigma_t \equiv \mathrm{Var}(\mu_t|y_t, Y^{t-1})$, we see that

$$R_t = G_t\Sigma_{t-1}G_{t-1} + W_t = \Sigma_{t-1}$$

$$K_t = R_t'F_t'(F_tR_tF_t' + V_t)^{-1} = \frac{\Sigma_{t-1}}{\Sigma_{t-1} + \sigma_v^2}$$

where

$$\Sigma_t = \Sigma_{t-1} - \frac{\Sigma_{t-1}^2}{\Sigma_{t-1} + \sigma_v^2} = \frac{\Sigma_{t-1}\sigma_v^2}{\Sigma_{t-1} + \sigma_v^2} = \frac{\sigma_v^2}{t + \frac{\sigma_v^2}{\Sigma_0}}, \tag{1.31}$$

which implies that the Kalman gains are equal to

$$K_t = \frac{\Sigma_{t-1}}{\Sigma_{t-1} + \sigma_v^2} = \frac{1}{t + \frac{\sigma_v^2}{\Sigma_0}}.$$

The expected value of the posterior distribution of the state is

$$\begin{aligned}\widehat{\mu}_t &= \widehat{\mu}_{t-1} + \nabla U_t + K_t(y_t - F_tG_t(\widehat{\mu}_{t-1} + \nabla U_{t-1}))\\ &= \widehat{\mu}_{t-1} + \nabla U_{t-1} + \frac{1}{t + \frac{\sigma_v^2}{\Sigma_0}}\left(y_t - (\widehat{\mu}_{t-1} + \nabla U_{t-1})\right).\end{aligned} \tag{1.32}$$

Therefore, the adjustment rule

$$\nabla U_{t-1} = -\widehat{\mu}_{t-1}$$

implies from Equation (1.32) that $\widehat{\mu}_t = K_t y_t$, and this in turn implies that

$$\nabla U_t = -K_t y_t = -\frac{1}{t + \frac{\sigma_v^2}{\Sigma_0}} y_t. \tag{1.33}$$

This is the same expression as the one derived by Grubbs [12] who solved the problem from a clever — but complicated — frequentist approach. If $\Sigma_0 \to \infty$, we get $\nabla U_t = -y_t/t$, Grubbs' elegant "harmonic" rule, which makes the adjustments progressively smaller following the harmonic series $\{1/t\}$.

Extending the previous development to the case when we have $p > 1$ responses of interest is straightforward, as the Kalman filter formulation is multivariate. This and other extensions, such as changing the objective function to include quadratic adjustment costs or assuming errors in the adjustments, is also relatively easy [9] compared with the frequentist approach, which results instead in intractable problems. For example, to include errors in the adjustments, all we have to do is to include a additive (normal) random error w_t to the state equation, and re-apply the Kalman filter recursive expressions. This shows the advantages of the Bayesian formulation.

Notice that in this application, the initial prior distribution is (μ_0, Σ_0) and has an easy interpretation: it is our guess of the setup error of the machine. Thus, for example, if we think *a priori* that on average no systematic error exists, we should set $\mu_0 = 0$ and set the variance Σ_0 to reflect our confidence in this guess. Using a frequentist point of view, Grubbs' indicates that what we call Σ_0 is simply the variance of the setup over many setup occurrences. Evidently, if historical information exists about previous batches produced, this can be used to estimate μ_0, Σ_0 in an "empirical Bayes form," and start the adjustment procedure accordingly. Related references to Bayesian setup adjustment are [5,20–22].

In Chapter 9, Pan discusses the setup adjustment problem in greater detail. Crowder's Ph.D. thesis [6] is an excellent reference for applications of Kalman filtering in process monitoring and control.

1.9 Bayesian Linear Regression and Process Optimization

Consider a normal linear regression model with p regressors $x_1, x_2, \ldots, x_p$, which themselves can be nonlinear transformations of underlying controllable factors. For example, $x_3 = x_1 x_2$, or $x_4 = \log(x_2)$, etc. We assume we have conducted N "experiments," an experiment consisting on observing the values of the p regressors together with the value of the response, y. In this section, we assume a single response is of interest. The N experimental conditions are gathered in a $N \times p$ matrix

$$\mathbf{X} = [x_{ij}],$$

which we will call the "design matrix" in what follows (this will include the actual $N \times k$ design matrix if k underlying factors are used). Put the observed response values in an $N \times 1$ vector $\boldsymbol{y}$. If the assumed normal model is valid, it should be valid for all N observations, so for each observation i, we can express the response as

$$y_i \,|\, \boldsymbol{\beta}, \mathbf{X} = \beta_1 x_{i1} + \beta_2 x_{i2} + \cdots + \beta_p x_{ip} + \varepsilon_i$$

where we assume $\varepsilon \sim \mathrm{N}(0, \sigma^2)$ and $\boldsymbol{\beta}$ denotes the $p \times 1$ vector of parameters. The response is then assumed to be the result of two effects: the first one, which we can explain, is due to the p regressors; the second one which we cannot attribute to any of the p factors, we thus model as a random variable with mean zero and constant variance. The two sources of uncertainty, due to not knowing the parameters and due to not knowing the intrinsic variability of the errors of the model, must be considered in any inference problem. This is achieved by using the Bayesian approach.

Based on this model, we have that

$$E[y_i \,|\, \boldsymbol{\beta}, \mathbf{X}] = \beta_1 x_{i1} + \beta_2 x_{i2} + \cdots + \beta_p x_{ip}$$

and

$$\mathrm{Var}(y_i|\boldsymbol{\beta}, \sigma^2, \boldsymbol{X}) = \sigma^2.$$

In many applications, x_1 is assumed to be 1, and the model has an intercept.[5] We can summarize the model by saying that

$$\boldsymbol{y}|\boldsymbol{\beta}, \sigma^2, \boldsymbol{X} \sim N(\boldsymbol{X\beta}, \sigma^2\boldsymbol{I})$$

where $\boldsymbol{y} = (y_1, \ldots, y_n)'$.

1.9.1 Analysis with a Non-Informative Prior

The non-informative prior for the parameters of this regression model is uniform in $(\boldsymbol{\beta}, \log\sigma)$, or

$$p(\boldsymbol{\beta}, \sigma^2|\boldsymbol{X}) \propto \frac{1}{\sigma^2}.$$

We assume both parameters are independent *a priori*. Note that we indicate that the design is given because we can design the experiment before deciding on the prior. Because of this, in regression cases it is also customary to define the prior taking advantage of our knowledge of the design. The resulting design is called a "g-prior" and was proposed by Zellner [34]. In this section we instead use non-informative priors.

The likelihood for N observations is

$$p(\boldsymbol{y}|\boldsymbol{\beta}, \sigma^2, \boldsymbol{X}) = \frac{1}{(2\pi)^{N/2}\sigma^N} e^{-\frac{1}{2\sigma^2}(\boldsymbol{y}-\boldsymbol{X\beta})'(\boldsymbol{y}-\boldsymbol{X\beta})},$$

so the posterior ($\propto$ prior $\times$ likelihood) is

$$p(\boldsymbol{\beta}, \sigma^2|\boldsymbol{y}, \boldsymbol{X}) = \frac{1}{(\sigma^2)^{N/2+1}} e^{-\frac{1}{2\sigma^2}(\boldsymbol{y}-\boldsymbol{X\beta})'(\boldsymbol{y}-\boldsymbol{X\beta})}. \tag{1.34}$$

The marginals for each parameter can be obtained by integration. For example, the marginal density for $\boldsymbol{\beta}$ is obtained from

$$\begin{aligned} p(\boldsymbol{\beta}|\boldsymbol{y}, \boldsymbol{X}) &\propto \int_0^\infty \frac{1}{(\sigma^2)^{N/2+1}} e^{-\frac{1}{2\sigma^2}(\boldsymbol{y}-\boldsymbol{X\beta})'(\boldsymbol{y}-\boldsymbol{X\beta})} d\sigma^2 \\ &\propto [(\boldsymbol{y}-\boldsymbol{X\beta})'(\boldsymbol{y}-\boldsymbol{X\beta})]^{-N/2}. \end{aligned}$$

[5] Models for mixture experiments usually have no intercept.

Add and subtract $X\widehat{\beta}$ in each parentheses, where $\widehat{\beta} = (X'X)^{-1}X'y$ makes the right-hand side equal to

$$
\begin{aligned}
&= [(y - X\widehat{\beta})'(y - X\widehat{\beta}) + (\beta - \widehat{\beta})'X'X(\beta - \widehat{\beta})]^{-N/2} \\
&= [(N-p)s^2 + (\beta - \widehat{\beta})'X'X(\beta - \widehat{\beta})]^{-N/2} \\
&= \left[1 + \frac{(\beta - \widehat{\beta})'X'X(\beta - \widehat{\beta})}{(N-p)s^2}\right]^{-N/2} \underbrace{[(N-p)s^2]^{N/2}}_{\text{a constant}} \\
&\propto \left[1 + \frac{(\beta - \widehat{\beta})'X'X(\beta - \widehat{\beta})}{(N-p)s^2}\right]^{-(N-p+p)/2} \underbrace{(s^2|X'X|^{-1})^{-1/2}}_{\text{another constant}},
\end{aligned}
$$

which has the form of a multivariate t distribution (see Appendix) with degrees of freedom $v = N - p$:

$$\beta|y, X \sim t_{N-p}(\widehat{\beta}, s^2(X'X)^{-1}).$$

Similarly, we can obtain the posterior marginal distribution of σ^2 by integrating Equation (1.34) over β, obtaining a scaled inverse chi-square distribution:

$$\sigma^2|y, X \sim Inv\ \chi^2(n-p, s^2).$$

The *posterior predictive density* can be shown to be equal [28] to

$$p(\tilde{y}|y) \propto \frac{1}{\left[1 + \frac{1}{N-p}\frac{(\tilde{y}-w'\widehat{\beta})^2}{s^2[1+w'(X'X)^{-1}w]}\right]^{(N+1-p)/2}},$$

which is a univariate Student t density (see Appendix):

$$\tilde{y}|y \sim t_{N-p}(w'\widehat{\beta}, s^2(1 + w'(X'X)^{-1}w)).$$

Note this implies that

$$\text{Var}(\tilde{y}|y) = \frac{N-p}{N-p-2}s^2(1 + w'(X'X)^{-1}w)$$

which is the sum of two components: the first, proportional to s^2, is due to intrinsic (sampling) variability; the second, proportional to $s^2(w'(X'X)^{-1}w)$, represents variance due to the uncertainty in the parameters.

Example

*Optimization of a single response process.*Consider now the use of the predictive density for optimization purposes. Peterson [27] first proposed this use in multiple response optimization and called it a Bayesian "reliability approach" to optimization (see also Peterson's contribution in Chapter 10 for a detailed presentation of this approach). The example data come from

TABLE 1.2

Data for a chemical experiment, from [28]

int.	x_1	x_2	x_1x_2	x_1^2	x_2^2	y_1 =yield	y_2 =viscosity
1	-1	-1	1	1	1	76.5	62
1	-1	1	-1	1	1	77	60
1	1	-1	-1	1	1	78	66
1	1	1	1	1	1	79.5	59
1	0	0	0	0	0	79.9	72
1	0	0	0	0	0	80.3	69
1	0	0	0	0	0	80	68
1	0	0	0	0	0	79.7	70
1	0	0	0	0	0	79.8	71
1	1.41421	0	0	2	0	78.4	68
1	−1.41421	0	0	2	0	75.6	71
1	0	1.41421	0	0	2	78.5	58
1	0	−1.41421	0	0	2	77	57

Montgomery [25], who gives a chemical experiment in which three responses are of interest: the yield of the process, the viscosity of the chemical, and the molecular weight. In this example, we will consider the viscosity response. Two underlying factors are controllable, the reaction time and the temperature, and we fit a quadratic polynomial. The $\boldsymbol{X}$ matrix containing the design (a rotatable CCD) and the observed responses are shown in Table 1.2. The process engineer wishes to know the operating conditions that would maximize the probability of the viscosity being between 62 and 68 units. To do this, we need to solve.

$$\max_{\sqrt{2} \leq x_1, x_2 \leq \sqrt{2}} \int_{62}^{68} p(\tilde{y}|y) d\tilde{y}$$

where the region over we wish to vary the two controllable factors ranges from $-\sqrt{2}$ to $\sqrt{2}$ in coded units, given the central composite design that was used. Because the objective function is not concave, MATLAB's `fmincon` nonlinear optimization solver is run from a set of initial points selected according to a random latin hypercube. For the viscosity response, we have that the highest probability found is 0.7305 at $w_1 = 0.0852$, $w_2 = 0.7845$, so the solution is well inside the experimental region.

Bayesian regression analysis based on a conjugate prior is discussed in Chapter 12 by Baba and Gilmour, who discuss the analysis of saturated experimental designs. The conjugate prior is similar to the Normal inverse Wishart described in the section on multivariate normal data. These authors conclude that the conjugate prior is quite inflexible to represent prior beliefs, and propose instead either the use of a finite mixture of densities as a prior or the use of non-conjugate priors and the corresponding MCMC analysis. Approaches to regression analysis with informative priors have the advantage of allowing optimization as in the previous example with little data, and to allow sequential optimization as more data is gathered from early on in an experiment. A successful commercial application of this idea is the Ultramax sequential optimizer software, developed by Moreno (see Moreno's description of this process optimization software in Chapter 11).

1.9.2 Comparison of the Bayesian Predictive Approach to Frequentist Predictions in Regression

In classical (frequentist) regression, a *prediction interval* is constructed from the pivot

$$\frac{e}{\sqrt{\widehat{\text{Var}}(e)}} = \frac{\tilde{y}|w - \widehat{y}|w}{s\sqrt{1 + w'(X'X)^{-1}w}} \sim t_{N-p}$$

where the point estimate for the next observation $\tilde{y}$ at point w is $\widehat{y}|w = w'\widehat{\beta} = \widehat{E}[\tilde{y}|w]$. That is, we take the prediction error, $\tilde{y}|w - \widehat{y}|w$, divided by its estimated standard error, which leads to a Student t distribution. This is used to set up a valid prediction interval on $\tilde{y}|w$, but notice that we still have that the distribution of the future observation is simply $\tilde{y}|w \sim N(w'\beta, \sigma^2)$. The t distribution so obtained is for the prediction error, where $\widehat{y}|w$ is random. The distribution of the pivot coincides with the predictive density of $\tilde{y}$ under a non-informative prior. However, the fundamental difference is that in the Bayesian approach, the probability measure is associated to $\tilde{y}|w$, not to $\widehat{y}|w = w'\widehat{\beta}$, which in the Bayesian setting is simply a constant. But a probability measure on $\tilde{y}$ that relates to our model is what we need to compute "reliabilities" (i.e., probabilities of conformance) of the form $P(\tilde{y}|\text{data}, w \in A)$, where A is a region defined by product specifications, as in the example. Therefore, no way exists to compute this type of probabilities from the prediction intervals (or regions, in case $\tilde{y}$ is multivariate), in the classical-frequentist approach.

1.10 Some General References in Bayesian Statistics

Several textbooks are devoted to Bayesian statistics. To close this chapter, we would like to provide our suggestions to general references that we have found particularly pedagogic (so this is indeed a subjective assessment). Three older but very clear expositions of Bayesian inference, which include most if not all of the topics discussed in this chapter, are the first edition of the book by Press [28], the book by Box and Tiao [3], and the book by Zellner [35]. The newer book by Press [29] includes modern developments, such as MCMC methods. One of the best recent references in this field is the book by Gelman et al. [11], which has a level of clarity not often encountered in textbooks. It is also, in our opinion, one of the best references on MCMC methods (see next chapter). The predictivist point of view, which we have emphasized in this chapter, is nicely discussed in the monograph by Geisser [10]. This book also includes an interesting chapter of Bayesian methods in feedback control and optimization problems.

The exposition in this chapter has not considered the important area of statistical decision theory, in which the Bayesian approach has played an important role. This is reflected in the books we suggest above. A now classic presentation of the decision theory approach in Bayesian statistics is the book by Berger [1], which is also an excellent reference on the foundations issues of Bayesian statistics compared to classical statistics.

Appendix

Transformation of Random Variables Theorem

Let X be a continuous random variable with density $p(x)$, and assume $Y = u(X)$ is a one-to-one transformation from $A = \{x : p(x) > 0\}$ to $B = \{y : p(y) > 0\}$ with inverse transformation $x = u^{-1}(y) = w(y)$. If the derivative $d/dy\ w(y)$ is continuous and nonzero in B, the density of Y is given by

$$p(y) = p(w(y)) \left| \frac{dw(y)}{dy} \right|.$$

(Scalar) Student t Density Function

$$p(t) = \frac{\Gamma((v+1)/2)}{\Gamma(v/2)\sqrt{v\pi}\sigma} \left(1 + \frac{1}{v}\left(\frac{t-\mu}{\sigma}\right)^2\right)^{-(v+1)/2}$$

where $E(t) = \mu = \text{mode}(t)$, $\text{Var}(t) = \frac{v}{v-2}\sigma^2\ (v > 2)$.

Multivariate Student t Density Function

A $q \times 1$ random vector $\boldsymbol{t}$ is distributed as a multivariate t if its density is

$$p(\boldsymbol{t}) = \frac{\Gamma((v+q)/2)}{\Gamma(v/2)v^{q/2}\pi^{q/2}} |\boldsymbol{\Sigma}|^{-1/2} \left(1 + \frac{1}{v}(\boldsymbol{t}-\boldsymbol{\mu})'\boldsymbol{\Sigma}^{-1}(\boldsymbol{t}-\boldsymbol{\mu})\right)^{-(v+q)/2}$$

where $E(\boldsymbol{t}) = \boldsymbol{\mu} = \text{mode}(\boldsymbol{t})$, $\text{Var}(\boldsymbol{t}) = \frac{v}{v-2}\boldsymbol{\Sigma}(v > 2)$.

Scaled Inverse χ^2 and Inverse Gamma Distributions

The Inv-$\chi^2(v_0, \sigma_0^2)$ (scaled inverse chi-squared) is the distribution of $\sigma_0^2 v_0^2 / \chi^2_{v_0}$, i.e., it is the inverse of a usual χ^2 distribution with v_0 degrees of freedom that is scaled by the quantity $\sigma_0^2 v_0^2$, hence its name. Its density is

$$p(\theta) = \frac{(\nu/2)^{\nu/2}}{\Gamma(\nu/2)} s^{\nu}\, \theta^{-(v/2+1)}\, e^{-\nu s^2/(2\theta)}, \quad \theta > 0,$$

which has mean $E(\theta) = \frac{\nu}{\nu-2}s^2$, $(\nu > 2)$, Mode$(\theta) = \frac{\nu}{\nu+2}s^2$ and Var$(\theta) = \frac{2\nu^2 s^4}{(\nu-2)^2(\nu-4)}$, $(\nu > 4)$.

The conjugate prior for the variance of a normal is the Inverse Gamma distribution ($IG(\alpha, \beta)$), whose density is

$$p(\theta) = \frac{\beta^{\alpha}}{\Gamma(\alpha)}\, \theta^{-(\alpha+1)}\, e^{-\beta/\theta}, \quad \theta > 0$$

with mean $E[\theta] = \frac{\beta}{\alpha-1}$, $(\alpha > 1)$, Mode$(\theta) = \frac{\beta}{\alpha+1}$, and Var$(\theta) = \frac{\beta^2}{(\alpha-1)^2(\alpha-2)}$, $(\alpha > 2)$.

The scaled inverse χ^2 distribution is then a particular IG$(\nu/2, \nu s^2/2)$ distribution, hence it is the conjugate prior distribution for the normal variance, as discussed in this chapter.

Inverse Wishart Distribution

This is the conjugate prior distribution for the covariance matrix of a multivariate normal distribution. It is the multivariate generalization of the scaled inverse χ^2. If $\mathbf{W}$ is a $p \times p$ positive definite matrix, then the IW density is

$$p(\mathbf{W}) = \left(2^{\nu p/2} \pi^{\,p(p-1)/4} \prod_{i=1}^{p} \Gamma\left(\frac{\nu+1-i}{2}\right)\right)^{-1} |\mathbf{S}|^{-\nu/2} |\mathbf{W}|^{(\nu-p-1)/2}\, e^{-\frac{1}{2}tr(S^{-1}\mathbf{W})}$$

with mean equal to $E[W] = \nu S$.

Key Results in Multivariate Statistics Used in Deriving the Kalman Filter

The following two results are used in deriving the Kalman filter expressions. A proof of them, which is not difficult, can be found in a book on multivariate statistics, such as Johnson and Wichern's [15].

Result 1. Let $\mathbf{X}_1$ be a q-dimensional random vector and $\mathbf{X}_2$ a $p-q$ dimensional random vector. If

$$\begin{pmatrix} \mathbf{X}_1 \\ \mathbf{X}_2 \end{pmatrix} \sim N_p\left(\begin{pmatrix} \boldsymbol{\mu}_1 \\ \boldsymbol{\mu}_2 \end{pmatrix}, \begin{pmatrix} \boldsymbol{\Sigma}_{11} & \boldsymbol{\Sigma}_{12} \\ \boldsymbol{\Sigma}_{21} & \boldsymbol{\Sigma}_{22} \end{pmatrix} \right)$$

and $|\boldsymbol{\Sigma}_{22}| > 0$, then the marginal distribution of $\mathbf{X}_1$ given $\mathbf{X}_2$ is given by

$$\mathbf{X}_1|\mathbf{X}_2 = x_2 \sim N_q\left(\boldsymbol{\mu}_1 + \boldsymbol{\Sigma}_{12}\boldsymbol{\Sigma}_{22}^{-1}(x_2 - \boldsymbol{\mu}_2), \boldsymbol{\Sigma}_{11} - \boldsymbol{\Sigma}_{12}\boldsymbol{\Sigma}_{22}^{-1}\boldsymbol{\Sigma}_{21}\right).$$

Result 2. Let $\mathbf{X}_t$ be a $q \times 1$ vector and $\mathbf{X}_2$ be a $(p-q) \times 1$ vector. If

$$\mathbf{X}_1|\mathbf{X}_2 = x_2 \sim N_q\left(\boldsymbol{\mu}_1 + \boldsymbol{\Sigma}_{12}\boldsymbol{\Sigma}_{22}^{-1}(x_2 - \boldsymbol{\mu}_2), \boldsymbol{\Sigma}_{11} - \boldsymbol{\Sigma}_{12}\boldsymbol{\Sigma}_{22}^{-1}\boldsymbol{\Sigma}_{21}\right) \tag{1.35}$$

and $\mathbf{X}_2 \sim N_{p-q}(\boldsymbol{\mu}_2, \boldsymbol{\Sigma}_{22})$, then we have that the joint distribution of $\mathbf{X}_1$ and $\mathbf{X}_2$ is

$$\begin{pmatrix}\mathbf{X}_1\\ \mathbf{X}_2\end{pmatrix} \sim N_p\left(\begin{pmatrix}\boldsymbol{\mu}_1\\ \boldsymbol{\mu}_2\end{pmatrix}, \begin{pmatrix}\boldsymbol{\Sigma}_{11} & \boldsymbol{\Sigma}_{12}\\ \boldsymbol{\Sigma}_{21} & \boldsymbol{\Sigma}_{22}\end{pmatrix}\right).$$

References

1. Berger, J.O., *Statistical Decision Theory and Bayesian Analysis*, 2nd ed., New York: Springer-Verlag, 1985.
2. Box, G.E.P., Jenkins, G.M., and Reinsel, G.C., *Time Series Analysis, Forecasting, and Control*, 3rd ed., Englewood Cliffs, NJ: Prentice Hall, 1994.
3. Box, G.E.P. and Tiao, G.C., *Bayesian Inference in Statistical Analysis*, Reading, MA: Addison Wesley, 1973.
4. Carlin, B.P., Gelfand, A.E., and Smith, A.F.M., "Hierarchical Bayesian analysis of changepoint problems," *Applied Statistics*, 41, 2, 389–405.
5. Colosimo, B.M., Pan, R., and del Castillo, E., "A sequential Markov Chain Monte Carlo approach to setup process adjustment over a set of lots," *Journal Applied Statistics*, 31(5), 499–520, 2004.
6. Crowder, S.V., *Kalman Filtering and Statistical Process Control*, Ph.D. Diss., Dept. of Statistics, Iowa State University, 1986.
7. Crowder, S.V. and Eshleman, L., "Small sample properties of an adaptive filter applied to low volume SPC," *Journal of Quality Technology*, 33, 1, 29–46, 2001.
8. Duncan, D.B. and Horn, S.D., "Linear dynamic recursive estimation from the point of view of regression analysis," *Journal of the American Statistical Association*, 67, 815–821, 1972.
9. Del Castillo, E., Pan, R., and Colosimo, B.M., "An unifying view of some process adjustment methods," *Journal of Quality Technology*, 35(3), 286–293, 2003.
10. Geisser, S., *Predictive Inference: An introduction*, New York: Chapman & Hall, 1993.
11. Gelman, A., Carlin, J.B., Stern, H.S., and Rubin, D.B., *Bayesian Data Analysis*, 2nd ed., Boca Raton FL: Chapman & Hall/CRC Press, 2004.
12. Grubbs, F.E., "An optimum procedure for setting machines or adjusting processes," *Industrial Quality Control*, July, reprinted in *Journal of Quality Technology*, 1983, 15(4), 186–189.

13. Hoadley, B., "The quality measurement plan," *Bell System Technical Journal*, 60, 215–273, 1981.
14. Jeffreys, H., *Theory of Probability*, 3rd ed., London: Oxford University Press, 1961.
15. Johnson, R.A. and Wichern, D.W., *Applied Multivariate Statistical Analysis*, 2nd ed., Englewood Cliffs, NJ: Prentice Hall, 1988.
16. Kadane, J.B., Dickey, J.M., Winkler, R.L., Smith, W.S., and Peters, S.C., "Intercative elicitation of opinion for a normal linear model," *Journal of the American Statistical Association*, 75, 845–854, 1980.
17. Kalman, R.E., "A new approach to linear filtering and prediction problems," *ASME Journal of Basic Engineering*, 35–45, March 1960.
18. Kass, R.E. and Wasserman, L., "The selection of prior distributions by formal rules," *Journal of the American Statistical Association*, 91, 435, 1343–1370, 1996.
19. Laplace, P.S. Marquis de, *A Philosophical Essay on Probabilities*, New York: Dover, 1951.
20. Lian, Z., Colosimo, B.M., and del Castillo, E., "Setup error adjustment: Sensitivity analysis and a new MCMC control rule," to appear in *Quality & Reliability Engineering International*, 2006.
21. Lian, Z. and del Castillo, E., "Setup adjustment under unknown process parameters and fixed adjustment cost," *Journal of Statistical Planning and Inference*, 136, 1039–1060, 2006.
22. Lian, Z., Colosimo, B.M., and del Castillo, E., "Setup adjustment of multiple lots using a sequential Monte Carlo method," to appear in *Technometrics*, 2006.
23. Meinhold, R.J. and Singpurwalla, N.D., "Understanding the Kalman filter," *The American Statistician*, 37, 2, 123–127, 1983.
24. Meinhold, R.J. and Singpurwalla, N.D., "Robustification of Kalman filter models," *Journal of the American Statistical Association*, 84(406), 479–486, 1989.
25. Montgomery, D.C., *Design and Analysis of Experiments*, 5th ed., New York: Wiley, 2001.
26. Myers, R.H. and Montgomery, D.C., *Response Surface Methodology*, 2nd ed., New York: Wiley, 2002.
27. Peterson, J.P. "A Bayesian reliability approach to multiple response surface optimization," *Journal of Quality Technology*, 36(2), 139–153, 2004.
28. Press, S.J., *Applied Multivariate Analysis: Using Bayesian and Frequentist Methods of Inference*, 2nd ed., Malabar, FL: Robert E. Krieger Publishing, 1982.
29. Press, S.J., *Subjective and Objective Bayesian Statistics*, 2nd ed., New York: John Wiley & Sons, 2003.
30. Robbins, H., "An empirical Bayes approach to statistics," in *Proceedings of the Thrid Berkeley Symposium on Mathematical Statistics and Probability*, 1, Berkeley, CA, University of California Press, 157–164, 1955.
31. Smith, J.Q., "A generalization of the Bayesian steady forecasting model," *Journal of the Royal Statistical Society*, B, 41, 375–387, 1979.
32. Stigler, S.M., *The History of Statistics: The Measurement of Uncertainty before 1900.* Cambridge, MA: Belknap Press, 1986.
33. West, M. and Harrison, J., *Bayesian Forecasting and Dynamic Models*, 2nd. ed., New York: Springer Verlag, 1997.

34. Zellner, A., "On assessing prior distributions and Bayesian regression analysis with g-prior distributions," *Bayesian Inference and Decision Techniques: Essays in Honour of Bruno de Finetti*, eds. Goel, P.K. and Zellner, A., Amsterdam: North-Holland, 1986.
35. Zellner, A., *An Introduction to Bayesian Inference in Econometrics*, New York: Wiley, 1971.
36. Zilong, L., and del Castillo, E., "Adaptive deadband control of a process with random drift," Technical paper, Engineering Statistics Laboratory, Penn State University, 2005.

2

Modern Numerical Methods in Bayesian Computation

Bianca M. Colosimo
Dipartimento di Meccanica, Politecnico di Milano

Enrique del Castillo
Department of Industrial & Manufacturing Engineering, Pennsylvania State University

CONTENTS

ABSTRACT This chapter presents a general overview of methods developed in the past few decades for performing Bayesian analysis via simulation. These methods are particularly useful when the posterior distribution is not analytically tractable as in hierarchical models and in models involving non-conjugate priors. After introducing the general approach for computing posterior distribution via simulation, the first part of this chapter is devoted to simulation-based approaches that make use of independent samples, i.e., the Rejection Sampling, the Importance Sampling, and Sampling Importance Resampling approaches and the Sequential Monte Carlo method. The chapter then gives particular attention to the most widely used Markov Chain Monte Carlo (MCMC) methods, with special focus on the Metropolis-Hastings and the Gibbs sampler algorithm and to methods for checking the convergence of MCMC simulations. The use of MCMC is further illustrated with reference to a classic hierarchical model, the variance component model, where MCMC computation is performed using available software packages (WinBUGS and CODA, which runs under R). An introductory tutorial on the WinBUGS package is presented in the Appendix.

2.1 Introduction

The previous chapter emphasized that the major conceptual challenge in Bayesian analysis involves the selection of an adequate prior. However, the main practical problem is related to computational issues involved in calculating the posterior distribution.

Let $y = (y_1, y_2, \ldots, y_n)$ denote observed data, characterized by the sampling distribution $p(y|\theta)$, where $\theta = (\theta_1, \theta_2, \ldots, \theta_D)$ are the parameters (as in Chapter 1, we use vector notation only when the data observations themselves are multivariate). Before observing the data, information on the parameters is modeled by a prior distribution, $p(\theta)$. The basic step in the Bayesian learning process is updating information on the parameters given observed data, i.e., computing the posterior distribution $p(\theta|y)$ through Bayes' theorem:

$$p(\theta|y) = \frac{p(y|\theta)\,p(\theta)}{\int p(y|\theta)\,p(\theta)d\theta} = \frac{q(\theta|y)}{m(y)}, \tag{2.1}$$

where $q(\theta|y)$ is the unnormalized density (also indicated as $q(\theta|y) \propto p(\theta|y)$), and $m(y) = \int q(\theta|y)d\theta$ is the marginal distribution that does not depend on parameter θ and is thus considered as a constant because it depends only on observed data y.

Given the posterior distribution, inference problems in Bayesian analysis can be solved by computing integrals involving the posterior distribution:

$$J = E(f(\theta)|y) = \int f(\theta)p(\theta|y)d\theta, \tag{2.2}$$

where, for example [2]:

- if $f(\theta) = \theta$, Equation (2.2) allows us to compute the posterior mean.
- if $f(\theta) = (\theta - \mu)(\theta - \mu)'$, with μ denoting the posterior mean, Equation (2.2) allows us to compute the posterior variance covariance matrix.
- if $f(\theta) = I_C(\theta)$, with $I_C(\theta)$ denoting the indicator function of θ in C, Equation (2.2) allows us to compute the posterior probability of the event $\theta \in C$ (this is used when computing Bayes' factors).
- if $f(\theta) = p(\tilde{y}|\theta, y)$, with $\tilde{y}$ denoting a future observation, Equation (2.2) allows us to compute the posterior predictive distribution.

As is clear from Equation (2.1) and Equation (2.2), the main computational task in Bayesian statistics is related to calculating complicated integrals that frequently are high-dimensional. Computation of these integrals can be skipped for simple (non-hierarchical) problems in which conjugate analysis is assumed. Thus, integration plays a major role in Bayesian analysis, substituting the role played by optimization in classical statistics.

In the last few decades, integration difficulties have been greatly reduced due to the development of simulation-based approaches, which aim to generate random samples and use these samples to solve inference problems via Equation (2.2). In particular, Section 2.1 of this chapter describes some well-known approaches based on simulating independent samples, namely, the Rejection Sampling, the Importance Sampling and Sampling Importance Resampling methods and the Sequential Importance Sampling approach. Section 2.2 deals with Markov Chain Monte Carlo (MCMC) methods, in which random samples used to compute integrals like Equation (2.2) are generated using a Markov chain. This section focuses on the most important approaches for MCMC simulation: the Metropolis-Hastings algorithm and the Gibbs sampler. As will be clear from Section 2.2, a major issue when performing MCMC simulations relates to the convergence of the algorithms. To make correct inferences, samples should be drawn once the steady-state distribution of a Markov chain has been achieved. Hence, the final part of Section 2.2 presents algorithms for convergence check. Section 2.3 then illustrates how powerful MCMC methods are to handle hierarchical models. In particular, we show the application of the Gibbs sampler for making inferences in a special type of hierarchical model, the variance components model, using available software for MCMC simulation and convergence checks. Finally, the Appendix to this chapter provides an introduction and mini-tutorial on the use of the Win BUGS software package.

2.2 Simulation-Based Approaches with Independent Samples

2.2.1 Rejection Sampling (RS)

Assume initially we are in the admittedly comfortable position of being able to generate independent identically distributed (i.i.d.) draws θ^k, $k = 1, 2, \ldots, K$ from the posterior distribution $p(\theta|y)$. In this case, the generic Equation (2.2) can be approximated using Monte Carlo integration as follows:

$$J = E(f(\theta)|y) = \int f(\theta)p(\theta|y)d\theta \approx \widehat{J} \approx \frac{1}{K}\sum_{k=1}^{K} f(\theta^k).$$

Thanks to Monte Carlo integration, computation of integrals required for Bayesian inference translates in generating i.i.d. samples from the posterior distribution, which is also called *target density*.

Assume now we are unable to generate samples from the posterior distribution $p(\theta|y)$, and suppose that we are able instead to draw samples from another density $g(\cdot)$ — called the "proposal density" — that obeys the following condition:

$$p(\theta|y) \leq Mg(\theta) \quad \forall\theta \text{ in the support of } p.$$

This condition requires that the ratio (or importance ratio) $\frac{p(\theta|y)}{g(\theta)}$ has a known bound M for any possible value of θ. Note that $p(\theta|y)$ is seen as a function of the parameters θ.

The *Rejection Sampling* (RS) algorithm consists of generating candidate samples from the proposal density $g(\cdot)$ and accepting them as drawn from the target density with probability $\frac{p(\theta|y)}{Mg(\theta)}$.

Figure 2.1 shows how the algorithm works. The curve on the top is the approximating function $Mg(\theta)$, whereas the curve on the bottom is the target density $p(\theta|y)$. As can be observed, the required condition $p(\theta|y) \leq Mg(\theta)$ is satisfied $\forall\theta$. Once the candidate draw $\tilde{\theta}$ is generated from the proposal density $g(\cdot)$, the probability of accepting this draw is the ratio of the height of the lower curve to the height of the upper curve in $\tilde{\theta}$. To accept the draw according to this probability, we can generate a random draw u from a uniform distribution $U(0, 1)$, compute the quantity $uMg(\tilde{\theta})$ — which will be uniformly distributed in the interval $[0; Mg(\tilde{\theta})]$ — and accept the draw if $uMg(\tilde{\theta}) \leq p(\tilde{\theta}|y)$.

The RS algorithm can be summarized in algorithmic form as follows:

Step 1: Set iteration counter $k = 1$.

Step 2: Draw a candidate sample $\tilde{\theta}_k$ from $g(\cdot)$ and generate u from a uniform distribution U(0, 1).

Step 3: Accept the candidate sample $\tilde{\theta_k}$, i.e. $\theta_k := \tilde{\theta}_k$, if $u \leq \frac{p(\tilde{\theta}_k|y)}{Mg(\tilde{\theta}_k)}$.

Step 4: if $k < K$, increase k, i.e., $k := k + 1$ and return to Step 2.

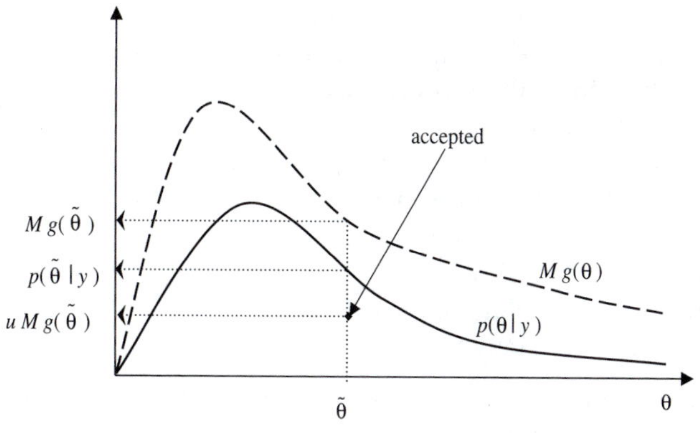

FIGURE 2.1
Illustration of the Rejection Sampling approach.

Following these steps, the accepted draws will be approximately distributed as the target density $p(\theta|y)$. In fact, we have that:

$$P(\theta \le c) = P\left(\tilde\theta \le c|U \le \frac{p(\tilde\theta|y)}{Mg(\tilde\theta)}\right) = \frac{P\left(\tilde\theta \le c, U \le \frac{p(\tilde\theta|y)}{Mg(\tilde\theta)}\right)}{P\left(U \le \frac{p(\tilde\theta|y)}{Mg(\tilde\theta)}\right)}$$

$$= \frac{\int_{-\infty}^{c}\int_{0}^{\frac{p(\tilde\theta|y)}{Mg(\tilde\theta)}} du g(\tilde\theta)d\tilde\theta}{\int_{-\infty}^{\infty}\int_{0}^{\frac{p(\tilde\theta|y)}{Mg(\tilde\theta)}} du g(\tilde\theta)d\tilde\theta} = \frac{\int_{-\infty}^{c}\frac{p(\tilde\theta|y)}{Mg(\tilde\theta)}g(\tilde\theta)d\tilde\theta}{\int_{-\infty}^{\infty}\frac{p(\tilde\theta|y)}{Mg(\tilde\theta)}g(\tilde\theta)d\tilde\theta}$$

$$= \frac{\frac{1}{M}\int_{-\infty}^{c} p(\tilde\theta|y)d\tilde\theta}{\frac{1}{M}\int_{-\infty}^{\infty} p(\tilde\theta|y)d\tilde\theta} = \int_{-\infty}^{c} p(\tilde\theta|y)d\tilde\theta.$$

The efficiency of this procedure directly depends on the bound M. The generated random draw will be accepted with probability $P(U \le \frac{p(\tilde\theta|y)}{g(\tilde\theta)}) = 1/M$, i.e., as a geometric distribution with mean $1/M$. As $M \to 1$, the efficiency of the algorithm increases.

An advantage of the RS algorithm is that it is a self-monitoring approach because the efficiency of the procedure can be directly monitored by computing the percentage of accepted draws. When this percentage is low, the bound M is not close to 1 and the resulting efficiency is poor.

A further advantage of the rejection sampling approach is that it does not require that the proposal density $g(\cdot)$ integrates to 1 (although it has to have a finite integral). Therefore, if the proposal density is chosen to be proportional to the posterior density, $g(\cdot) \propto p(\cdot|y)$, the algorithm can work very well accepting every draw with probability 1, given that the bound M is properly selected.

A last comment concerns the role of the posterior in the algorithm. In the illustrated version of the approach, it is assumed that we are unable to generate samples from the posterior distribution $p(\theta|y)$ but we are able to compute the posterior for a given value of θ in Step 3. This can appear as a limiting drawback. Fortunately, the algorithm can work by using the unnormalized density $q(\theta|y) \propto p(\theta|y)$, i.e., $p(\theta|y) = q(\theta|y) \int q(\theta|y)$, in place of the the target density $p(\theta|y)$.

2.2.2 Importance Sampling (IS) and Sampling Importance Resampling (SIR)

For many Bayesian problems in which the bound required by the Rejection Sampling approach can be difficult to find, an alternative approach is *Importance Sampling* (IS). To make posterior inferences through Equation (2.2), the Importance Sampling method is based on the following simple equalities:

$$J = E(f(\theta)|y) = \int f(\theta)p(\theta|y)d\theta$$

$$= \int \frac{f(\theta)p(\theta|y)}{g(\theta)} g(\theta)d\theta = \int f(\theta)w(\theta)g(\theta)d\theta, \tag{2.3}$$

where $g(\theta)$ is a proposal density that has to be chosen and $w(\theta) = \frac{p(\theta|y)}{g(\theta)}$ represents the *importance weight function*. Equation (2.3) can be computed through Monte Carlo integration as:

$$J = \int f(\theta)w(\theta)g(\theta)d\theta \approx \widehat{J} \approx \frac{1}{K}\sum_{k=1}^{K} f(\theta^k)w(\theta^k), \tag{2.4}$$

where $\theta^k, k = 1, 2, \ldots, K$ are i.i.d. draws generated from the proposal density $g(\theta)$, and the importance weights $w(\theta^k)$ can be computed as:

$$w(\theta^k) = \frac{p(\theta^k|y)}{g(\theta^k)} \quad \forall k = 1, 2 \ldots, K.$$

The Importance Sampling algorithm can be hence summarized in algorithmic form as follows.

Step 1: Generate K draws $\theta^1, \theta^2, \ldots, \theta^K$ from the proposal density $g(\theta)$.

Step 2: Compute the importance weights $w(\theta^k) = \frac{p(\theta^k|y)}{g(\theta^k)}$ for $k = 1, \ldots, K$.

Step 3: Use Equation (2.4) to compute the required posterior estimate.

Unlike the Rejection Sampling, the approximate or proposal density $g(\theta)$ must be normalized and should have a support that is included in the support of $p(\theta|y)$. The selection of a proper proposal density should also induce a "nice" behavior of weights. In fact, the approach will produce poor estimates if many weights assume small values and just a few weights have very high values.

If the target density $p(\theta|y)$ is only known up to a normalizing constant (as often happens in Bayesian analysis), the Importance Sampling algorithm can be used by slightly changing the described approach. Given a draw θ^k generated from the proposal density $g(\theta)$, assume we are able to compute $q(\theta^k|y)$ instead of $p(\theta^k|y)$, where $q(\theta|y) \propto p(\theta|y)$. In this case, the Importance Sampling algorithm works as before by simply substituting Equation (2.4) with

$$J = E(f(\theta)|y) \approx \widehat{J} \approx \frac{\frac{1}{K}\sum_{k=1}^{K} f(\theta^k)w(\theta^k)}{\frac{1}{K}\sum_{k=1}^{K} w(\theta^k)}, \tag{2.5}$$

where:

$$w(\theta^k) = \frac{q(\theta^k|y)}{g(\theta^k)} \quad \forall k = 1, 2 \ldots, K. \tag{2.6}$$

Rubin [22] suggests adding a resampling step to the procedure, thus proposing the *Sampling Importance Resampling* (SIR) approach. This further step is performed after Step 1 and Step 2 of the original procedure, starting from the K draws $\theta^1, \ldots, \theta^k, \ldots, \theta^K$ and their associated weights $w(\theta^1), \ldots, w(\theta^k), \ldots, w(\theta^K)$. The resampling step consists in selecting L draws from the discrete distribution $(\theta^1, \ldots, \theta^k, \ldots, \theta^K)$ with probability given by the importance weights. In algorithmic form, the SIR method consists of substituting only the third step of the original IS approach with the following substeps.

Step 3a: Sample a value in $\theta^1, \theta^2, \ldots, \theta^K$ where the probability of sampling θ^k is proportional to the weight $w(\theta^k)$.

Step 3b: Repeat sampling (Step 3a) without replacement $L - 1$ more times.

Step 3c: Compute the required posterior estimate, according to Equation (2.4), or Equation (2.5) if an unnormalized proposal density is being used.

As outlined by Gelman et al. [14], the resampling step is performed without replacement to improve the accuracy of the estimates. In fact, in the worst case of many small importance weights and very few large values, if we sample with replacement, we will end up selecting the same values of θ^k again and again.

Unlike RS, IS's and SIR's accuracy cannot be easily monitored. However, observing the histogram of the weights can help to outline situations in which the distribution of weights is bad behaved.

2.2.3 Sequential Importance Sampling (SIS) or Sequential Monte Carlo (SMC)

The *Sequential Importance Sampling* (SIS) or *Sequential Monte Carlo* (SMC) approach [8] is a sequential variant of the SIR algorithm. As SIR, the SIS method consists of generating a set of draws (also called "particles" in the context of sequential methods) from the proposal distribution of the unknown parameters and associating a weight to each set of particles. Unlike IR or SIR, the SIS method allows us to deal with data that becomes available sequentially, as in many real-life applications related with process control or process monitoring (for an application in data mining, see [1]). To better explain the motivation behind SIR, let $Y^t = (y_1, y_2, \ldots, y_t)$ denote all the data collected until time t. Assume that posterior estimates are required each time new data is available. At time t, the SIR approach requires us to perform all the mentioned steps (generating draws and computing their associated weights, resampling, performing the required estimates). When the new data y_{t+1} is collected, the SIR algorithm should start over from the beginning without using information (draws, weights, or estimates) obtained in the previous step. Unlike SIR, the SIS approach allows us to sequentially update weights each time a new observation is available.

To show how SIS works, consider the special case in which the importance sampling function $g(\theta)$ is the prior distribution $p(\theta)$. As showed by Smith and Gelfand [23], this resampling strategy simply means that more weight is given to prior samples that are more "likely" to happen. Because the unnormalized density $q(\theta|Y^t)$ is the product of the likelihood times the prior — see Equation (2.1) — at time t, the kth weight in Equation (2.6) can be rewritten as

$$w(\theta^k)_t = \frac{q(\theta^k|Y^t)}{g(\theta^k)} = \frac{p(Y^t|\theta^k)\,p(\theta^k)}{p(\theta^k)} = p(Y^t|\theta^k), \quad \forall k = 1, 2, \ldots, K.$$

In this case, the unnormalized weight associated with the kth particle can be computed as

$$w(\theta^k)_t = p(Y^t|\theta^k) = p(y_t|\theta^k)\,p(Y^{t-1}|\theta^k) = p(y_t|\theta^k)\,w(\theta^k)_{t-1}.$$

Therefore, the weight associated with the kth particle changes accordingly to the likelihood of observing y_t evaluated at each particle θ^k. This expression allows us to sequentially update the weights each time a new data is observed and is thus applicable in situations in which data arise sequentially.

Despite the benefits of sequential updating, this method poses some implementation difficulties in practice. The problem arises if the generated particles remain the same throughout all iterations and only the weights or the frequencies associated with the particles change. In this case, a phenomenon known as the "degeneracy" of the sample can arise. Degeneracy means that after some iterations of SMC, just a few of the K original particles will have weights greater than zero. In fact, when the K particles are generated from noninformative priors, most of these particles (the ones that are less likely, given the data observed) will have weights equal to zero after few iterations. In these cases, particles associated with weights greater than zero will be very few, say $K^* << K$. In other words, the sequential algorithm will be based on an "impoverished" sample, thus inducing biased estimates of unknown parameters. This phenomenon is particularly severe when the original prior distributions have large variances or high dimension, when the number of particles is small, or when the observed data set is large (i.e., when the number of iterations of the SMC method is large).

The degree of degeneracy due to an "impoverished" sample [1] can be monitored via the Effective Sample Size (ESS) [18], [21], which at time index t, is given by

$$ESS_t = \frac{K}{1 + var(w(\theta^k)_t)}, \tag{2.7}$$

where K is the original number of particles and $w(\theta^k)_t$ is the weight associated to particle k at time t. In Equation (2.7), the degree of degeneracy is described by the variance of weights $w(\theta^k)_t$ at time t. If at time t just few particles have associated weights greater than zero, the variance $var(w(\theta^k)_t)$ will be large and the ESS_t small.

To overcome the degeneracy problem, a "rejuvenation" step of the K particles can be performed using the One-pass Particle Filtering (1PFS) algorithm described in [1]. The rejuvenation step disperses the particles when ESS drops below a specific level, thus reducing the degeneracy problem. The rejuvenation step involves approximating the posterior densities of particles with a shrinkage kernel smoothing method.

2.3 Markov Chain Monte Carlo (MCMC)

Until now, all the simulation-based approaches presented (RS, IS, and SIR) assume that generating independent draws from the proposal density is possible.

The *Markov Chain Monte Carlo* (MCMC) methods [17] are instead based on drawing dependent samples that admit as stationary distribution the target density. In particular, the type of dependence considered is a Markov chain, i.e., the next state of the sample will depend just on the current state of the chain. As clear from the name, the approach is based on performing Monte Carlo integration on samples generated by a Markov chain. In fact, MCMC allows us to compute posterior inference of the type given in Equation (2.2) using the following approximation:

$$J = E(f(\theta)|y) \approx \widehat{J} \approx \frac{1}{K - k_0} \sum_{k=k_0+1}^{K} f(\theta^k), \tag{2.8}$$

where θ^k are dependent samples generated by a Markov chain whose stationary distribution is the target posterior density, and k_0 represents the end of the transient period (also called the "burn-in" period) at which the Markov chain has eventually reached its steady-state (i.e., draws generated after the k_0^{th} step can be assumed to be generated by the stationary distribution). As clear from Equation (2.8), the Monte Carlo integration step must be performed discarding samples generated from the transient state because these samples are not yet drawn from the target density. Therefore, one of the main concerns in MCMC is related to convergence issues and to methods aimed at deciding the iteration step at which convergence of the Markov chain can be considered achieved.

To perform MCMC, the generated Markov chain should satisfy the ergodicity property. *Ergodicity* is achieved when the chain is irreducible and aperiodic. Irreducibility means that any state can be reached from any other state in a finite number of moves. This property is required to let the chain forget the starting point, as any state can be reached from any starting point. The chain should be also aperiodic, which means that no cyclic pattern arises in the chain. This property allows us to avoid cyclic behavior of the generated states.

Therefore, the general steps required for computing posterior estimates by using MCMC can be summarized as follows:

Step 1: Start from arbitrary values for unknown parameters θ^0.

Step 2: Generate K draws from a Markov chain whose transition kernel is $T(\theta^k|\theta^{k-1})$ and which admits as stationary distribution the target density (i.e., the posterior density).

Step 3: Skip k_0 samples from the burn-in period (transient period) of the chain, and compute required estimates by using Equation (2.8).

As clear from this brief description, the core of any MCMC approach is the *transition kernel* or *transition density* $T(\theta^k|\theta^{k-1})$, which represents the probability of moving from one state θ^{k-1} of the chain to the next θ^k, thus defining the way in which a new state of the chain is obtained given the actual state. All the MCMC algorithms use the basic algorithmic structure described by Step 1 to Step 3 but basically change in the way in which the transition kernel in Step 2 is defined. In the following section the two most popular MCMC methods, the Metropolis-Hastings approach and the Gibbs Sampler, will be described.

2.3.1 Metropolis-Hastings (M-H) Algorithm

As outlined by Chib and Greemberg [6], traditional Markov chain theory starts from a specific transition kernel and focuses on determining conditions under which the invariant or stationary distribution exists and conditions under which the transition kernel converges to this invariant distribution. In MCMC methods, the problem is turned upside down: the stationary distribution is known (at least up to a constant multiple) and coincides with the posterior density we want to find, while the main problem is selecting a suitable transition kernel that converges to this stationary distribution. With reference to the Bayesian problem, the target density $p(\cdot)$ we want to sample from is the posterior distribution $p(\cdot|y)$; however, to keep notation simpler, we will skip dependence on data , i.e., $|y$, from this point on.

The problem of finding an appropriate transition kernel can be solved starting from the "detailed balance equation" or "reversibility condition." A Markov chain with transition kernel $T(\theta^{k+1}|\theta^k)$ satisfies this condition if a distribution $p(\cdot)$ exists such that:

$$T(\theta^k|\theta^{k+1})p(\theta^{k+1}) = T(\theta^{k+1}|\theta^k)p(\theta^k) \quad \forall(\theta^k, \theta^{k+1}). \tag{2.9}$$

If this result holds, the chain is reversible and the distribution $p(\cdot)$ is the stationary or invariant distribution because

$$\begin{aligned}\int T(\theta^{k+1}|\theta^k)p(\theta^k)d\theta^k &= \int T(\theta^k|\theta^{k+1})p(\theta^{k+1})d\theta^k \\ &= p(\theta^{k+1})\int T(\theta^k|\theta^{k+1})d\theta^k = p(\theta^{k+1}),\end{aligned}$$

which is the definition of an invariant distribution. The balance Equation (2.9) intuitively says that if two samples θ^k and θ^{k+1} are taken from the stationary distribution, the probability of moving from θ^k to θ^{k+1} is equal to the probability of moving in the opposite way, i.e., the chain is reversible. Therefore, if the transition kernel is selected to satisfy the reversibility condition in Equation (2.9), the stationary distribution of the chain will be the desired target density. Unfortunately, finding a chain that exactly satisfies the reversibility condition can be difficult: if an arbitrary proposal kernel density T' is selected, the balance equation might not hold. Without loss of generality, suppose that the selected proposal density T' is such that

$$T'(\theta^{k+1}|\theta^k)p(\theta^k) > T'(\theta^k|\theta^{k+1})p(\theta^{k+1}). \tag{2.10}$$

In this case, the chain moves from θ^k to θ^{k+1} too often and from θ^{k+1} to θ^k too rarely. To correct this asymmetry, a factor $\alpha(\theta^k, \theta^{k+1}) < 1$ can be used to reduce the number of moves from θ^k to θ^{k+1}, changing the balance equation as follows:

$$T'(\theta^{k+1}|\theta^k)\alpha(\theta^k, \theta^{k+1})p(\theta^k) = T'(\theta^k|\theta^{k+1})p(\theta^{k+1}). \tag{2.11}$$

Obviously, a symmetric factor $\alpha(\theta^{k+1}, \theta^k) = 1$ is implicitly considered on the right-hand side of Equation (2.11). The factor $\alpha(\theta^k, \theta^{k+1})$ can thus represent the probability of moving from θ^k to θ^{k+1} and can be derived from the previous equation as $\alpha(\theta^k, \theta^{k+1}) = \frac{T'(\theta^k|\theta^{k+1})p(\theta^{k+1})}{T'(\theta^{k+1}|\theta^k)p(\theta^k)}$. If the inequality in Equation (2.10) is reversed, i.e., the chain moves from θ^k to θ^{k+1} too rarely, we should set $\alpha(\theta^k, \theta^{k+1}) = 1$, i.e., allow with probability 1 this move, and derive $\alpha(\theta^{k+1}, \theta^k) < 1$ as before. Therefore, to build a kernel that satisfies the reversibility condition by construction, we should accept a move from θ^k to θ^{k+1} with probability

$$\alpha(\theta^k, \theta^{k+1}) = min\left\{\frac{T'(\theta^k|\theta^{k+1})p(\theta^{k+1})}{T'(\theta^{k+1}|\theta^k)p(\theta^k)}, 1\right\}. \tag{2.12}$$

This factor is thus playing the role of an acceptance ratio in the algorithm: given the actual state of the chain θ^k, a candidate sample $\tilde{\theta}^{k+1}$ is generated and accepted as the next state of the chain with probability $\alpha(\theta^k, \tilde{\theta}^{k+1})$, computed using Equation (2.12). If the candidate sample is not accepted, the chain will remain in the current state θ^k. Note also that this acceptance ratio resembles the acceptance probability used in the Rejection Sampling algorithm described in Section 2.2.1. Note that in that algorithm when a candidate sample is rejected, a new sample is drawn, whereas in the *Metropolis-Hastings* (M-H) algorithm, the current sample is taken as the next one. Given the simplified notation adopted in this section, the target density $p(\cdot)$ in Equation (2.12) is actually the posterior density, i.e., $p(\cdot|y)$. Fortunately, the unnormalized density $q(\cdot|y)$ introduced in Equation (2.1) can be used here in place of the target density $p(\cdot|y)$. In fact, the normalizing constant $-m(y)$ in Equation 2.1 - appears both in the numerator and in the denominator of Equation (2.12).

Back to the basic problem of generating draws from the posterior density $p(\theta|y)$, the M-H algorithm can be summarized in algorithmic form as follows:

Step 1: Initialize the chain by selecting an (arbitrary) starting point θ^0, and set the stage $k = 0$.

Step 2: At step k, draw a proposal sample $\tilde{\theta}^{k+1}$ from $T'(\tilde{\theta}^{k+1}|\theta^k)$.

Step 3: Accept the sample, i.e. $\theta^{k+1} := \tilde{\theta}^{k+1}$, with probability

$$\alpha(\theta^k, \tilde{\theta}^{k+1}) = min\left\{\frac{T'(\theta^k|\tilde{\theta}^{k+1})\,p(\tilde{\theta}^{k+1})}{T'(\tilde{\theta}^{k+1}|\theta^k)\,p(\theta^k)}, 1\right\};$$

otherwise, set $\theta^{k+1} := \theta^k$. To perform this acceptance step, generate a random draw u from U(0,1) if $u < \alpha(\theta^k, \tilde{\theta}^{k+1})$, then set $\theta^{k+1} := \tilde{\theta}^{k+1}$; otherwise, set $\theta^{k+1} := \theta^k$.

Step 4: If $k \leq K$, increase k, i.e., $k := k + 1$, and return to Step 2.

Step 5: Skip samples from the burn-in period of the chain, and compute the required estimates by using Equation (2.8).

As previously mentioned, the MCMC approach works well if the chain is irreducible and aperiodic. These conditions are usually verified if the proposal kernel density has a support included in the support of the target density. This simple requirement explains why the M-H algorithm has been so extensively applied. However, this condition does not assure a good efficiency (fast convergence) of the adopted algorithm, which depends on the selection of the transition kernel $T'(a, b)$. In the following, the more commonly used types of transition kernels to adopt in the M-H algorithm are described.

1. *Symmetric kernel*, for which $T'(a, b) = T'(b, a)$. In this case, computation of the acceptance ratio simplifies to $\alpha(a, b) = min\{\frac{p(b)}{p(a)}, 1\}$. A common selection in this case is $T'(a, b) = f(b - a)$, where f is a symmetric density, i.e., $f(-c) = f(c)$, such as a multivariate uniform centered at 0, a multivariate-t, or a multivariate normal with

mean equal to 0. In this last case, the candidate sample is given by $\boldsymbol{b} = N(\boldsymbol{a}, \Sigma)$, or equivalently by $\boldsymbol{b} = \boldsymbol{a} + N(\boldsymbol{0}, \Sigma)$, i.e., the candidate is given by the current sample plus random noise. This case is called a *Metropolis Random Walk.*

2. *Independent chain.* In this case, the candidate sample is generated independently from the actual state of the chain, i.e., $T'(a, b) = f(b)$. Again, f can be a multivariate normal or multivariate-t but we need to choose both the location and the spread of this density.

Further possible choices for the M-H algorithm are described in [6].

Example 1:

Simulating a bivariate normal using the M-H algorithm (adapted from [6] and [14]). To illustrate how the M-H algorithm works, consider a single observation $\boldsymbol{y} = (y_1\ y_2)'$ is observed from a bivariate normal distribution with unknown mean $\boldsymbol{\theta} = (\theta_1\ \theta_2)'$ and known covariance matrix $\Sigma = \begin{pmatrix} 1 & \rho \\ \rho & 1 \end{pmatrix}$. Using a uniform prior distribution on $\boldsymbol{\theta}$, the posterior distribution is is a bivariate normal given by

$$\begin{pmatrix} \theta_1 \\ \theta_2 \end{pmatrix} \Big| \mathbf{y} \sim N\left(\begin{pmatrix} y_1 \\ y_2 \end{pmatrix}, \begin{pmatrix} 1 & \rho \\ \rho & 1 \end{pmatrix} \right).$$

Although this example is trivial (simulating draws from the bivariate normal is quite an easy task), it can clearly show how the M-H algorithm can be implemented. Assume $\mathbf{y} = (1\ 2)'$ and $\rho = 0.9$. Using a symmetric transition kernel, the acceptance ratio in this case is given by

$$\alpha(\boldsymbol{a}, \boldsymbol{b}) = min\left\{ \frac{exp[-0.5(\boldsymbol{b} - \boldsymbol{y})'\Sigma^{-1}(\boldsymbol{b} - \boldsymbol{y})]}{exp[-0.5(\boldsymbol{a} - \boldsymbol{y})'\Sigma^{-1}(\boldsymbol{b} - \boldsymbol{y})]}, 1 \right\}.$$

Consider a Metropolis Random Walk transition density, $\boldsymbol{b} = \boldsymbol{a} + \boldsymbol{c}$, where $\boldsymbol{c}$ is a bivariate uniform, i.e., a bivariate distribution in which the ith component is uniformly distributed in $(-\delta_i, \delta_i)$, for $i = 1, 2$. In the performed simulations, we set $\delta_i = 1$ for $i = 1, 2$. The two plots in Figure 2.2 illustrate the behavior of four chains simulated with the M-H algorithm for 30 and 1000 iterations. Each chain is obtained from a different starting point. As can be observed, with just 30 iterations the chain has not reached the steady-state behavior because each chain has not yet "forgotten" the starting point. However, after 1000 iterations the chains seem to have reached the steady-state point because samples are drawn from the target distribution. The left plot in Figure 2.3 shows the final behavior of the M-H chain obtained using an initial starting point at (–5,–5), 10,000 iterations, and an initial burn-in equal to 2000 iterations (i.e., the plot is obtained considering just the last 8000 iterations of the chain). The plot on the right of Figure 2.3 shows the accepted (dot) and the rejected (cross) draws obtained with the M-H algorithm after burn-in has been already skipped (the acceptance probability obtained in this

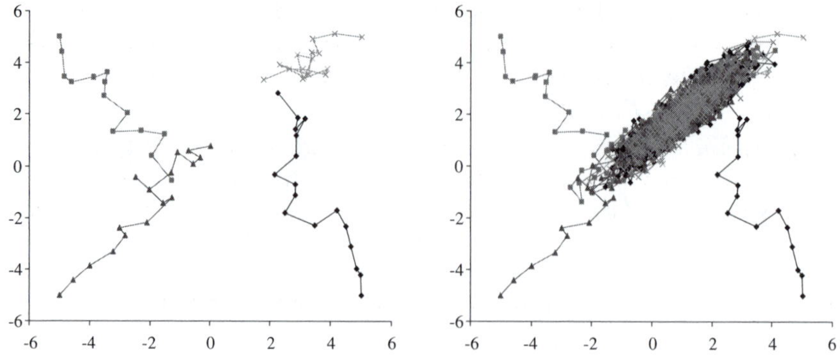

FIGURE 2.2
Four chains obtained with the M-H algorithm to simulate a bivariate normal using starting points (−5,−5), (−5,5), (5,−5) and (5,5); the plots show the first 30 (left) and the first 1000 (right) iterations.

simulation is around 47%). As can be observed, the M-H algorithm properly rejects candidate samples drawn far from the region in which the actual bivariate density has a higher likelihood.

2.3.2 Gibbs Sampling (GS)

One of the most powerful MCMC methods is the *Gibbs sampling* (GS), which has the relevant advantage of being almost independent of the number of parameters (and model's stages when hierarchical models as the one described in Section 2.4 are considered). As with any MCMC approach, the Gibbs sampler adopts the general algorithmic structure described at the end

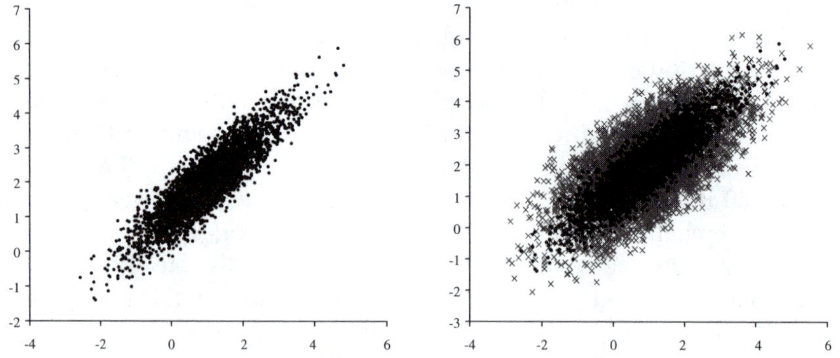

FIGURE 2.3
On the left: Scatter plot of samples obtained to simulate the bivariate normal with 10,000 iterations of the M-H algorithm and a burn-in equal to 2000. On the right: Scatter plots of accepted (dot) and rejected (cross) samples in the chain showed on the left.

of Section 2.3. The main difference consists in the specific transition kernel that allows us to move from one state to the next of the chain. To show how the Gibbs sampler works, consider the general problem of estimating D parameters $\theta_1, \theta_2, \ldots, \theta_D$ and assume that the "full conditional" (posterior) distributions,

$$p(\theta_d|y, \theta_{l\neq d}) = p(\theta_d|y, \theta_1, \theta_2, \ldots, \theta_{d-1}, \theta_{d+1}, \ldots, \theta_D) \quad d = 1, \ldots, D,$$

are all given. Given an arbitrary set of starting values $(\theta_1^0, \theta_2^0, \ldots, \theta_D^0)$, the first iteration in Gibbs sampling is performed as follows:

$$\begin{aligned} &\text{Draw} \quad \theta_1^1 \sim p(\theta_1|y, \theta_2^0, \ldots, \theta_D^0) \\ &\text{Draw} \quad \theta_2^1 \sim p(\theta_2|y, \theta_1^1, \theta_3^0, \ldots, \theta_D^0) \\ &\qquad \ldots \\ &\text{Draw} \quad \theta_D^1 \sim p(\theta_D|y, \theta_1^1, \ldots, \theta_{D-1}^1), \end{aligned}$$

thus obtaining as a result a set of points $(\theta_1^1, \theta_2^1, \ldots, \theta_D^1)$ that represent the starting values for the next step. Iterating the process and denoting with l the generic iteration of the algorithm, Geman and Geman [15] showed that the simulated draws can be viewed as generated from the joint posterior density as $l \to \infty$, i.e.,

$$(\theta_1^l, \theta_2^l, \ldots, \theta_D^l) \xrightarrow{d} (\theta_1, \theta_2, \ldots, \theta_D) \sim p(\theta_1, \theta_2, \ldots, \theta_D|y),$$

and hence draws obtained for each parameter θ_d ($d = 1, \ldots, D$) can be viewed as generated from the marginal posterior density as $1 \to \infty$ (see [9]), i.e.,

$$\theta_d^l \xrightarrow{d} \theta_d \sim p(\theta_d|y).$$

Therefore, the Gibbs sampling can be described in algorithmic form as follows:

Step 1: Initialize the chain by selecting an (arbitrary) starting point $(\theta_1^0, \theta_2^0, \ldots, \theta_D^0)$, and set the stage $k = 1$.

Step 2: At iteration k, performs the following D steps:

- Draw the sample θ_1^k from $p(\theta_1|y, \theta_2^{k-1}, \ldots, \theta_D^{k-1})$, i.e., the full conditional posterior.
- Draw the sample θ_2^k from $p(\theta_2|y, \theta_1^k, \theta_3^{k-1}, \ldots, \theta_D^{k-1})$.

  ...
- Draw the sample θ_D^k from $p(\theta_D|y, \theta_1^k, \ldots, \theta_{D-1}^k)$.

Step 3: If $k < K$, increase k, i.e., $k := k + 1$, and return to Step 2.

Step 4: Skip k_0 samples from the burn-in period of the chain, and compute required estimates by using Equation (2.8).

With reference to the general structure of the MCMC algorithms, the Gibbs sampler can be seen as a specific instance that uses as its transition kernel [2]:

$$T(\theta^k|\theta^{k-1}) = \prod_{d=1}^{D} p(\theta_D|y, \theta_j^k, \theta_l^{k-1};\ j < d, l > d)$$

To show why the Gibbs sampler works, let us start from a simple case in which the vector of parameters is composed by $D = 2$ parameters (θ_1, θ_2) [9]. The idea behind the Gibbs sampler is to extract the marginal posterior distributions $p(\theta_d|y)$ $(d = 1, 2)$ from the full conditional distributions, $p(\theta_1|y, \theta_2)$ and $p(\theta_2|y, \theta_1)$. To derive the marginal posterior distribution, we can compute two integrals:

$$p(\theta_1|y) = \int p(\theta_1|y, \theta_2)p(\theta_2|y)d\theta_2 \tag{2.13}$$

$$p(\theta_2|y) = \int p(\theta_2|y, \theta_1)p(\theta_1|y)d\theta_1 \tag{2.14}$$

The solution of each integral requires the solution of the other one, thus determining a system of two linear integral equations, given by Equation (2.13) and Equation (2.14). Substituting the last equation in the first one and for simplicity skipping the conditional dependence on the data (i.e., $|y$) in the notation, we obtain:

$$p(\theta_1) = \int p(\theta_1|\theta_2) \int p(\theta_2|\theta_1')p(\theta_1')d\theta_1' d\theta_2 = \int h(\theta_1, \theta_1')p(\theta_1')d\theta_1', \tag{2.15}$$

where $h(\theta_1, \theta_1') = \int p(\theta_1|\theta_2)p(\theta_2|\theta_1')d\theta_2$. Equation (2.15) is called a *fixed point integral equation*, because substituting for $p(\theta_1')$ in the right-hand side the correct density of the marginal posterior, we obtain it once again on the left, as θ_1' is just a "dummy" argument. If instead we substitute on the right-hand side a density $p_0(\theta_1)$ that is close to the real one, we will obtain on the left a different density that can be denoted as $p_1(\theta_1)$. Iterating this substitution process, Equation (2.15) can be rewritten as

$$p_{i+1}(\theta_1) = \int h(\theta_1, \theta_1')p_i(\theta_1')d\theta_1' \text{fori} = 0, 1, \ldots \tag{2.16}$$

Tanner and Wong [26] proved that under a mild regularity condition, the sequence thus generated, $\{p_i(\theta_1)\}$, converges monotonically to $p(\theta_1)$ and therefore an iterative approach to derive the true marginal density is provided. Unfortunately, the remaining difficulty consists in computing the integral in Equation (2.15) at each step of the iterative approach, which is often impossible to evaluate (otherwise no iterative approach will be needed). The problem can be overcome by replacing the analytical solution with a sampling-based substitution at each step of the iterative approach. In this case, at the first step we draw $\theta_1^0 \sim p_0(\theta_1)$ and then $\theta_2^1 \sim p(\theta_2|\theta_1^0)$, which is characterized by the marginal distribution

$$p_1(\theta_2) = \int p(\theta_2|\theta_1)p_0(\theta_1)d\theta_1. \tag{2.17}$$

The next draw $\theta_1^1 \sim p(\theta_1|\theta_2^1)$ has thus marginal distribution

$$p_1(\theta_1) = \int p(\theta_1|\theta_2)p_1(\theta_2)d\theta_2. \tag{2.18}$$

Substituting Equation (2.17) on the right-hand side of Equation (2.18), the last can be rewritten as

$$p_1(\theta_1) = \int p(\theta_1|\theta_2) \int p(\theta_2|\theta_1')p_0(\theta_1')d\theta_1'd\theta_2 = h(\theta_1, \theta_1')p_0(\theta_1')d\theta_1',$$

which is the recursive Equation (2.16) for $i = 0$. Therefore, iterating the approach, the sequences $\{\theta_1^i\}$ and $\{\theta_2^i\}$ are such that

$$\theta_1^i \xrightarrow{d} \theta_1 \sim p(\theta_1)\text{and}\,\theta_2^i \xrightarrow{d} \theta_2 \sim p(\theta_2).$$

The approach just presented is the "data augmentation" algorithm developed by Tanner and Wong [26] in the context of missing data and is closely related to the Gibbs sampler approach (as outlined in [9], the two approaches are identical if $D = 2$ parameters and are slightly different if the number of parameters is greater than 2).

As clear from the previous description, the main elements needed for implementing Gibbs sampling are the full (posterior) conditionals. Starting from the joint posterior distribution, the full conditional can be easily found by considering that

$$p(\theta_d|y, \theta_{r\neq d}) = \frac{p(\theta_d, \theta_{r\neq d}|y)}{\int p(\theta_d|y, \theta_{r\neq d})d\theta_d} \propto p(\theta_d, \theta_{r\neq d}|y).$$

Therefore, without considering the normalizing constant, the full conditional can be derived as the expression of the joint posterior distribution that contains the specific parameter.

Example 2:
Simulating a bivariate normal using Gibbs sampling. Back to the bivariate problem used to show the M-H algorithm in Example 1, assume we want to simulate samples from a bivariate normal $N(\boldsymbol{y}, \Sigma)$ where $\boldsymbol{y} = (y_1\ y_2)'$ and $\Sigma = \begin{pmatrix} 1 & \rho \\ \rho & 1 \end{pmatrix}$. As showed by Gelman et al. [14], this distribution represents the posterior when one sample $\boldsymbol{y}$ is observed from a bivariate normal $N(\boldsymbol{\theta}, \Sigma)$ with known Σ and with a uniform prior assumed for $\boldsymbol{\theta}$. As already mentioned, this is a trivial example as simulating draws from the bivariate (posterior) normal is quite easy. However, the example is useful to show how the Gibbs sampler can be implemented. To apply the Gibbs sampler to $\boldsymbol{\theta} = (\theta_1, \theta_2)'$, Gelman et al. [14] show that the conditional posterior distribution can be derived from properties of multivariate normal distributions (discussed in Chapter 1 in the derivation of Kalman filters) and are given by

$$\theta_1|\theta_2, \boldsymbol{y} \sim N(y_1 + \rho(\theta_2 - y_2), 1 - \rho^2)$$
$$\theta_2|\theta_1, \boldsymbol{y} \sim N(y_2 + \rho(\theta_1 - y_1), 1 - \rho^2).$$

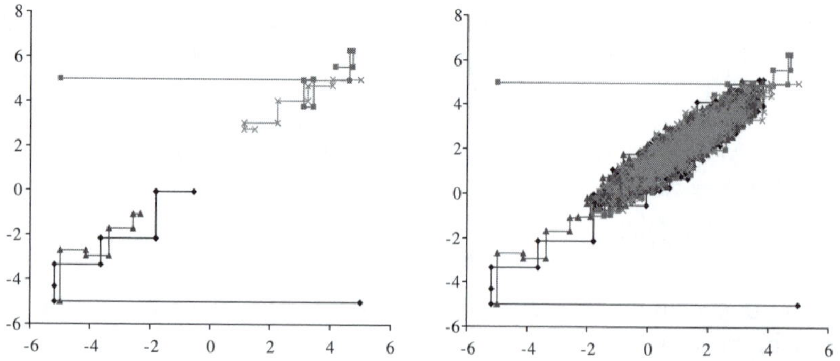

FIGURE 2.4
Draws obtained for the bivariate normal example using four independent chains of the Gibbs sampler with 5 (left) and 500 (right) iterations of the algorithm (each iteration is composed by two moves of the chain to update the two components of θ) and starting points (−5,−5), (−5,5), (5,−5) and (5,5).

Therefore, the Gibbs sampler can be simply implemented and consists of drawing samples alternatively from the conditional posterior densities. Assuming $\boldsymbol{y} = (1\ 2)'$ and $\rho = 0.9$, results obtained using the Gibbs sampler are shown in Figure 2.4 and Figure 2.5. In particular, Figure 2.4 reports samples obtained with 5 (left plot) and 500 (right plot) iterations of the algorithm. Note that each iteration consists in two steps: The first step is used to update the first component of $\boldsymbol{\theta}$ and the second one to update the second component of $\boldsymbol{\theta}$.

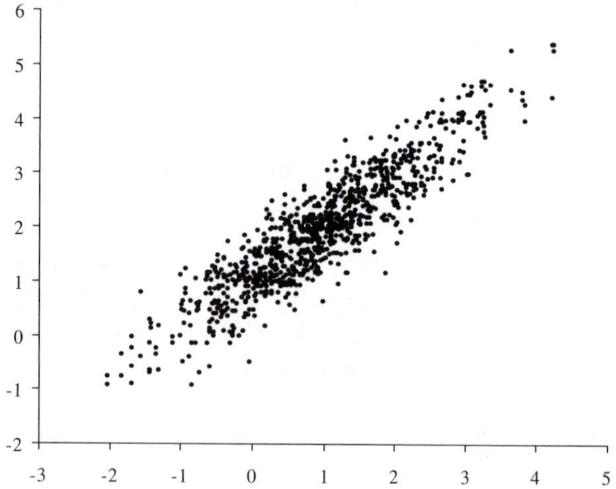

FIGURE 2.5
Samples drawn form the posterior distribution using the Gibbs sampler for the bivariate normal example (obtained by skipping the first 500 iterations as burn-in and using a further set of 4000 iterations).

Figure 2.5 reports samples obtained at the steady state of the Markov chain (in this case, obtained in 4000 iterations of the Gibbs sampler after an initial burn-in of 1000 iterations has been discarded).

2.3.3 Convergence Diagnostics in MCMC Approaches

A complete implementation of a "MCMC method requires dealing with convergence issues. In particular, specifying the k_0 (truncation point) and K (stopping point) in Equation (2.8) is necessary. The k_0 variable, often called the burn-in period, is constituted of observations that must be discarded before computing moments of the posterior densities. This burn-in period represents the transient period in which the Markov chain has not yet reached stability and is thus highly influenced by starting points (initial values of the parameters that must be estimated).

The second variable, K, is the total number of iterations and is required to determine the additional number of samples $K - k_0$ that must be drawn after convergence has been reached to compute all relevant moments of the posterior distribution of Equation 2.8. This variable is critical due to the Markov nature of the algorithm. If convergence has been reached, all the samples are identically distributed. However, because the samples are autocorrelated, the slower the simulation algorithm is moving within the sample space, the higher the number of samples required to efficiently obtain the required estimates.

Although typical choices adopted in the literature ($k_0 = 1000$ and $K = 10000$) have shown to be appropriate for many applications, convergence diagnostic algorithms should be used to evaluate if these values are satisfactory for the specific problem under study. Convergence diagnostic tools analyze outputs produced by MCMC simulations to decide whether convergence has been reached with the adopted settings. Numerous studies [4, 7] reviewed methods proposed in the literature for convergence assessment. At this time, no diagnostic algorithm seems globally superior to all others, and the recommended approach is applying simultaneously more than one diagnostic method to assess convergence. In the following the most commonly used approaches will be described. These algorithms are implemented in the CODA library [19], which runs under R, the freely available version of S-plus.

2.3.3.1 Raftery and Lewis' Convergence Diagnostic

The algorithm proposed by Raftery and Lewis [20] is aimed at calculating the length of the burn-in period and the whole number of iterations required to estimate a given quantile of the posterior distribution of parameters of interest. This convergence approach is based on a single run of a Markov chain. In particular, the method requires as input the quantile Q that must be estimated, the accuracy of this estimate r, and a probability s, i.e., the estimated quantile should lie within $\pm r$ of the true value with probability s. Default values suggested by Raftery and Lewis are $Q = 0.025$, $r = 0.005$ and $s = 0.95$.

2.3.3.2 Geweke's Convergence Diagnostic

Geweke [16] proposed a convergence diagnostic that is based on considering two subsets of a single chain (usually the first 10% and the last 50%) in which burn-in has already been skipped. Given the two subsequences of samples, the method is based on testing the assumption that the mean of samples drawn from the first subset of the chain is equal to the mean of samples drawn from the last part of the chain. If the assumption of equality of the means can not be rejected, samples should be drawn from the same (stationary) distribution and this means that convergence has already been achieved. To test the assumption, Geweke suggests using the following statistics:

$$Z = \frac{\bar{\theta}_A - \bar{\theta}_B}{\sqrt{\frac{1}{n_A}\hat{S}_\theta^A(0) + \frac{1}{n_B}\hat{S}_\theta^B(0)}},$$

where n_A and n_B denote the number of samples used in the first and in the last part of the chain, $\bar{\theta}_A$ and $\bar{\theta}_B$ are the sample means of θ obtained using samples drawn from the first and the last part of the chain; and $S_\theta^A(0)$ and $S_\theta^B(0)$ denote estimates of the spectral density at zero (i.e., variance estimates). The Z statistic is asymptotically distributed as a standard normal.

2.3.3.3 Gelman and Rubin's Convergence Diagnostic

Methods based on a single simulated chain can have the main drawback of masking excessively slow convergence of Markov chain simulation. Looking at one single chain, convergence could seem achieved although the chain is stuck in one place of the target distribution. To overcome this problem, Gelman and Rubin [13] (see also [11]) suggest performing multiple simulations, starting from different initial values for parameters that must be estimated. These values must be chosen to be "overdispersed" with reference to the target density. Because the target density is not known in advance, this technique has been criticized for the lack of guidelines about selecting the starting points. However, this choice can be performed considering basic knowledge of the problem under study. For example, a rough idea of the process capability can give information on the range of variation (at least its magnitude) of the quality characteristic, and this information can be easily used in determining overdispersed initial values for parameters derived through MCMC. Once starting values are selected, the Gelman and Rubin algorithm is based on mixing simulated chains and comparing the variance within each chain to the total variance of the mixture of chains. These (estimated) variances permit us to derive the "estimated potential scale reduction" factor, or "shrink factor," $\hat{R}$. As simulations converge, $\hat{R}$ declines to 1, thus assessing that parallel chains are essentially overlapping. The rule of thumb suggested when performing this diagnostic is to continue simulation until $\hat{R}$ is close to 1, for example, lower than 1.2.

2.4 Bayesian Computation in Hierarchical Models

Many real problems are characterized by different unknown parameters often related to each other by a "structure," depending on the specific problem addressed. *Hierarchical models* (HM) are suitable to deal with statistical analysis for this type of complex multivariate models. Some typical problems that can be studied with a HM framework are:

1. Hierarchically designed experiments, when observed data are drawn from clusters that are nested on each other as in random effects (variance components) or mixed models analysis. As an example, consider the case in which parts are produced in small lots.
2. Meta-analysis, where the objective is to combine results coming from different studies that share the same inferential problem. The statistical analysis should therefore account for extra variability due to the different conditions in which each experiment has been performed while simultaneously trying to merge evidence arising from each single experiment. Most of the time, ideal conditions that should characterize experiment execution cannot be respected (all the experiments performed on the same machine, in the same conditions, and soon). Therefore, a meta-analysis can result in a significant reduction in experimental costs and time.
3. When the assumption of (conditional) independence can be considered an unrealistic simplification and therefore all the data cannot be considered to be drawn from a single population. Most of the time, heterogeneity describes real data. In this case, using mixture models and latent variables in a hierarchical structure is common practice.
4. Model selection and representation of model uncertainty. In principle, when different models can represent the collected data, a further stage in a hierarchical structure can be added to represent models under study. In this case, the statistical analysis can be conducted for all the models at once.

Hierarchical models can be seen as a 'multi-stage' extension of Bayes' Theorem. Assume the observed data $y = (y_1, y_2, \ldots, y_t)$ are characterized by a probability distribution $f(y|\theta)$, where $\theta = (\theta_1, \theta_2, \ldots, \theta_K)$ is the vector of unknown parameters. Instead of considering these parameters as a set of unknown constants, as in a frequentist approach, the Bayesian point of view is based on modeling these quantities as random variables. The unknown parameters are *a priori* supposed to be distributed according to a prior distribution $p(\theta|\eta)$, where η is a vector of unknown *hyperparameters*. Inference on

θ is then based on the posterior distribution, given by

$$p(\theta|\eta, y) = \frac{p(y|\theta)p(\theta|\eta)}{\int p(y|u)p(u|\eta)du}.$$

Considering each conditional model (data given the parameters and parameters given the hyperparameters) as a stage in a hierarchy, the basic approach to Bayesian inference can be seen as a two-stage hierarchical model in which:

- *First stage:* composed by the distribution of data given the parameters $p(y|\theta)$.
- *Second stage:* composed by the distribution of the parameters given the hyperparameters $p(\theta|\eta)$.

Because the prior itself depends on a set of hyperparameters η, a further stage can be added assuming that this set of unknown quantities is itself a vector of random variables, characterized by an *hyperprior* (or more generally a *second stage prior*) distribution, thus deriving a third stage in the hierarchy:

- *Third stage:* Distribution of the hyperparameters $p(\eta)$.

This practice of specifying a model over more than one level in a hierarchy can be theoretically extended to a desired number of levels. The choice of the number of levels is strictly related to the problem addressed, although most of the applications require a three-level model [5].

Despite their flexibility, hierarchical models have been applied just to simple problems because of computational difficulties associated to posterior estimates. The advent of modern computational methods, first of all MCMC approaches, allowed us to overcome these difficulties, as will be shown in the next section.

2.4.1 Variance Components Model

A simple hierarchical model is the well-known and frequently adopted *random effects* or *variance components* model. As an example, suppose the data are obtained in lots or batches (or different experiments) and the objective of the analysis is to find the variance components, i.e., the amount of variation imputable to changes from lot to lot (between-lots or between-batches variability) versus the one occurring within lots (within-lot or within-batch variability). As an example, the first variance component (between-lots) can be related to changes in raw materials or processing conditions, whereas the second one (within-lot) is mainly due to the inner (or natural) variability related with the process and the measuring system. In this case, the model characterizing data observed is given by

$$y_{ij} = \theta_i + \varepsilon_{ij} \quad (i = 1, \ldots, I; \quad j = 1, \ldots, J),$$

where, assuming conditional independence throughout:

- $i = 1, \ldots, I$ is the index denoting lots and $j = 1, \ldots, J$ is the index of parts in each lot;
- y_{ij} is the characteristic observed in the jth part of the ith lot;
- $\boldsymbol{\theta} = (\theta_i,\ i = 1, \ldots, I)$ is a vector of normally distributed random variables,

$$\theta_i | \mu, \sigma_\theta^2 \sim N(\mu, \sigma_\theta^2), \tag{2.19}$$

where in the "pure" random effect model μ is assumed equal to 0;
- ε_{ij} are normally distributed independent random variables,

$$\varepsilon_{ij} | \sigma_\varepsilon^2 \sim N(0, \sigma_\varepsilon^2).$$

To show the modeling flexibility offered by hierarchical models, assume that a further objective of the analysis is to distinguish between variability due to the production process and variability due to the measuring system. By performing repeated measures of the quality characteristic on parts processed, a stage can be added in the hierarchy and a further variance component can be estimated. As outlined by the example, a strict connection offer exists between the "design" of the experiment (how many measures must be taken on the same characteristic, how many parts in a lot, and so on.) and the following statistical analysis.

The model specified until now is quite general and is independent of the adoption of a frequentist or a Bayesian point of view. It is characterized by the problem of estimating the variance components σ_θ^2 *and* σ_ε^2 from the data. Box and Tiao [3], Gelman et al. [14], and Carlin and Louis [5] contrasted random effects model estimation using classical and Bayesian analyses of variance. Using a classical approach, the variance between lots σ_θ^2 is obtained as the difference of the between- and within-batch mean squares divided by the number of batches. In cases where the between-batch mean square is lower than the within-batch mean square, this leads to the unsatisfactory situation of a negative variance estimate. Furthermore, the classical approach induces difficulties in computation of the confidence interval for this variance component σ_θ^2 because of the complicated sampling distribution. Bayesian estimation of variance components allows us to overcome these difficulties [3].

To estimate unknown parameters in the random effects model, the Bayesian approach requires us to assume a set of prior distributions. A common choice for the random effect model [9, 10] is adopting conjugacy at each step of the hierarchical model, thus assuming:

$$\begin{aligned}
\mu | \mu_0, \sigma_0^2 &\sim N(\mu_0, \sigma_0^2) \\
\sigma_\theta^2 | a_1, b_1 &\sim IG(a_1, b_1) \\
\sigma_\varepsilon^2 | a_2, b_2 &\sim IG(a_2, b_2),
\end{aligned}$$

where $\mu_0, \sigma_0^2, a_1, b_1, a_2$, and b_2 are assumed known and IG represents an Inverse-Gamma distribution. A vague prior for μ can be assumed by setting σ_0^2 sufficiently great, e.g., $\sigma_0^2 \geq 10^4$. Parameters of the Inverse-Gamma

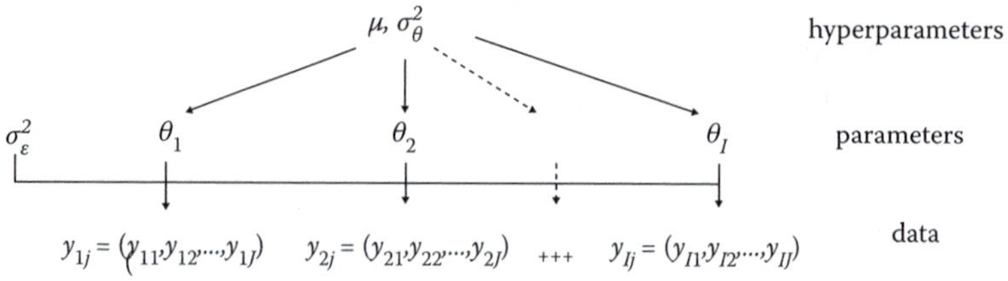

FIGURE 2.6
The random effects hierarchical model.

a_1, b_1 (or a_2, b_2) can be assumed equal to 0 to model the usual improper prior for σ_θ^2 (σ_ε^2). A common choice for a_1, b_1 (or a_2, b_2) is to set both equal to 0.001 to model "vague" prior information [24]. Here we point out that this usual "vague" IG prior has been recently criticized as actually being informative, and recommendations have been given for using some other distribution family as a prior in the variance components [12, 25]. These alternative choices can lead to better results (i.e., a smaller posterior uncertainty concerning the variance components). However, in this chapter we will describe the standard approach adopting Inverse-Gamma priors for both variance components [9, 10].

The random effects model can be described by three, partially 'nested' stages: data, parameters, and hyperparameters as described in Figure 2.6. After the observations $y = (y_{ij},\ i = 1, \ldots, I;\ j = 1, \ldots, J)$ have been collected, inference on the unknown parameters must be performed. Due to conjugacy adopted at each step, all the (full) conditional posterior required to perform Gibbs sampling can be derived as follows [10]:

$$p(\theta_i | y, \mu, \sigma_\theta^2, \sigma_\varepsilon^2) = N\left(\frac{J\sigma_\theta^2}{J\sigma_\theta^2 + \sigma_\varepsilon^2}\bar{y}_i + \frac{\sigma_\varepsilon^2}{J\sigma_\theta^2 + \sigma_\varepsilon^2}\mu, \frac{\sigma_\theta^2\sigma_\varepsilon^2}{J\sigma_\theta^2 + \sigma_\varepsilon^2}\right), i = 1, 2, \ldots, I$$

$$p(\mu | y, \boldsymbol{\theta}, \sigma_\theta^2, \sigma_\varepsilon^2) = N\left(\frac{\sigma_\theta^2\mu_0 + \sigma_0^2\sum_i \theta_i}{\sigma_\theta^2 + I\sigma_0^2}, \frac{\sigma_\theta^2\sigma_0^2}{\sigma_\theta^2 + I\sigma_0^2}\right)$$

$$p(\sigma_\theta^2 | y, \boldsymbol{\theta}, \mu, \sigma_\varepsilon^2) = IG\left(a_1 + \frac{1}{2}I, b_1 + \frac{1}{2}\sum_i(\theta_i - \mu)^2\right)$$

$$p(\sigma_\varepsilon^2 | y, \boldsymbol{\theta}, \mu, \sigma_\theta^2) = IG\left(a_2 + \frac{1}{2}IJ, b_2 + \frac{1}{2}\sum_i\sum_j(y_{ij} - \theta_i)^2\right)$$

where $\bar{y}_i = \frac{\sum_j y_{ij}}{J}$ and $\boldsymbol{\theta} = (\theta_1, \theta_2, \ldots, \theta_I)$.

To show how Gibbs sampling works in the variance component models, consider the data reported in Table 2.1. This data were presented in [3] in the

TABLE 2.1

Dyestuff Data y_{ij} for $i = 1, \ldots, 6$ and $j = 1, \ldots, 5$ (Yield of Dyestuff in Grams of Standard Color Computed with Respect to a Target Value of 1400 g) [3]

Batch	Dyestuff data				
1	1545	1440	1440	1520	1580
2	1540	1555	1490	1560	1495
3	1595	1550	1605	1510	1560
4	1445	1440	1595	1465	1545
5	1595	1630	1515	1635	1625
6	1520	1455	1450	1480	1445

framework of Bayesian estimation in one-way random effects model. Data in Table 2.1 refers to the yield of dyestuff measured in five samples taken from six batches of raw material.

The MCMC simulation is coded in the WinBUGS (**B**ayesian inference **U**sing **G**ibbs **S**ampling) language [24], which we introduce in the Appendix to this chapter. The random effects model is one of the examples in Volume 1 of the Bugs and WinBUGS manuals [24]. The convergence diagnostic algorithms used (Raftery and Lewis', Geweke's, and Gelman and Rubin's convergence diagnostics) are run under R — the freely available version of S-plus —using the CODA library [19]. Two MCMC chains are simulated starting from different initial values of the unknown parameters (μ, σ_θ^2, σ_ε^2, and θ_i, where i=1,...,6). Among these parameters, the θ_i's can be randomly generated by the software, using Equation (2.19). Initial values used in these examples are reported in Table 2.2. To see how the effect of the starting values disappears as the MCMC simulation proceeds, consider the plots shown in Figure 2.7. They represent the first 2000 iterations of the MCMC chains estimating μ, σ_θ^2 and σ_ε^2, given data reported in Table 2.1. As can be observed, the two chains seem to have almost forgotten the starting point after 1000 iterations. Therefore, a first guess for the burn-in can be set to $k_0 = 2000$. The total number of runs K is set to 100, 000 because of the high autocorrelation observed within each chain. Given this first attempt for the burn-in and the total length of the Markov chain, a convergence check can be performed using all the methods described.

Figure 2.8 shows the Z-scores obtained using the Geweke statistics for the two chains starting from the overdispersed starting points in Table 2.2. Z-scores outside the two bands at ± 2 denote that the test of equality

TABLE 2.2

Initial Values Adopted in the Two MCMC Chains for the Random Effects Model

Chain	μ	σ_θ^2	σ_ε^2
First	2000	1	1
Second	1000	0.01	0.01

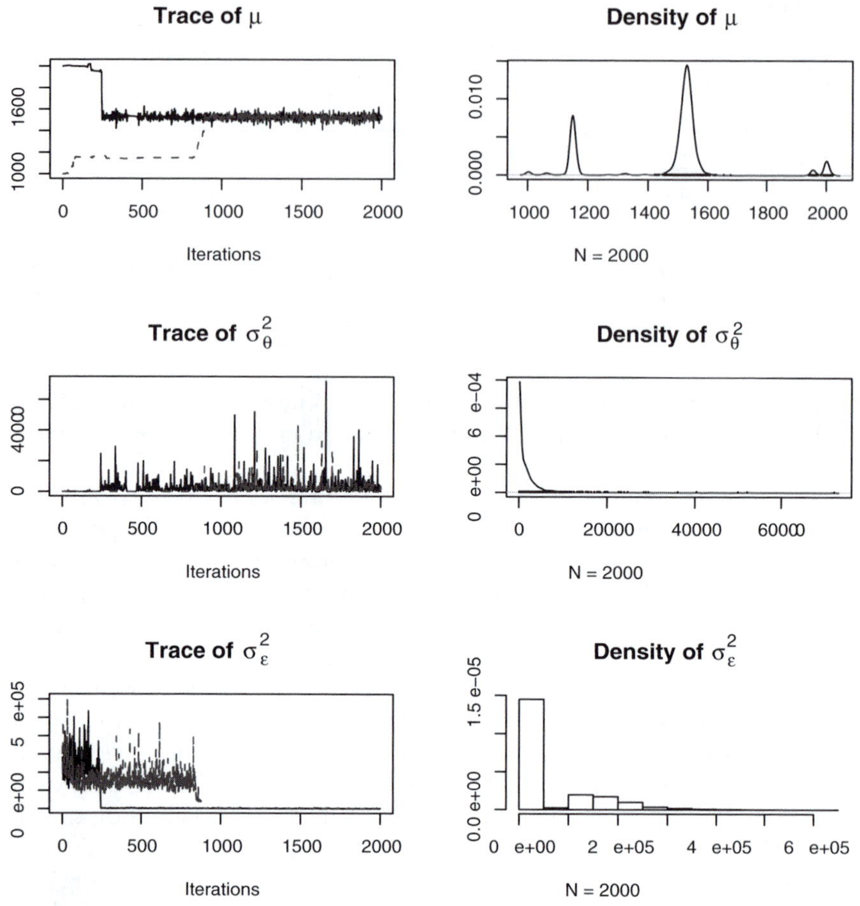

FIGURE 2.7
First 2000 iterations of the two MCMC chains simulating μ, σ_θ^2, and σ_ε^2 in the random effect model, given data in Table 2.1.

of means failed when the corresponding burn-in (reported on the abscissa) is adopted. These Z-scores show that burn-in lower than 20, 000 can cause problems, especially for convergence of the variance components σ_ε^2 and σ_θ^2.

Table 2.3 shows results obtained using the Raftery and Lewis diagnostic for the two chains used for the random effect model. As can be observed, values reported for the burn-in period are quite small. Therefore, convergence is basically achieved using the current burn-in truncation point of $k_0 = 2000$, according to this diagnostic check. However, the total number of draws suggested for estimating σ_θ^2 is almost double than the one adopted ($K = 10, 0000$). The requirement for longer chains is due to the high autocorrelation (i.e., the

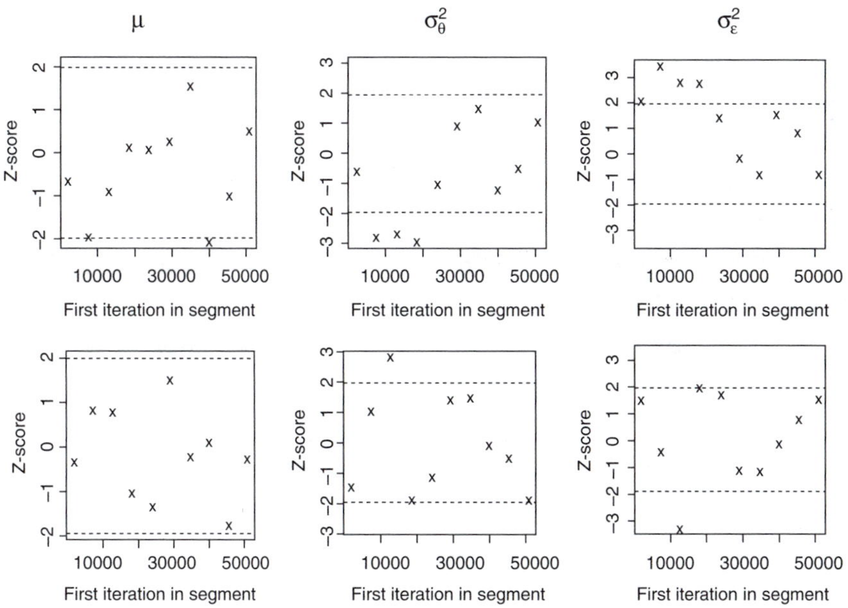

FIGURE 2.8
Z-scores (Geweke's convergence diagnostic) for the two simulated Markov chains (first chain, first row; second chain, second row), random effects model. From left to right, Z scores corresponding to μ, σ_θ^2 and σ_ε^2. The burn-in time was $k_0 = 2000$ and total number of random draws was $K = 10,0000$.

"dependence factor" show in Table 2.3) that the Markov chain shows for this parameter.

Figure 2.9 shows the Gelman and Rubin' shrink factor $\hat{R}$ as a function of the burn-in truncation point. In all the plots the median and the 97.5% shrink factor is below the critical value 1.2. Therefore, according to the Gelman and Rubin diagnostic, convergence has been reached with the adopted burn-in truncated at $k_0 = 2000$.

One additional Markov chain was simulated using the parameters inferred from Geweke's diagnostic statistic (i.e., a burn-in truncation point of

TABLE 2.3

Results Obtained Using the Raftery and Lewis Diagnostic for the Two Chains with Burn-In $k_0 = 2000$ and Whole Number of Draws $K = 10,0000$

	First chain			Second chain		
Parameter	**Burn-in**	**Total**	**Dependence factor**	**Burn-in**	**Total**	**Dependence factor**
μ	4	8000	2.14	6	7952	2.12
σ_θ^2	148	211,492	56.50	160	213,824	57.10
σ_ε^2	4	8024	2.14	3	4111	1.10

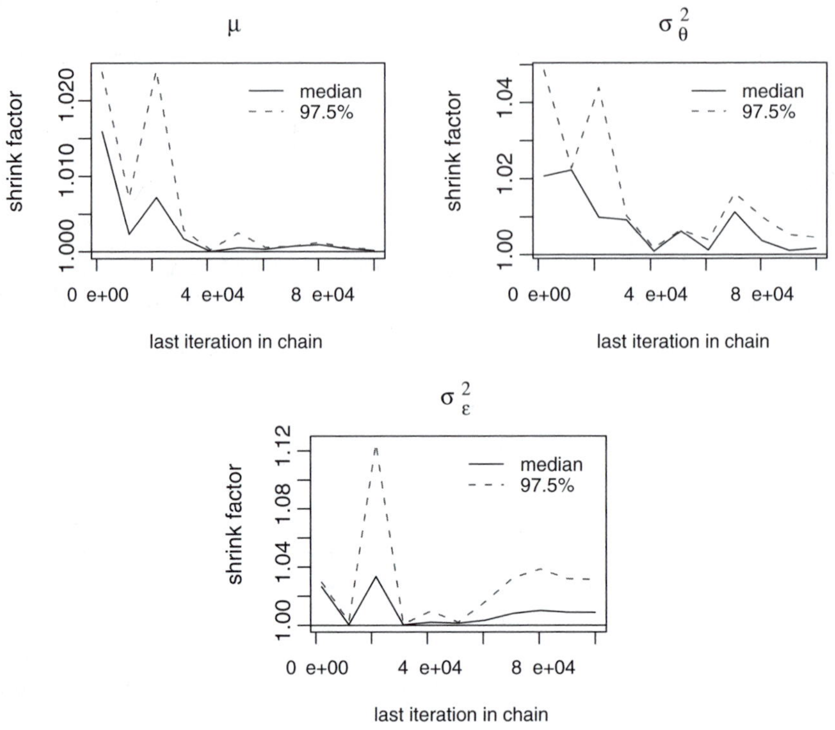

FIGURE 2.9
Gelman and Rubin shrink factor $\hat{R}$ as a function of the burn-in truncation point, random effects model. From left to right: shrink factors for μ, σ_ε^2, and σ_θ^2. The burn-in period was $k_0 = 2000$ and the total number of draws was $K = 100,000$.

$k_0 = 25,000$) and from Raftery and Lewis' diagnostic statistic (i.e., a total number of draws of $K = 200,000$). Figure 2.10 shows that the Z-scores are now always within the bound at ± 2 and hence convergence seems achieved according to this diagnostic approach. Results obtained with the Raftery and Lewis' diagnostic (showed in Table 2.4) confirm that with these new settings, the burn-in and the total number of samples are satisfactory.

TABLE 2.4

Results Obtained Using the Raftery and Lewis Diagnostic for the Chain with Burn-in Truncation at $k_0 = 25,000$ and a Total Number of Draws of $K = 200,000$

Parameter	Burn-in	Total	Dependence factor
μ	6	8296	2.21
σ_θ^2	168	182,252	48.70
σ_ε^2	4	7830	2.09

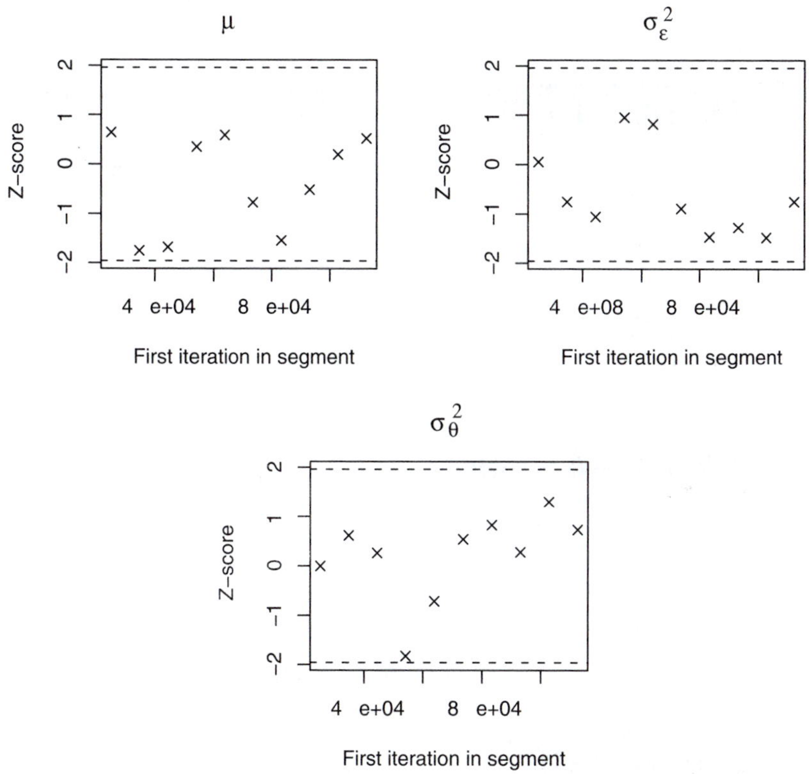

FIGURE 2.10
Z-scores (Geweke's convergence diagnostic) for μ, σ_{ε}^2, and σ_{θ}^2, using a burn-in truncation point of $k_0 = 25,000$ and a total number of draws of $K = 200,000$.

Results obtained with this last chain are reported in Figure 2.11. Table 2.5 shows the final estimates obtained with the Gibbs sampling for unknown parameters. As can be observed in Figure 2.11, the posterior distribution for σ_{θ}^2 has a very long upper tail, and hence the posterior mean for this parameter is considerably larger than the median in Table 2.5.

TABLE 2.5
Final Estimates Obtained for the Random Effects Model with Burn-In Truncation Point of 25,000 and Total Number of Draws of K=200,000

Node	Mean	SD	2.5%	Median	97.5%	Start	Sample
μ	1527.0	21.49	1484.0	1527.0	1571.0	25,000	175,001
σ_{θ}^2	2192.0	4126.0	0.006919	1279.0	10130.0	25,000	175,001
σ_{ε}^2	3038.0	1115.0	1557.0	2806.0	5794.0	25,000	175,001

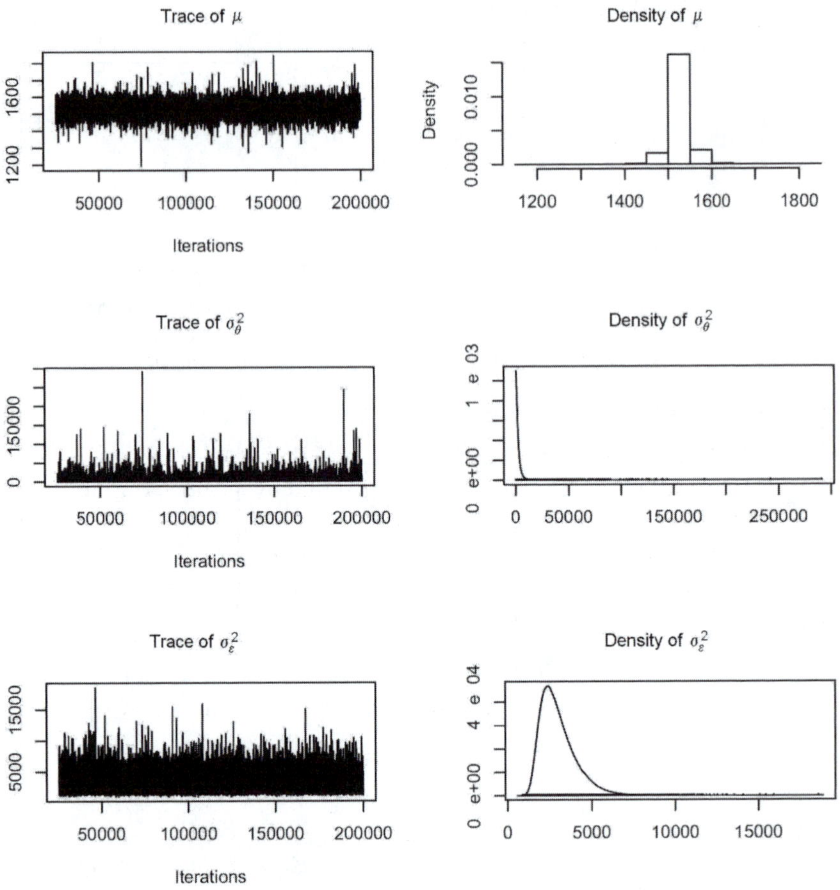

FIGURE 2.11
Steady-state behavior of the two MCMC chains when simulating μ, σ_θ^2, and σ_ε^2 in the random effects model, given the data in Table 2.1.

Appendix: Mini-Tutorial on WinBUGS 1.4

WinBUGS is a statistical software for Bayesian analysis of statistical models using MCMC. It was developed under the BUGS (Bayesian inference Using Gibbs Sampling) project that began in 1989 in the MRC Biostatistics Unit of the University of Cambridge (UK) and that was further developed jointly with the Imperial College School of Medicine at St Mary's, London.

The current version of the software as of this writing (WinBUGS 1.4.1) is available at http://www.mrc-bsu.cam.ac.uk/bugs/welcome.shtml. In this short tutorial, the main steps required to run MCMC using WinBUGS will be described with reference to the variance components model described in

Section 2.4.1 and included as an examples in Volume 1 in the WinBUGS help documentation. We will refer to this example as the "Dyes example." To run a MCMC simulation using WinBUGS, two main steps are required:

1. Specify the model in the WinBUGS language and provide all the data required to make inferences;
2. Perform the MCMC simulation and compute all the desired statistics.

In the following, these two steps will be briefly described. Further details can be found in the WinBUGS User Manual.

Model Specification and Data Format

WinBUGS allows us to specify the model in two alternative forms: a graphical representation, in which the model is represented by a directed graph, and a text-based description, in which the model is defined in the BUGS language. In the following, the second type of representation will be briefly described. In BUGS language, two basic relationships can be used:

- Stochastic or probabilistic relationships, which are denoted by the tilde symbol, ~;
- Deterministic or logical relationships, which are denoted by the left-arrow symbol, <-.

Stochastic relationships are used to assign a distribution to variables. As an example, `y dorm(mu,tau)` specifies that `y` is a stochastic variable that is normally distributed with mean `mu` and precision (i.e., the inverse of the variance) `tau`. Logical relationships are used to define variables as functions of other variables or constants. As an example: `sigma2<-1/tau` allows us to define the variance `sigma2` as a function of the precision `tau`. To handle arrays, indexes can be used to specify different elements in the array and a "for-loop" can be used as follows:

```
for(i in 1:5){
            mu[i] ~ dnorm(mu.all, tau.btw)
}
```

where `mu` is a vector of five elements that are normally distributed. Hence, the random effects model for the Dyes example described in Section 2.4.1 can be specified in the BUGS language as follows:

```
model {
        for( i in 1 : batches ) {
            mu[i] ~ dnorm(mu.all, tau.btw)
            for( j in 1 : samples ) {
                y[i , j] ~ dnorm(mu[i], tau.with)
```

```
            }
        }
        mu.all ~ dnorm(0.0, 1.0E-10)
        # prior for within-variation
        tau.with ~ dgamma(0.001, 0.001)
        sigma2.with <- 1 / tau.with
        # prior for between-variation
        tau.btw ~ dgamma(0.001, 0.001)
        sigma2.btw <- 1 / tau.btw
}
```

where # allows us to comment out the expression which follows and `dgamma` refers to the Gamma distribution. As we mentioned, WinBUGS allows us to describe a model both in the BUGS language (text-based description) and in a graphical form. The graphical representation of the model described is reported in Figure 2.12. As can be observed, the model is characterized by two constants `batches` and `samples`, an array of data denoted with `y[i,j]` and unknown parameters `mu.all`, `tau.with`, `tau.btw`, and `mu[i]`. To perform the MCMC simulation, we must specify values for the constants and the observed data. With reference to the Dyes example, `batches` = 6, `samples` = 5, and data `y[,]` are reported in Table 2.1. These values can be given in the following S-plus format:

```
list(batches = 6, samples = 5,
       y = structure(
       .Data = c(1545, 1440, 1440, 1520, 1580,
        1540, 1555, 1490, 1560, 1495,
        1595, 1550, 1605, 1510, 1560,
        1445, 1440, 1595, 1465, 1545,
        1595, 1630, 1515, 1635, 1625,
        1520, 1455, 1450, 1480, 1445), .Dim = c(6, 5))).
```

To start performing the MCMC simulation, initial values for all the unknown parameters must be defined for each Markov chain that must be simulated. To assign initial values reported in Table 2.2, the same S-plus format can be used for the first chain:

```
list(mu.all=2000, tau.with=1, tau.btw=1);
```

and for the second chain:

```
list(mu.all=1000, tau.with=100, tau.btw=100).
```

Note that initial values for unknown parameters `mu[i]` can be directly generated by WinBUGS, and that this is a preferable choice for random effects model (for fixed effects model, the WinBUGS manual suggests to define initial values for all the parameters involved).

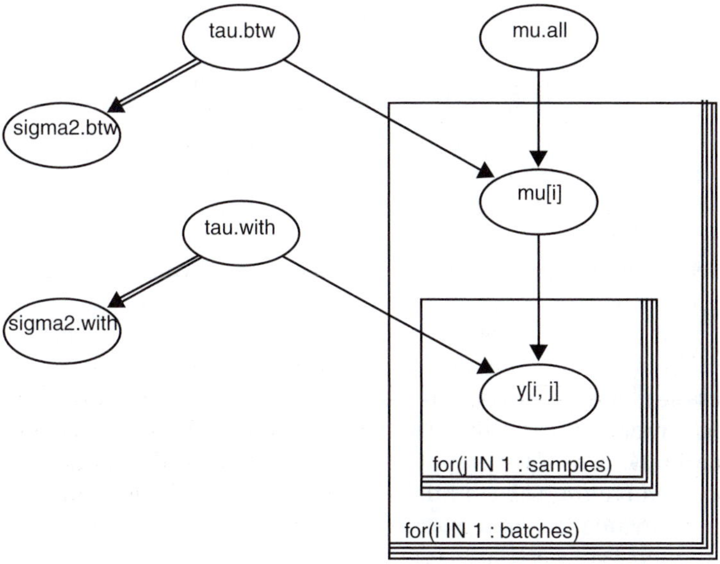

FIGURE 2.12
Graphical representation of the variance components model in WinBUGS.

Performing the MCMC Simulation and Obtaining the Required Posterior Statistics

Given the specified model, the first step consists in checking whether the model syntax is correct. To perform this step, we should point to Model in the tool bar and highlight the Specification... option. The Specification Tool window will appear (Figure 2.13). In the window containing the model code, the word `model` must be selected and highlighted and the check model button in the Specification Tool window clicked. If the model is correct, the message "model is syntactically correct" should appear in the bottom left of the WinBUGS program window.

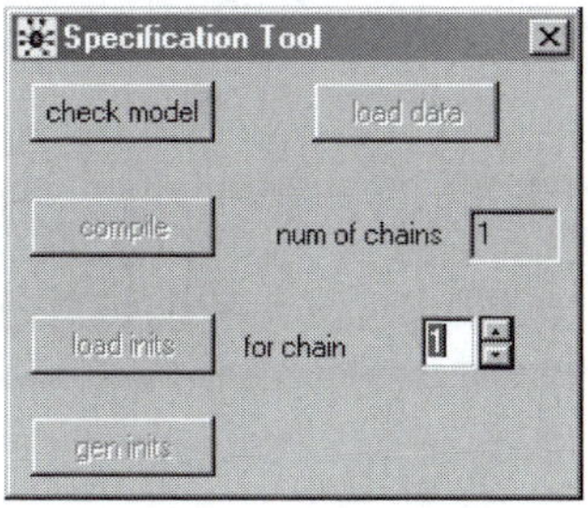

FIGURE 2.13
The Specification Tool dialog box in WinBUGS.

The second step consists of loading the data and can be performed by highlighting the word `list` at the beginning of the data file and clicking the load data button in the Specification Tool. The message "data loaded" should appear in the bottom left of the WinBUGS program window.

The final step consists of specifying the number of chains we want to run and the initial values for each chain. To specify the number of chains, the desired number should be inserted in the box labelled num of chains in the Specification Tool. After clicking the compile button, the message "model compiled" should appear. For each chain, the initial values can be loaded by highlighting the word `list` at the beginning of each set of initial values and clicking the load inits button in the Specification Tool. When some parameters have no initial values (as in the random effects model we are referring to), the message "this chain contains uninitialized variables" should appear. By clicking the gen inits button, the software will generate initial values by sampling from the prior distributions for all the uninitialized variable in all the chains. The final message "initial values loaded, model initialized" testifies that the software is ready to start running the simulation.

To properly collect results of the simulations, we need to specify to WinBUGS which samples should be stored before running the MCMC simulation. This step can be performed by selecting the Samples... option from the Inference tool bar. In the Sample Monitor Tool window (reported in Figure 2.14), we can type the name of the parameter to be monitored in the box node, select the "burn-in samples" that need to be discarded in the box beg, and the whole number of samples we want to simulate in the box end, and then click the button labelled set. This procedure must be repeated for all the parameters that we want to monitor. If we want to compute statistics based just on samples from every kth iteration, we need to type k in the box thin.

The Sample Monitor Tool window contains some further commands that can be of interest: the trace and history buttons allow us to obtain plots of

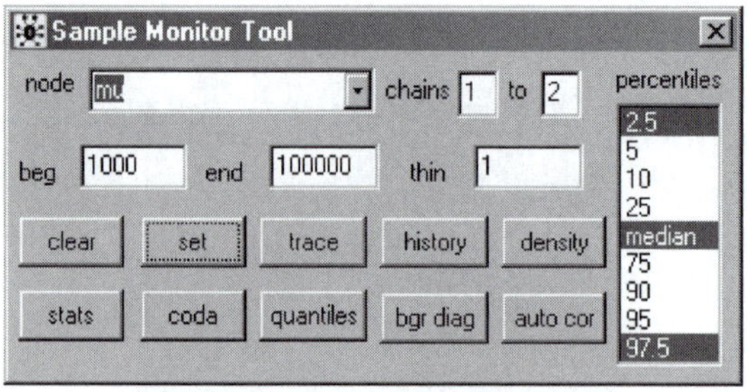

FIGURE 2.14
The Sample Monitor Tool dialog box in WinBUGS.

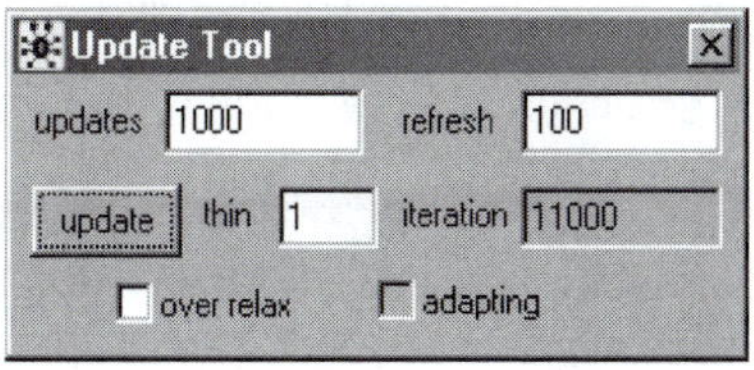

FIGURE 2.15
The Update Tool dialog box in WinBUGS.

the values of each parameter monitored against the iteration number. The only difference between the two commands is that trace generates a dynamic plot showing the last iterations of the simulation, whereas the history command generates a plot that shows the whole behavior of the chain from the beginning of the simulation to the current time. Both the commands are useful to support convergence check. In fact, when multiple chains are simulated, these commands show plots in which each chain is represented with a different color on the same graph. Convergence can be qualitatively considered achieved when the different chains overlap. Further plots as smoothed kernel density or autocorrelation functions can be obtained with the corresponding buttons.

To run the simulation, we must point to Model in the tool bar and highlight the Update... option. The Update tool window will appear (Figure 2.15). In this window, we can specify the number of MCMC updates to be carried out in the updates box and the number of updates between redrawing the screen (and hence redrawing the trace plots) in the refresh box. By setting the box thin equal to $k > 1$, we can specify that just samples from every k^{th} iteration must be stored. Note that this thinning option is different from the one appearing in the Sample Monitor Tool dialog box. When we specify thinning in the Update tool window, we are permanently discarding samples. When thinning is specified via the Sample Monitor Tool, we are temporarily discarding some samples to perform the required statistics but we are not reducing the number of samples stored (and hence we can not use this command to free-up memory). In the Update tool dialog box, the two fields over relax and adapting are specifically related with options concerning the MCMC method used (for further details, refer to the WinBUGS manual in the "Update options..." section).

Once the desired number of samples have been simulated through the Update tool, we can eventually obtain all the summaries from the posterior distribution by typing * in the node box of the Sample Monitor Tool and clicking the button marked stats. Smoothed kernel density plots based on the posterior samples can be also obtained by clicking the button labelled density. The button coda allows us to generate *ASCII* files containing all monitored values. These files can then be read with the CODA diagnostic package (running under R or S-plus) to perform the desired convergence checks.

References

1. Balakrishnan, S. and Madigan, D., "A one-pass sequential Monte Carlo method for Bayesian analysis of massive datasets," Technical Paper, http://www.stat.rutgers.edu/~madigan/papers, 2004.
2. Barbieri, M.M., *Metodi MCMC nell'inferenza statistica*. Societa' Italiana di Statistica, monografie, 1996.
3. Box, G.E.P. and Tiao, G.C., *Bayesian Inference in Statistical Analysis*, New York: John Wiley and Sons, 1973.
4. Brooks, S.P. and Roberts, G.O., "Assessing convergence of Markov Chain Monte Carlo algorithms," *Statistics and Computing*, 8, pp. 319–335, 1998.
5. Carlin, B.P. and Louis, T.A., *Bayes and Empirical Bayes Methods for Data Analysis*, 2nd ed., Boca Raton FL: Chapman & Hall/CRC, 2000.
6. Chib, S. and Greemberg E., "Understanding the Metropolis-Hastings algorithm" *The American Statistician*, 49, 4, 327–335, 1995.
7. Cowles, M.K. and Carlin, B.P. "Markov Chain Monte Carlo convergence diagnostics: A comparative review," *Journal of the American Statistical Association*, 91, 883–904, 1995.
8. Doucet, A., de Freitas, N., Gordon, N. (Eds.) *Sequential Monte Carlo Methods in Practice*, New York: Springer-Verlag, 2001.
9. Gelfand, A.E. and Smith, A.F.M., "Sampling-based approaches to calculating marginal densities," *Journal of the American Statistical Association*, 85, 410, 398–409, 1990.
10. Gelfand, A.E., Hills, S.E., Racine-Poon, A., and Smith, A.F.M., "Illustration of Bayesian inference in normal data models using Gibbs sampling," *Journal of the American Statistical Association*, 85, 412, 972–985, 1990.
11. Gelman, A., "Inference and monitoring convergence," in Gilks, W.R., Richardson, S. and Spiegelhalter, D.J. (Eds.), *Markov Chain Monte Carlo in Practice*, London: Chapman & Hall, 1996.
12. Gelman, A., "Prior distributions for variance parameters in hierarchical models," to appear in *Bayesian Analysis*, http://www.stat.columbia.edu/∼gelman, 2005.
13. Gelman, A. and Rubin, D., "Inference from iterative simulation using multiple sequences," *Statistical Science*, 7, 457–511, 1992.
14. Gelman, A., Carlin, J.B., Stern, H.S., and Rubin, D.B., *Bayesian Data Analysis*, 2nd ed., Boca Raton FL: Chapman & Hall/CRC, 2004.
15. Geman, S. and Geman, D., "Stochastic relaxation, Gibbs distribution and Bayesian restoration of images," *IEEE Transactions on Pattern Analysis and Machine Intelligence* 6, 721–741, 1984.
16. Geweke, J., "Evaluating the accuracy of sampling-based approaches to the calculation of posterior moments," in Bernardo, J.M., Smith, A.F.M., Dawid, A.P., and Berger, J.O. (Eds.), *Bayesian Statistics* 4, New York: Oxford University Press. 1992.
17. Gilks, W.R., Richardson, S., and Spiegelhalter, D.J. (Eds.), *Markov Chain Monte Carlo in Practice*, London: Chapman & Hall, 1996.
18. Kong, A., Liu, J., and Wong, W., "Sequential imputation and Bayesian missing data problems," *Journal of the American Statistical Association*, 89, 278–288, 1994.
19. Plummer, M., Best, N.G., Cowles, M.K., and Vines, S.K., *CODA Manual Version 0.9-5*, http://www-fis.iarc.fr/coda, 2005.

20. Raftery, A.E. and Lewis, S.M., "Implementing MCMC," in Gilks, W.R., Richardson, S., and Spiegelhalter, D.J., (Eds.), *Markov Chain Monte Carlo in Practice,* London: Chapman & Hall, 1996.
21. Ridgeway, G. and Madigan, D., "A sequential Monte Carlo Method for Bayesian analysis of massive datasets," *Journal of Knowledge Discovery and Data Mining*, 7, 301–319, 2002.
22. Rubin, D.B., "Using the SIR algorithm to simulate posterior distributions," in Bernardo, J.M., DeGroot, M.H., Lindley, D.V., and Smith, A.F.M., (Eds.), *Bayesian Statistics* 3, Cambridge, MA: Oxford University Press, 1988.
23. Smith, A.F.M. and Gelfand, A.E., "Bayesian statistics without tears: A sampling-resampling perspective," *The American Statistician*, 46, 2, 84–88, 1992.
24. Spiegelhalter, D.J., Thomas, A., Best, N.G., Gilks, W.R., and Lunn, D., "BUGS: Bayesian inference using Gibbs sampling," MRC Biostatistics Unit, Cambridge, UK, http://www.mrc-bsu.cam.ac.uk/bugs, 1994, 2003.
25. Spiegelhalter, D.J., Abrams, K.R., and Myles, J.P., *Bayesian Approaches to Clinical Trials and Health-Care Evaluation*, New York: John Wiley & Sons, 2004.
26. Tanner, M.A. and Wong, W.H., "The calculation of posterior distributions by data augmentation," *Journal of the American Statistical Association*, 82, 528–550, 1987.

Part II

Process Monitoring

3

A Bayesian Approach to Statistical Process Control

Panagiotis Tsiamyrtzis
Department of Statistics, Athens University of Economics and Business

Douglas M. Hawkins
School of Statistics, University of Minnesota

CONTENTS

ABSTRACT The frequentist Shewhart charts have proved valuable for the first stage of quality improvement in many manufacturing settings. However, their statistical foundation is on a model with exactly known process parameters and independent identically distributed process readings. One or more aspects of this foundation are often lacking in real problems. A Bayesian framework allowing an escape from the independence and the known-parameter assumptions provides a conceptually sounder and more effective approach for process control when one moves away from this first idealization of a process.

3.1 Introduction

The Shewhart X-bar chart [12] is the most widely used statistical process control tool. While many practitioners downplay statistical models for the chart, we believe these are essential to fully appreciate its capabilities and limitations. In the idealized situation that all process readings follow a common normal distribution with known parameters and are independent, all the properties of the chart are well-known. Its run length (both in- and out-of-control) is geometric with a known probability of signal. With independence and known parameters, one can advance to more sensitive tools such as the cumulative sum [10] and the exponentially weighted moving average charts [11], and again the independent known-parameter normal model allows for exact calculation of all statistical properties of the charts.

However, increasing realization exists that process parameters are hardly ever known to an adequate precision for these theoretical calculations to be plausible. Furthermore, many process drift. The most familiar is the tool wear problem, but other settings are also found. Food safety is one such problem. Food may start out safe, but it deteriorates with time — ideally steadily, but perhaps (for example, if refrigeration fails) suddenly and catastrophically. In clinical chemistry, standards change slowly over time, and they too can deteriorate suddenly in case of contamination or environmental change. Other problems do not have systematic drift but show varying degrees of serial association — blending problems are an example of this.

These thoughts motivate the search for methods that escape the known-parameter straitjacket; that allow for serial dependence [17], and that can model sharp shifts [2]. Bayesian methods in *statistical process control* (SPC) are particularly attractive because one can easily convert the inference part into a decision theory problem [16], but mainly because they allow for optimal use of partial information on the process [4, 9]. In what follows, we will develop one such framework, concentrating on the monitoring of the mean of a normally distributed process measurement.

3.2 Statistical Modeling

We consider the problem of control of the mean of a Normal distribution model when the mean is unknown and not necessarily constant and the variance is unknown. Furthermore, the mean of the process will be modeled as a random walk that is subject to random jumps. More precisely, writing θ_n for the process true mean at stage n, when moving between successive stages of the process, we model:

$$\theta_{n+1}|(\theta_n, \sigma^2, p, \delta) \sim \left\{ \begin{array}{ll} N(\theta_n, \sigma^2) & \text{with probability} \quad p \\ N(\theta_n + \delta, \sigma^2) & \text{with probability } 1-p \end{array} \right\},$$

where σ^2 represents the random drift of the mean. The mean can also undergo a jump of size δ with probability $1-p$. This model does not give rise to a steady state; rather, θ_n increases (or decreases) without bound. Despite the absence of an asymptotic steady state, the model is still attractive for settings that do not have steady states. Examples are the tool wear problem; spoilage problems (of food, vaccine, perishable raw materials) and chemical assay standards, in each of which at some point the process will become unacceptable and must be stopped. Our model allows this to occur through either or both of drifting and step changes. We are mainly interested in the mean of the process but also have nuisance parameters $\phi = (\sigma^2, p, \delta)$.

Finally, we assume that θ_n is measured only with some measurement error variability, i.e.,

$$X_{n+1}|\theta_{n+1} \sim N(\theta_{n+1}, c\sigma^2),$$

where $c > 0$ is a known constant expressing the ratio of the measurement to the drift random variability. (We do not need to know the measurement error variance, just the ratio of these two variances.)

The case where the vector of nuisance parameters ϕ was known or could be estimated from historical data has been treated in [14], where the inference for the mean based on the posterior distribution of the parameter of interest was derived and its relation to the Kalman filter [8] approach was studied.

In what follows, we will treat the vector ϕ of nuisance parameters in a Bayesian fashion, using a prior distribution and obtaining the joint posterior. Integrating out the nuisance parameters, we will obtain the posterior for the parameters of interest.

3.2.1 Prior Settings

For the random walk model with occasional jumps we studied here, we adopted a sequential updating scheme where the posterior distribution at each stage of the process will be the prior of the next stage. To have a closed form updating mechanism, using natural conjugates for the prior is best. More precisely, our model is:

$$\theta_{n+1}|(\theta_n, \sigma^2, p, \delta) \sim pN(\theta_n, \sigma^2) + (1-p)N(\theta_n + \delta, \sigma^2).$$

The natural conjugate prior for the pair (θ, σ^2) is the Normal-Inverted Gamma (N-IG) distribution. With this choice, we get independently:

$$\left\{ \begin{array}{ll} (\theta_n, \sigma^2) & \sim N - IG \\ p & \sim Beta(\alpha, \beta) \\ \delta & \sim Normal(\delta_0, \lambda\sigma^2) \end{array} \right\}.$$

The Beta distribution for p is quite convenient because appropriate choice of the parameters α and β can lead to a member of a wide range of possible a priori beliefs. These include prior ignorance (uniform distribution) or more informative beliefs (symmetric, U-shaped, strictly increasing or decreasing, and so on).

The size of the jump is set by δ. We allow jumps of random size where prior knowledge will allow us to decide on the predetermined value of the mean δ_0, whereas the prior confidence we have for the size of δ will be expressed via an appropriate choice of λ (a multiple of the model's variability, σ^2), where small values will lead to informative cases and large values express ignorance. Providing that λ is quite small (indicating very informative cases), the type of processes we are interested in will determine the sign of δ_0. For processes where the mean gradually increases (like the tool wear setting) and we are concerned with not exceeding an upper limit, δ_0 will be positive. Alternately, δ_0 will be negative for the type of processes where we are interested in guarding against a lower control limit. The case where $\delta_0 = 0$ (and $\lambda > 0$) will not affect the mean of the process but its variance because the unconditional mean will take the form of a contaminated Normal distribution. This special case refers to the processes where it is not the mean that can undergo sudden jumps but the variance, which might be elevated from one stage to the next.

3.2.2 Model Evolution

Before the process starts to operate, we have an initial prior distribution:

$$\pi(\theta_0, \sigma^2) \sim N(\mu^{(0)}, k^{(0)}\sigma^2) IG(A^{(0)}, B^{(0)})$$

where $\mu^{(0)}, k^{(0)}, A^{(0)}$, and $B^{(0)}$ are hyperparameters whose values we will assume known. Due to the fact that our model allows the process mean to either jump or not jump at every stage of the process, the posterior distribution after n steps will be a mixture of 2^n components. More precisely, once the nuisance parameters p and δ are integrated out, we will have the prior for the stage $n+1$, (i.e., before we will observe x_{n+1}) $(\theta_{n+1}, \sigma^2)|\mathbf{X}_n$ to be a mixture of 2^{n+1} N-1G distributions, where by $\mathbf{X}_n = \{x_1, x_2, \ldots, x_n\}$ we will denote the data available from time 1 until n. Then combining the likelihood and the prior in Bayes theorem, we will obtain the posterior distribution of $(\theta_{n+1}, \sigma^2)|\mathbf{X}_{n+1}$. More precisely, we have the following:

THEOREM 3.1
At time $n+1$ the posterior distribution of (θ_{n+1}, σ^2) will be a mixture of 2^{n+1} Normal-Inverted Gamma distributions:

$$p(\theta_{n+1}, \sigma^2|\mathbf{X}_{n+1}) \sim \sum_{j=0}^{2^{n+1}-1} w_j^{(n+1)} N\left(\mu_j^{(n+1)},\ k_j^{(n+1)}\sigma^2\right) IG\left(A^{(n+1)},\ B_j^{(n+1)}\right),$$

where the weights and the parameters of the posterior obey the following recursive rules

$$A^{(n+1)} = A^{(n)} + \frac{1}{2}$$

and for $i = 0, 1, \ldots, 2^n - 1$,

$$\mu_{2i}^{(n+1)} = \frac{(k_i^{(n)} + 1)x_{n+1} + c\mu_i^{(n)}}{c + k_i^{(n)} + 1}$$

$$\mu_{2i+1}^{(n+1)} = \frac{(\lambda + k_i^{(n)} + 1)x_{n+1} + c(\mu_i^{(n)} + \delta_0)}{c + \lambda + k_i^{(n)} + 1}$$

$$k_{2i}^{(n+1)} = \frac{(k_i^{(n)} + 1)c}{c + k_i^{(n)} + 1}$$

$$k_{2i+1}^{(n+1)} = \frac{(\lambda + k_i^{(n)} + 1)c}{c + \lambda + k_i^{(n)} + 1}$$

$$B_{2i}^{(n+1)} = \left[\frac{(x_{n+1} - \mu_i^{(n)})^2}{2(c + k_i^{(n)} + 1)} + \frac{1}{B_i^{(n)}}\right]^{-1}$$

$$B_{2i+1}^{(n+1)} = \left[\frac{(x_{n+1} - \mu_i^{(n)} - \delta_0)^2}{2(c + \lambda + k_i^{(n)} + 1)} + \frac{1}{B_i^{(n)}}\right]^{-1}$$

$$w_{2i}^{(n+1)} = \frac{\left(\frac{\alpha}{\alpha+\beta}\right) w_i^{(n)} m_i(x_{n+1})}{NC}$$

$$w_{2i+1}^{(n+1)} = \frac{\left(\frac{\beta}{\alpha+\beta}\right) w_i^{(n)} m_i^*(x_{n+1})}{NC},$$

where

$$m_i(x_{n+1}) = \frac{\Gamma\left(A^{(n)} + \frac{1}{2}\right)}{\Gamma(A^{(n)})} \sqrt{\frac{B_i^{(n)}}{2\pi\left(c + k_i^{(n)} + 1\right)}} \times \left[1 + \frac{B_i^{(n)}(x_{n+1} - \mu_i^{(n)})^2}{2(c + k_i^{(n)} + 1)}\right]^{-(A^{(n)} + \frac{1}{2})}$$

$$m_i^*(x_{n+1}) = \frac{\Gamma\left(A^{(n)} + \frac{1}{2}\right)}{\Gamma(A^{(n)})} \sqrt{\frac{B_i^{(n)}}{2\pi\left(c + \lambda + k_i^{(n)} + 1\right)}} \times \left[1 + \frac{B_i^{(n)}(x_{n+1} - \mu_i^{(n)} - \delta_0)^2}{2(c + \lambda + k_i^{(n)} + 1)}\right]^{-(A^{(n)} + \frac{1}{2})}$$

and

$$NC = \sum_{i=0}^{2^n-1} \left[\left(\frac{\alpha}{\alpha+\beta}\right) w_i^{(n)} m_i(x_{n+1}) + \left(\frac{\beta}{\alpha+\beta}\right) w_i^{(n)} m_i^*(x_{n+1}) \right].$$

The proof of the theorem is given in the Appendix.

As we mentioned in the prior settings section, in very informative cases of δ, appropriate choice of δ_0 provides for processes that we are interested in upward shifts ($\delta_0 > 0$), downward shifts ($\delta_0 < 0$), or no jumps but elevated variance ($\delta_0 = 0$) between successive stages. If we are interested for cases where the mean can undergo both positive and negative jumps, the conditional distribution of the mean $\theta_{n+1}|(\theta_n, \sigma^2, p, \delta)$ would then be given as a mixture of three components, including two unidirectional jump components of opposite direction. This would provide a posterior distribution with 3^n components [13].

A binary representation is helpful in identifying the terms of the mixture. At stage n, the index of any of the 2^n components in the mixture can be used as a "state vector" defining that term's provenance. We do this by writing the term number/state vector as an n-bit binary integer. Bit i is 1 if the process jumped at stage i and is 0 otherwise. Each component of the mixture has as weight its mixture (posterior) probability, $w_i^{(n)}$. The state vector $j = 0$ in the sum refers to the length n string of zeros that represents the no-jump case. As we can see in the proof of the theorem, the even-numbered components represent the case of not having a jump at the nth stage of the process (with all 2^{n-1} possible scenarios for the history), whereas the odd-numbered components refer to the case of having a jump at the current stage (again, with all possible history).

3.2.3 Approximating the Posterior

In the posterior distribution, the number of components in the mixture grows exponentially with n, the number of stages in the process. However, after a few stages of the increasingly large number of terms, most will have small posterior probabilities and so make small contributions to the overall mixture distribution. This fact motivates us to use an approximation, replacing the exact distribution with another mixture having far fewer components. This can be done in several ways. In [14], the authors proposed prunine all the components that have "small" weights $w_i^{(n)}$, because all their descendants will have even smaller weights in the updated prior distribution. This is done because the pruned components represent the most unlikely scenarios of model evolution, and thus we end up keeping the "active" components whose weights are non-negligible. Instead of pruning the low-probability components, we could (as suggested in [15]) pool neighboring components, replacing them with a single approximating normal distribution.

Either approximation will reduce the number of terms in the model and make it able to handle longer sequences. If we adopt the "pooling" approach, however, we lose with it the ability to identify a term's history from the binary representation of its number.

3.3 Inference

As a result of the updating mechanism at the end of each stage, we obtain the joint posterior distribution of the mean θ_n and the variance σ^2 of the model. The marginal posterior distribution of each parameter can be obtained easily by integrating out the other. In particular, if we integrate out the mean θ_n, we obtain the posterior of $\sigma^2|\mathbf{X}_n$ as a mixture of 2^n Inverted Gamma distributions. Alternetely, if we integrate out σ^2, then we get the posterior distribution of $\theta_n|\mathbf{X}_n$ as a mixture of 2^n Student t distributions. This leads to the following:

LEMMA 3.1
The posterior distribution of $\sigma^2|\mathbf{X}_n$ will be

$$p(\sigma^2|\mathbf{X}_n) \sim \sum_{i=0}^{2^n-1} w_i^{(n)} IG\left(A^{(n)}, B_i^{(n)}\right),$$

whereas the marginal posterior mean distribution will be

$$p(\theta_n|\mathbf{X}_n) \sim \sum_{i=0}^{2^n-1} w_i^{(n)}\, Student\left(\theta_n \mid \mu_i^{(n)}, \frac{A^{(n)} B_i^{(n)}}{k_i^{(n)}}, 2A^{(n)}\right)$$
$$= \sum_{i=0}^{2^n-1} w_i^{(n)} \frac{\Gamma\left(A^{(n)} + \frac{1}{2}\right)}{\Gamma\left(A^{(n)}\right)} \sqrt{\frac{B_i^{(n)}}{2\pi k_i^{(n)}}} \left[1 + \frac{B_i^{(n)}\left(\theta_n - \mu_i^{(n)}\right)^2}{2k_i^{(n)}}\right]^{-\left(\frac{2A^{(n)}+1}{2}\right)}.$$

In what follows, we will apply some of the standard decision theory tools in drawing inference for the mean of the process (inference on the model variability σ^2 proceeds analogously). In the framework of point estimation regarding the mean of the process, we have that under squared error loss, the Bayes rule is the mean of the posterior distribution, i.e.,

$$\hat{\theta}_n = \sum_{i=0}^{2^n-1} w_i^{(n)} \mu_i^{(n)}.$$

One situation of interest is where the process has an upper limit on the tolerable θ_n and, correspondingly, the jumps are positive. If we wish to signal when the mean is thought to have crossed some upper threshold value M, we can set up a sequence of hypothesis tests:

$$\left\{\begin{array}{ll} H_0: & \theta_n \leq M \\ H_1: & \theta_n > M \end{array}\right\}.$$

If we are concerned for processes with negative jumps, the symmetric hypothesis test can be applied. At each stage of the process, we perform the hypothesis test and allow the process to continue operating when H_0 is not rejected. Alternately, rejection of H_0 at some stage will stop the process and some corrective action will be taken (like an intervention and initialization of the whole process).

Deciding on whether H_0 will be rejected or not can be done in several different ways. For instance, one might use the posterior coverage probabilities of the region $(-\infty, M]$. At each stage n of the process, we can estimate the probability

$$p_n = p(\theta_n \leq M|\mathbf{X}_n) = \sum_{i=0}^{2^n-1} w_i^{(n)} \mathcal{F}_i(M),$$

where $\mathcal{F}_i$ is the Student t cumulative distribution function (cdf) corresponding to component i. We will reject H_0 when $p_n < c^*$, where c^* is some prespecified constant. For instance, if we use the "generalized $0-1$" loss function, with c_I and c_{II} being the costs of type I and II errors, respectively, the value of $c^* = c_{II}/(c_I + c_{II})$ provides a Bayes test [1]. A time series plot of the p_n probabilities will offer further insight into the behavior of the mean.

A different alternative for doing the hypothesis testing is to use the ratio of the posterior to prior odds of H_0, which is known as Bayes factor [7]. Depending on the value of this ratio, we will get evidence for or against H_0 (in [7], a table of cutoff values to be used is provided). More generally, in [3] one can find a variety of sequential decision problem approaches that can be adapted for our setup.

3.4 Forecasting

A particularly interesting feature in the Bayesian paradigm is forecasting. Namely, one can use the available data $\mathbf{X}_n = \{x_1, \ldots, x_n\}$ to derive the predictive distribution of the next (unseen) observation X_{n+1}. From [6], the predictive distribution $X_{n+1}|\mathbf{X}_n$ in our model setting will be given by

$$P(X_{n+1}|\mathbf{X}_n) = \iint f(X_{n+1}|\theta_{n+1}, \sigma^2)\pi(\theta_{n+1}, \sigma^2|\mathbf{X}_n)d\theta_{n+1}d\sigma^2,$$

where $f(X_{n+1}|\theta_{n+1}, \sigma^2)$ is the likelihood and $\pi(\theta_{n+1}, \sigma^2|\mathbf{X}_n)$ is the prior at stage $n+1$. From the proof of Theorem 3.1 we have that

$$P(X_{n+1}|\mathbf{X}_n) = \sum_{i=0}^{2^n-1} \left[\left(\frac{\alpha w_i^{(n)}}{\alpha + \beta} \right) m_i(x_{n+1}) + \left(\frac{\beta w_i^{(n)}}{\alpha + \beta} \right) m_i^*(x_{n+1}) \right],$$

where the form of $m_i(x_{n+1})$ and $m_i^*(x_{n+1})$ were provided in the Theorem 3.1. Thus, we have the following:

LEMMA 3.2
The predictive distribution of $X_{n+1}|\mathbf{X}_n$ will be given as a mixture of 2^{n+1} Student t distributions:

$$P(X_{n+1}|\mathbf{X}_n) = \frac{\alpha}{\alpha+\beta}\sum_{i=0}^{2^n-1} w_i^{(n)} Student\left(x_{n+1} \mid \mu_i^{(n)}, \frac{A^{(n)}B_i^{(n)}}{c+k_i^{(n)}+1}, 2A^{(n)}\right)$$
$$+\frac{\beta}{\alpha+\beta}\sum_{i=0}^{2^n-1} w_i^{(n)} Student\left(x_{n+1} \mid \mu_i^{(n)}+\delta_0, \frac{A^{(n)}B_i^{(n)}}{\lambda+c+k_i^{(n)}+1}, 2A^{(n)}\right).$$

This predictive distribution leads to, for instance, means and predictive intervals for the upcoming observation, which can be useful in the common tolerance interval settings.

3.5 Making the Model Robust

In this section, we study the behavior of our model in the presence of outliers. In [5], two types of time series outliers were identified, innovative and additive. Innovative outliers persist, whereas additive outliers affect a single process reading but leave no after-effect. A heavy-tailed distribution for the measurement error would lead to additive outliers, whereas the jump included in the modeling of the mean offers protection against innovative outliers.

If a possibility of additive outliers exists, making the SPC scheme against them make robust would be helpful. A convenient method of doing so (that preserves conjugacy) is the γ scale-contaminated normal distribution:

$$X_n|(\theta_n, \sigma^2) \sim \gamma\ N(\theta_n,\ c\sigma^2) + (1-\gamma)\ N(\theta_n,\ Lc\sigma^2),$$

where $0 < \gamma < 1$ and $L > 1$. The choice of specific values for γ and L depend on the application.

Using a mixture of two components for both the mean and the measurement equation will cause our model's posterior distribution of $(\theta_n, \sigma^2)|\mathbf{X}_n$ to be a mixture of 4^n different components.

3.6 Example

The developed methodology will be applied to a data set (Table 3.1), provided to us by Dr. Daniel Schultz of the Rogasin Institute, that provides a usual laboratory quality control setting. Clinical chemistry laboratories carry out assays for a variety of biological markers in the blood (triglycerides, in our data). As each day's run involves an instrument calibration, having quality controls on

TABLE 3.1

The 13 Consecutive Measurements of the Riglycerides (in mg/dL)

Time	1	2	3	4	5	6	7	8	9	10	11	12	13
Trigl.	112	109	108	107	110	109	114	113	111	112	113	119	120

the day's runs is important. These are commonly provided by using quality control samples— specimens. These are then assayed along with the patient unknowns, and the reading checked against the previous assays and any prior information on the specimen. The specimens are commonly a large pool of blood serum stored under conditions where the lipids remain quite stable. However, some day-to-day change exists and if the storage conditions are lost, the pool can spoil. These control assays are therefore natural candidates for our model.

The fact that the lab has a long history of previous pools, and pools used for different blood components, means that quite good prior information is available when a new pool is introduced. From several series of historic data, we have elicited the following prior settings:

$$\left\{ \begin{array}{lcl} (\theta_0, \sigma^2) & \sim & N(110, \sigma^2) IG\,(3, 1/6) \\ p & \sim & Beta(8, 2) \\ \delta & \sim & Normal(4, \sigma^2/2) \end{array} \right\}.$$

In Figures 3.1 and Figure 3.2, we plot the prior distributions of p and σ^2, respectively, where the prior mean of p (the probability of not having a jump) has an average value $E(p) = 0.8$, whereas for the marginal of σ^2, we put a prior that has $E(\sigma^2) = 3$ and $V(\sigma^2) = 9$.

Assembling the information on three different analysts in three different lipid assays, we selected $c = 1$ (the sensitivity analysis regarding the choice c and the rest of hyperparameters will be carried over). Finally, with respect to the upper specification limit, this was specified to be $M = 118$ mg/DL.

Next, we applied the proposed methodology and we obtained the full posterior probability for each stage of the process. (A Matlab code that runs the proposed model and reproduces this example's results can be downloaded at http://www.stat-athens.aueb.gr/~pt/software.) The data along with the posterior means at each stage (Bayes rule for the parameter θ_n under squared error loss) can be seen in Figure 3.3.

Performing the sequence of hypothesis testing of whether the mean has exceeded the upper threshold value M, or not we obtain the posterior coverage probabilities of H_0 (p_n) and the Bayes factors shown in Tables 3.2 and Table 3.3 respectively.

Based on these posterior coverage probabilities, we are very confident that the mean is well below the threshold value for the first 11 observations. At the 12th observation (which, based on the sequential view of the data, looks to be

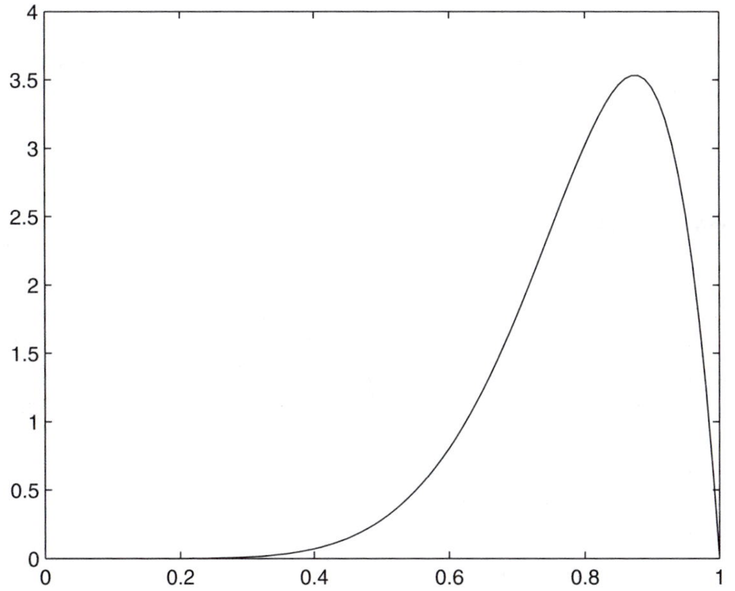

FIGURE 3.1
The Beta prior distribution for the probability of not having a jump p.

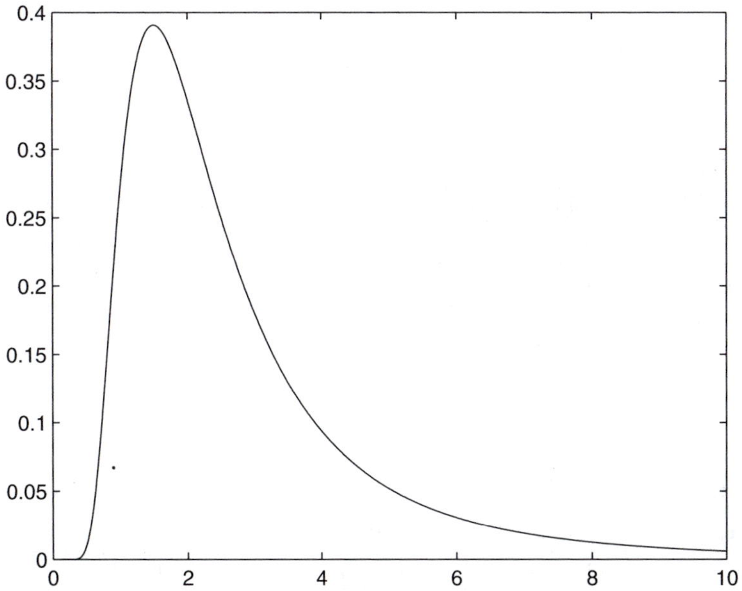

FIGURE 3.2
The marginal prior distribution for the parameter σ^2.

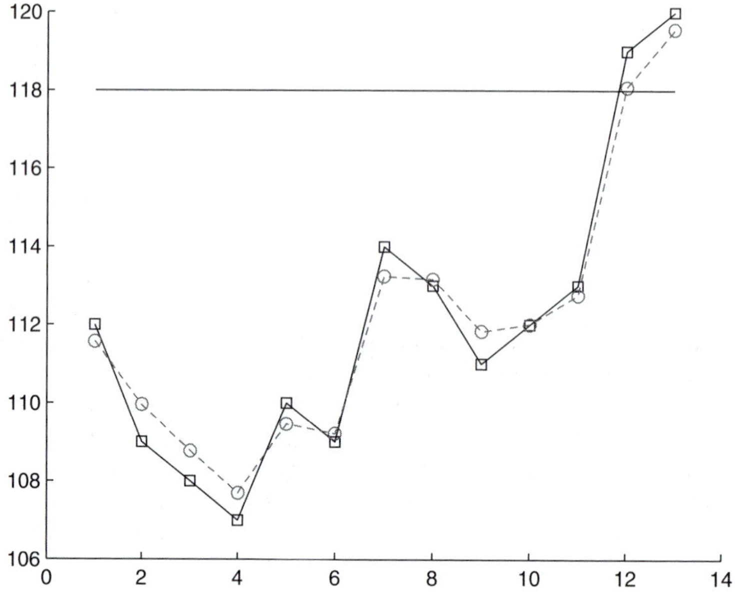

FIGURE 3.3
The solid line represents the data and the dashed line refers to the Bayes posterior mean for the marginal posterior of θ_n.

TABLE 3.2

The Posterior Probabilities $P(\theta_n > M|\mathbf{X}_n)$ for Each Stage of the Process

Time	1	2	3	4	5	6	7	8	9	10	11	12	13
p_n	0.999	0.999	1.000	1.000	1.000	1.000	0.999	0.999	1.000	1.000	0.999	0.454	0.088

TABLE 3.3

The Bayes Factor for Each Stage of the Process

Time	1	2	3	4	5	6	7	8	9	10	11	12	13
Bayes factor	33.868	499.46	765.11	1429.2	165.76	2068.6	5.5456	219.33	2069	693.67	247.5	0.046681	0.15597

TABLE 3.4

The Marginal Posterior Probability of Having a Jump at Each Stage of the Process (Independent of the History)

Time	1	2	3	4	5	6	7	8	9	10	11	12	13
Prob. of jump	0.1958	0.0270	0.0290	0.0283	0.2385	0.0481	0.6666	0.0540	0.0199	0.0573	0.0963	0.8544	0.1774

either an outlier or a result of a mean jump), the posterior coverage of H_0 falls below 0.5, and with the 13th observation, drops to a very low value, indicating that the mean has exceeded the upper threshold. This is an indication that either the pool has deteriorated to the point that a new pool might be advisable or that some change in the assay itself has occurred (possibly a calibration issue or an instrument change) that should be investigated and remedied. The similar story is provided to us from the Bayes factors as well, where we observe that at stage 12 of the process, we have a value well below 1, indicating strong evidence against H_0 according to the guidelines given in [7].

Based on the binary representation of the weights (which are the posterior probabilities of the respective components), we can calculate the marginal probability of a jump occurring at each stage (irrespective of the earlier history) by simply summing all the odd number weights. These marginal probabilities can be seen in Table 3.4.

From these posterior probabilities, the mean of the processes appears to have underwent two jumps, one at stage 7 and a second at stage 12. Combing these values with the posterior coverage probabilities, we observe that the second jump was decisive in setting the mean above the threshold value.

A sensitivity analysis showed that the results were quite robust with respect to the choice of the hyperparameters, as long as we are not becoming extremely informative. In this spirit, the value of c (ratio of the model to measurement variance) hardly alters the posterior coverage probabilities unless it becomes very small —which is the equivalent of directly observing the mean of the process— or very large —where so much error contaminates the actual measurement that the underlying pattern is lost.

3.7 Conclusions

Statistical process control has been immensely valuable as a first line of attack in quality improvement. However, many real-world settings are not accommodated by the traditional Shewhart or the newer, more sensitive tools. Bayesian methods make a good bridge between the frequentist extremes of total knowledge and total ignorance of process parameters. We have sketched a Bayesian formulation that allows for serial correlation and jumps in the process mean. It is particularly suitable for short-run processes, which is a setting that is not covered well by existing methodologies.

Appendix

The proof of the theorem will be given via induction. Proving that it holds for $n = 1$ is easy. Assume this is also true for n, i.e.,

$$p(\theta_n, \sigma^2|\mathbf{X}_n) \sim \sum_{i=0}^{2^n-1} w_i^{(n)} N(\mu_i^{(n)},\ k_i^{(n)}\sigma^2) IG(A^{(n)},\ B_i^{(n)}).$$

We then show that it is true for $n + 1$. We have

$$\left.\begin{array}{ll} \theta_{n+1}|\theta_n, \sigma^2, p, \delta \sim & pN(\theta_n,\ \sigma^2) + (1-p)N(\theta_n+\delta,\ \sigma^2) \\ \pi(\theta_n, \sigma^2|\mathbf{X}_n) & \sim \sum_{i=0}^{2^n-1} w_i^{(n)} N(\mu_i^{(n)},\ k_i^{(n)}\sigma^2) IG(A^{(n)},\ B_i^{(n)}) \\ \pi(p) & \sim Beta(\alpha,\ \beta) \\ \pi(\delta) & \sim N(\delta_0,\ \lambda\sigma^2) \end{array}\right\}.$$

From the above, we are interested in deriving the distribution of $(\theta_{n+1}, \sigma^2)|\mathbf{X}_n$, which will play the role of the updated prior in the Bayes theorem. We have

$$\begin{aligned} \pi(\theta_{n+1}, \sigma^2|\mathbf{X}_n) &= \iiint \pi(\theta_{n+1}, \sigma^2, p, \delta) dp\, d\delta\, d\theta_n \\ &= \iiint \pi(\theta_{n+1}|\theta_n, \sigma^2, p, \delta)\pi(\theta_n, \sigma^2|\mathbf{X}_n)\pi(p)\pi(\delta) dp\, d\delta\, d\theta_n \\ &= \int \pi(\theta_n, \sigma^2|\mathbf{X}_n) \\ &\quad \times \left[\int \pi(\delta) \left[\int \pi(\theta_{n+1}|\theta_n, \sigma^2, p, \delta)\pi(p)dp\right] d\delta\right] d\theta_n = (\mathbf{I}). \end{aligned}$$

For the inner integral, we have

$$\begin{aligned} I_1 &= \int \pi(\theta_{n+1}|\theta_n, \sigma^2, p, \delta)\pi(p)dp \\ &= \int \left[\frac{p}{\sqrt{2\pi\sigma^2}} exp\left\{-\frac{(\theta_{n+1}-\theta_n)^2}{2\sigma^2}\right\} + \frac{1-p}{\sqrt{2\pi\sigma^2}} exp\left\{-\frac{(\theta_{n+1}-\theta_n-\delta)^2}{2\sigma^2}\right\}\right] \\ &\quad \times \left[\frac{1}{Be(\alpha, \beta)} p^{\alpha-1}(1-p)^{\beta-1}\right] dp \\ &= \frac{Be(\alpha+1, \beta)}{Be(\alpha, \beta)} \frac{1}{\sqrt{2\pi\sigma^2}} exp\left\{-\frac{(\theta_{n+1}-\theta_n)^2}{2\sigma^2}\right\} \\ &\quad + \frac{Be(\alpha, \beta+1)}{Be(\alpha, \beta)} \frac{1}{\sqrt{2\pi\sigma^2}} exp\left\{-\frac{(\theta_{n+1}-\theta_n-\delta)^2}{2\sigma^2}\right\} \end{aligned}$$

$$= \left(\frac{\alpha}{\alpha+\beta}\right)\frac{1}{\sqrt{2\pi\sigma^2}}\,exp\left\{-\frac{(\theta_{n+1}-\theta_n)^2}{2\sigma^2}\right\}$$
$$+\left(\frac{\beta}{\alpha+\beta}\right)\frac{1}{\sqrt{2\pi\sigma^2}}\,exp\left\{-\frac{(\theta_{n+1}-\theta_n-\delta)^2}{2\sigma^2}\right\}$$

Then the second integral becomes

$$I_2 = \int \pi(\delta)\times I_1 d\delta$$
$$= \int\left[\frac{1}{\sqrt{2\pi\lambda\sigma^2}}\,exp\left\{-\frac{(\delta-\delta_0)^2}{2\lambda\sigma^2}\right\}\right]\times\left[\frac{\alpha}{(\alpha+\beta)\sqrt{2\pi\sigma^2}}\,exp\left\{-\frac{(\theta_{n+1}-\theta_n)^2}{2\sigma^2}\right\}\right.$$
$$\left.+\frac{\beta}{(\alpha+\beta)\sqrt{2\pi\sigma^2}}\,exp\left\{-\frac{(\theta_{n+1}-\theta_n-\delta)^2}{2\sigma^2}\right\}\right]d\delta$$
$$= \frac{\alpha}{(\alpha+\beta)\sqrt{2\pi\sigma^2}}\,exp\left\{-\frac{(\theta_{n+1}-\theta_n)^2}{2\sigma^2}\right\}+\frac{\beta}{(\alpha+\beta)\sqrt{2\pi\sigma^2}\sqrt{2\pi\lambda\sigma^2}}$$
$$\int exp\left\{-\frac{(\delta-\delta_0)^2}{2\lambda\sigma^2}-\frac{(\theta_{n+1}-\theta_n-\delta)^2}{2\sigma^2}\right\}d\delta$$
$$= \frac{\alpha}{(\alpha+\beta)\sqrt{2\pi\sigma^2}}\,exp\left\{-\frac{(\theta_{n+1}-\theta_n)^2}{2\sigma^2}\right\}$$
$$+\frac{\beta}{(\alpha+\beta)\sqrt{2\pi\sigma^2}\sqrt{\lambda+1}}\,exp\left\{-\frac{(\theta_{n+1}-\theta_n-\delta_0)^2}{2(\lambda+1)\sigma^2}\right\},$$

and thus we have

$$(\mathbf{I}) = \pi(\theta_{n+1},\sigma^2|\mathbf{X}_n) = \int\pi(\theta_n,\sigma^2|\mathbf{X}_n)\times I_2 d\theta_n$$
$$= \int\left[\sum_{i=0}^{2^n-1}\frac{w_i^{(n)}}{\sqrt{2\pi k_i^{(n)}\sigma^2}}\,exp\left\{-\frac{\left(\theta_n-\mu_i^{(n)}\right)^2}{2k_i^{(n)}\sigma^2}\right\}\right.$$
$$\left.\times\frac{1}{\Gamma(A^{(n)})\left[B_i^{(n)}\right]^{A^{(n)}}}\frac{1}{(\sigma^2)^{A^{(n)}+1}}\,exp\left\{-\frac{1}{B_i^{(n)}\sigma^2}\right\}\right]$$
$$\times\left[\frac{\alpha}{(\alpha+\beta)\sqrt{2\pi\sigma^2}}\,exp\left\{-\frac{(\theta_{n+1}-\theta_n)^2}{2\sigma^2}\right\}\right.$$
$$\left.+\frac{\beta}{(\alpha+\beta)\sqrt{2\pi\sigma^2}\sqrt{\lambda+1}}\,exp\left\{-\frac{(\theta_{n+1}-\theta_n-\delta_0)^2}{2(\lambda+1)\sigma^2}\right\}\right]d\theta_n$$

$$= \frac{1}{2\pi(\alpha+\beta)\Gamma(A^{(n)})(\sigma^2)^{A^{(n)}+2}} \sum_{i=0}^{2^n-1} \frac{w_i^{(n)}}{\sqrt{k_i^{(n)}}} \frac{1}{\left[B_i^{(n)}\right]^{A^{(n)}}} exp\left\{-\frac{1}{B_i^{(n)}\sigma^2}\right\}$$

$$\times \left[\alpha \int exp\left\{-\frac{(\theta_n-\theta_{n+1})^2}{2\sigma^2} - \frac{\left(\theta_n-\mu_i^{(n)}\right)^2}{2k_i^{(n)}\sigma^2}\right\} d\theta_n\right.$$

$$\left. + \frac{\beta}{\sqrt{\lambda+1}} \int exp\left\{-\frac{(\theta_n-(\theta_{n+1}-\delta_0))^2}{2(\lambda+1)\sigma^2} - \frac{\left(\theta_n-\mu_i^{(n)}\right)^2}{2k_i^{(n)}\sigma^2}\right\} d\theta_n\right]$$

$$= \frac{1}{2\pi(\alpha+\beta)\Gamma(A^{(n)})(\sigma^2)^{A^{(n)}+2}} \sum_{i=0}^{2^n-1} \frac{w_i^{(n)}}{\sqrt{k_i^{(n)}}} \frac{1}{\left[B_i^{(n)}\right]^{A^{(n)}}} exp\left\{-\frac{1}{B_i^{(n)}\sigma^2}\right\}$$

$$\times \left[\alpha \frac{\sqrt{2\pi k_i^{(n)}\sigma^2}}{\sqrt{k_i^{(n)}+1}} exp\left\{-\frac{\left(\theta_{n+1}-\mu_i^{(n)}\right)^2}{2(k_i^{(n)}+1)\sigma^2}\right\}\right.$$

$$\left. + \frac{\beta}{\sqrt{\lambda+1}} \frac{\sqrt{\lambda+1}\sqrt{2\pi k_i^{(n)}\sigma^2}}{\sqrt{\lambda+1+k_i^{(n)}}} exp\left\{-\frac{\left(\theta_{n+1}-\delta_0-\mu_i^{(n)}\right)^2}{2(\lambda+1+k_i^{(n)})\sigma^2}\right\}\right]$$

$$= \sum_{i=0}^{2^n-1} \frac{w_i^{(n)}}{(\alpha+\beta)} \frac{1}{\Gamma(A^{(n)})\left[B_i^{(n)}\right]^{A^{(n)}}} \frac{1}{(\sigma^2)^{A^{(n)}+1}} exp\left\{-\frac{1}{B_i^{(n)}\sigma^2}\right\}$$

$$\times \left[\frac{\alpha}{\sqrt{2\pi\left(k_i^{(n)}+1\right)\sigma^2}} exp\left\{-\frac{\left(\theta_{n+1}-\mu_i^{(n)}\right)^2}{2(k_i^{(n)}+1)\sigma^2}\right\}\right.$$

$$\left. + \frac{\beta}{\sqrt{2\pi\left(\lambda+k_i^{(n)}+1\right)\sigma^2}} exp\left\{-\frac{\left(\theta_{n+1}-\mu_i^{(n)}-\delta_0\right)^2}{2(\lambda+k_i^{(n)}+1)\sigma^2}\right\}\right].$$

Therefore, the prior of (θ_{n+1}, σ^2) at stage $n+1$ of the process will be

$$\pi(\theta_{n+1}, \sigma^2|\mathbf{X}_n) \sim \sum_{i=0}^{2^n-1} \left[\frac{\alpha w_i^{(n)}}{\alpha+\beta} N(\mu_i^{(n)}, (k_i^{(n)}+1)\sigma^2) IG(A^{(n)}, B_i^{(n)})\right.$$

$$\left. + \frac{\beta w_i^{(n)}}{\alpha+\beta} N(\mu_i^{(n)}+\delta_0, (\lambda+k_i^{(n)}+1)\sigma^2) IG(A^{(n)}, B_i^{(n)})\right]$$

$$\sim \sum_{i=0}^{2^n-1} \left[\frac{\alpha w_i^{(n)}}{\alpha+\beta} \pi_i(\theta_{n+1}, \sigma^2|\mathbf{X}_n) + \frac{\beta w_i^{(n)}}{\alpha+\beta} \pi_i^*(\theta_{n+1}, \sigma^2|\mathbf{X}_n)\right],$$

where

$$\pi_i(\theta_{n+1}, \sigma^2|\mathbf{X}_n) \equiv N\big(\mu_i^{(n)}, (k_i^{(n)}+1)\sigma^2\big)IG\big(A^{(n)}, B_i^{(n)}\big)$$

and

$$\pi_i^*(\theta_{n+1}, \sigma^2|\mathbf{X}_n) \equiv N\big(\mu_i^{(n)}+\delta_0, (\lambda + k_i^{(n)}+1)\sigma^2\big)IG\big(A^{(n)}, B_i^{(n)}\big).$$

At time $n+1$, the observation x_{n+1} will become available from the likelihood:

$$f(x_{n+1}|\theta_{n+1}, \sigma^2) \sim N(\theta_{n+1}, c\sigma^2).$$

Then for the posterior of interest, we will have

$$p(\theta_{n+1}, \sigma^2|\mathbf{X}_{n+1}) \propto f(x_{n+1}|\theta_{n+1}, \sigma^2)\pi(\theta_{n+1}, \sigma^2|\mathbf{X}_n)$$

$$= \sum_{i=0}^{2^n-1}\left[\frac{\alpha w_i^{(n)}}{\alpha+\beta} f(x_{n+1}|\theta_{n+1}, \sigma^2)\pi_i(\theta_{n+1}, \sigma^2|\mathbf{X}_n) + \frac{\beta w_i^{(n)}}{\alpha+\beta} f(x_{n+1}|\theta_{n+1}, \sigma^2)\pi_i^*(\theta_{n+1}, \sigma^2|\mathbf{X}_n)\right].$$

We will call

$$m_i(x_{n+1}) = \iint f(x_{n+1}|\theta_{n+1}, \sigma^2)\pi_i(\theta_{n+1}, \sigma^2|\mathbf{X}_n)d\theta_{n+1}d\sigma^2$$

$$= \iint\left[\frac{1}{\sqrt{2\pi c\sigma^2}}\exp\left\{-\frac{(\theta_{n+1}-x_{n+1})^2}{2c\sigma^2}\right\}\frac{1}{\sqrt{2\pi\,(k_i^{(n)}+1)\sigma^2}} \times exp\left\{-\frac{(\theta_{n+1}-\mu_i^{(n)})^2}{2(k_i^{(n)}+1)\sigma^2}\right\} \times \frac{1}{\Gamma(A^{(n)})\big[B_i^{(n)}\big]^{A^{(n)}}}\frac{1}{(\sigma^2)^{A^{(n)}+1}} \times exp\left\{-\frac{1}{B_i^{(n)}\sigma^2}\right\}\right]d\theta_{n+1}d\sigma^2$$

$$= \int\frac{1}{\sqrt{2\pi c\sigma^2}}\frac{1}{\sqrt{2\pi\,(k_i^{(n)}+1)\sigma^2}}\frac{1}{\Gamma(A^{(n)})\big[B_i^{(n)}\big]^{A^{(n)}}}\frac{1}{(\sigma^2)^{A^{(n)}+1}} \times exp\left\{-\frac{1}{B_i^{(n)}\sigma^2}\right\} \times \left[\int exp\left\{-\frac{(\theta_{n+1}-\mu_i^{(n)})^2}{2(k_i^{(n)}+1)\sigma^2}\right\} \times exp\left\{-\frac{(\theta_{n+1}-x_{n+1})^2}{2c\sigma^2}\right\}d\theta_{n+1}\right]d\sigma^2 = \textbf{(II)},$$

but we know that

$$exp\left\{-\frac{(\theta-x)^2}{2c\sigma^2}-\frac{(\theta-\mu)^2}{2k\sigma^2}\right\}=exp\left\{-\frac{\left[\theta-\frac{kx+c\mu}{k+c}\right]^2}{2\left(\frac{kc}{k+c}\right)\sigma^2}\right\}\times exp\left\{-\frac{(x-\mu)^2}{2(k+c)\sigma^2}\right\}$$

and thus for the inner integral I_3, we will have

$$I_3=exp\left\{-\frac{\left(x_{n+1}-\mu_i^{(n)}\right)^2}{2(c+k_i^{(n)}+1)\sigma^2}\right\}\int exp\left\{-\frac{\left[\theta_{n+1}-\frac{\left(k_i^{(n)}+1\right)x_{n+1}+c\mu_i^{(n)}}{\left(c+k_i^{(n)}+1\right)}\right]^2}{2\left[\frac{\left(k_i^{(n)}+1\right)c}{c+k_i^{(n)}+1}\right]\sigma^2}\right\}d\theta_{n+1}$$

$$=\sqrt{2\pi\left[\frac{(k_i^{(n)}+1)c}{c+k_i^{(n)}+1}\right]\sigma^2}\;exp\left\{-\frac{\left(x_{n+1}-\mu_i^{(n)}\right)^2}{2(c+k_i^{(n)}+1)\sigma^2}\right\},$$

then

$$(\mathbf{II})=\int\frac{1}{\sqrt{2\pi c\sigma^2}}\frac{1}{\sqrt{2\pi\,(k_i^{(n)}+1)\sigma^2}}\frac{1}{\Gamma(A^{(n)})\left[B_i^{(n)}\right]^{A^{(n)}}}\frac{1}{(\sigma^2)^{A^{(n)}+1}}$$

$$\times exp\left\{-\frac{1}{B_i^{(n)}\sigma^2}\right\}\times\sqrt{2\pi\left[\frac{(k_i^{(n)}+1)c}{c+k_i^{(n)}+1}\right]\sigma^2}$$

$$\times exp\left\{-\frac{\left(x_{n+1}-\mu_i^{(n)}\right)^2}{2(c+k_i^{(n)}+1)\sigma^2}\right\}d\sigma^2$$

$$=\frac{1}{\sqrt{2\pi\,(c+k_i^{(n)}+1)}}\frac{1}{\Gamma(A^{(n)})\left[B_i^{(n)}\right]^{A^{(n)}}}$$

$$\times\int\frac{1}{(\sigma^2)^{A^{(n)}+\frac{1}{2}+1}}\,exp\left\{-\left[\frac{\left(x_{n+1}-\mu_i^{(n)}\right)^2}{2(c+k_i^{(n)}+1)}+\frac{1}{B_i^{(n)}}\right]\frac{1}{\sigma^2}\right\}d\sigma^2,$$

where the integral is the kernel of an Inverted Gamma distribution. Thus we have

$$I_4=\int\frac{1}{(\sigma^2)^{A^{(n)}+\frac{1}{2}+1}}\,exp\left\{-\left[\frac{\left(x_{n+1}-\mu_i^{(n)}\right)^2}{2(c+k_i^{(n)}+1)}+\frac{1}{B_i^{(n)}}\right]\frac{1}{\sigma^2}\right\}d\sigma^2$$

$$=\Gamma\left(A^{(n)}+\frac{1}{2}\right)\left[\frac{\left(x_{n+1}-\mu_i^{(n)}\right)^2}{2(c+k_i^{(n)}+1)}+\frac{1}{B_i^{(n)}}\right]^{-\left(A^{(n)}+\frac{1}{2}\right)}$$

and therefore

$$m_i(x_{n+1}) = \frac{\Gamma\left(A^{(n)} + \frac{1}{2}\right)}{\Gamma(A^{(n)})} \sqrt{\frac{B_i^{(n)}}{2\pi\left(c + k_i^{(n)} + 1\right)}} \left[1 + \frac{B_i^{(n)}\left(x_{n+1} - \mu_i^{(n)}\right)^2}{2(c + k_i^{(n)} + 1)}\right]^{-\left(A^{(n)} + \frac{1}{2}\right)}.$$

Similarly,

$$m_i^*(x_{n+1}) = \iint f(x_{n+1}|\theta_{n+1}, \sigma^2)\pi_i^*(\theta_{n+1}, \sigma^2|\mathbf{X}_n)d\theta_{n+1}d\sigma^2$$

$$= \frac{\Gamma(A^{(n)} + \frac{1}{2})}{\Gamma(A^{(n)})} \sqrt{\frac{B_i^{(n)}}{2\pi\left(c + \lambda + k_i^{(n)} + 1\right)}}$$

$$\times \left[1 + \frac{B_i^{(n)}\left(x_{n+1} - \mu_i^{(n)} - \delta_0\right)^2}{2(c + \lambda + k_i^{(n)} + 1)}\right]^{-\left(A^{(n)} + \frac{1}{2}\right)}.$$

We then have

$$p(\theta_{n+1}, \sigma^2|\mathbf{X}_{n+1}) \propto \sum_{i=0}^{2^n-1} \left[\frac{\alpha w_i^{(n)}}{\alpha + \beta} m_i(x_{n+1}) p_i(\theta_{n+1}, \sigma^2|\mathbf{X}_{n+1}) + \frac{\beta w_i^{(n)}}{\alpha + \beta} m_i^*(x_{n+1}) p_i^*(\theta_{n+1}, \sigma^2|\mathbf{X}_{n+1})\right]$$

where $p_i(\cdot)$ and $p_i^*(\cdot)$ will be the posterior distributions, which (based on Bayes theorem) will be Normal-Inverted Gamma. If we call NC the normalizing constant of the posterior distribution, i.e.,

$$NC = \sum_{i=0}^{2^n-1} \left[\left(\frac{\alpha}{\alpha + \beta}\right) w_i^{(n)} m_i(x_{n+1}) + \left(\frac{\beta}{\alpha + \beta}\right) w_i^{(n)} m_i^*(x_{n+1})\right],$$

we then get

$$p(\theta_{n+1}, \sigma^2|\mathbf{X}_{n+1}) = \sum_{i=0}^{2^n-1} \left[\frac{\left(\frac{\alpha}{\alpha+\beta}\right) w_i^{(n)} m_i(x_{n+1})}{NC} p_i(\theta_{n+1}, \sigma^2|\mathbf{X}_{n+1}) + \frac{\left(\frac{\beta}{\alpha+\beta}\right) w_i^{(n)} m_i^*(x_{n+1})}{NC} p_i^*(\theta_{n+1}, \sigma^2|\mathbf{X}_{n+1})\right]$$

$$= \sum_{i=0}^{2^n-1} \left[w_{2i}^{(n+1)} p_i(\theta_{n+1}, \sigma^2|\mathbf{X}_{n+1}) + w_{2i+1}^{(n+1)} p_i^*(\theta_{n+1}, \sigma^2|\mathbf{X}_{n+1})\right]$$

where

$$w_{2i}^{(n+1)} = \frac{\left(\frac{\alpha}{\alpha+\beta}\right) w_i^{(n)} m_i(x_{n+1})}{NC}$$

and

$$w_{2i+1}^{(n+1)} = \frac{\left(\frac{\beta}{\alpha+\beta}\right) w_i^{(n)} m_i^*(x_{n+1})}{NC}.$$

For the posterior distribution, applying standard Bayes theory we have

$$p_i(\theta_{n+1}, \sigma^2|\mathbf{X}_{n+1}) \sim N\left(\mu_{2i}^{(n+1)},\ (k_{2i}^{(n+1)}+1)\sigma^2\right) IG\left(A^{(n+1)},\ B_{2i}^{(n+1)}\right)$$

$$p_i^*(\theta_{n+1}, \sigma^2|\mathbf{X}_{n+1}) \sim N\left(\mu_{2i+1}^{(n+1)},\ (k_{2i+1}^{(n+1)}+1)\sigma^2\right) IG\left(A^{(n+1)},\ B_{2i+1}^{(n+1)}\right),$$

where

$$A^{(n+1)} = A^{(n)} + \frac{1}{2}$$

$$\mu_{2i}^{(n+1)} = \frac{(k_i^{(n)}+1)x_{n+1} + c\mu_i^{(n)}}{c + k_i^{(n)} + 1}$$

$$\mu_{2i+1}^{(n+1)} = \frac{(\lambda + k_i^{(n)}+1)x_{n+1} + c\left(\mu_i^{(n)} + \delta_0\right)}{c + \lambda + k_i^{(n)} + 1}$$

$$k_{2i}^{(n+1)} = \frac{(k_i^{(n)}+1)c}{c + k_i^{(n)} + 1}$$

$$k_{2i+1}^{(n+1)} = \frac{(\lambda + k_i^{(n)}+1)c}{c + \lambda + k_i^{(n)} + 1}$$

$$B_{2i}^{(n+1)} = \left[\frac{\left(x_{n+1} - \mu_i^{(n)}\right)^2}{2(c + k_i^{(n)} + 1)} + \frac{1}{B_i^{(n)}}\right]^{-1}$$

$$B_{2i+1}^{(n+1)} = \left[\frac{\left(x_{n+1} - \mu_i^{(n)} - \delta_0\right)^2}{2(c + \lambda + k_i^{(n)} + 1)} + \frac{1}{B_i^{(n)}}\right]^{-1}.$$

References

1. Casella, G. and Berger, R.L., *Statistical Inference*, Pacific Grove, CA: Wadsworth & Brooks-Cole, 1990.
2. Chang, J. T. and Fricker, R.D., "Detecting when a monotonically increasing mean has crossed a threshold," *Journal of Quality Technology*, 31, 217–234, 1999.

3. De Groot, M.H., *Optimal Statistical Decision*, New York: McGraw-Hill, 1970.
4. Harvey, A.C., *Forecasting, Structural Time Series Models and the Kalman Filter*, New York: Cambridge University Press, 1989.
5. Fox, A.J., "Outliers in time series," *Journal of the Royal Statistical Society, Series B*, 34, 350–363 1972.
6. Geisser, S., *Predictive Inference: An Introduction*, London: Chapman & Hall, 1993.
7. Jeffreys, H., *Theory of Probability*, 2nd Ed., Oxford UK: University Press, 1948.
8. Kalman, R.E., "A new approach to linear filtering and prediction problems," *Journal of Basic Engineering*, 82, 35–45 1960.
9. Kirkendall, N.J., "The relationship between certain Kalman filter models and exponential smoothing models," in *Statistical Process Control in Automated Manufacturing*, Keats, J.B. and Hubele, N.F. (Eds.) New York: Marcel Dekker, 1989.
10. Page, E.S., "Continuous inspection schemes," *Biometrika*, 41, 100–115, 1954.
11. Roberts, S.W., "Control chart tests based on geomrtric moving averages," *Technometrics*, 1, 239–250, 1959.
12. Shewhart, W.A., *Economic Control of Quality of Manufactured Product*, New York: Van Nostrand, 1931.
13. Tsiamyrtzis, P., "A Bayesian approach to quality control problems," *Ph.D. Dissertation*, School of Statistics, University of Minnesota 2000.
14. Tsiamyrtzis, P. and Hawkins, D.M., "A Bayesian scheme to detect changes in the mean of a short run process," *Technometrics*, 47, 446–456, 2005.
15. West, M., "Approximating posterior distributions by mixtures," *Journal of Royal Statistical Society, Series B*, 55, 2, 409–422, 1993.
16. Woodward, P.W., and Naylor, J.C., "An application of Bayesian methods in SPC," *The Statistician*, 42, 461–469, 1993.
17. Wright, C.M., Booth, D.E., and Hu, M.Y., "Joint estimation: SPC method for short-run autocorellated data," *Journal of Quality Technology*, 33, 365–378, 2001.

4

Empirical Bayes Process Monitoring Techniques

Jyh-Jen H. Shiau
Institute of Statistics, National Chiao Tung University

Carol J. Feltz
Division of Statistics, Northern Illinois University

CONTENTS

ABSTRACT In this chapter, some empirical Bayes monitoring techniques are presented for statistical process control (SPC). In particular, techniques are presented for monitoring univariate and multivariate continuous measurements, as well as yield/defect data (pass/fail data) and polytomous data. For each type of data, a Bayesian model is assumed for the process data, a prior with unknown hyperparameters is chosen, and empirical Bayes estimators are accordingly given for each of the parameters in the Bayesian model. In addition to estimating where the current process is, the estimators provide a distribution for the process parameter of interest, such as yield. Furthermore, by combining the empirical Bayes techniques with exponential smoothing, one can see where the process has been in the past, as well as where it is currently. Recursive equations for the estimators are provided for efficient computation, which is essential for successful implementation of real-time analysis of processes in factories.

4.1 Introduction

Many *statistical process control* (SPC) techniques have been proven useful in quality and productivity improvement of products and processes. Among them, monitoring processes by control charts has become a standard or even a required procedure in many industries, especially in manufacturing. Most of the control charts designed so far were derived based on statistical inferences from frequentist points of view. Only a few researchers have developed process monitoring schemes from Bayesian perspectives.

In this chapter, we present an empirical Bayes methodology for monitoring process data, including univariate/multivariate and continuous/discrete measurements, in which the process parameters of interest such as mean or yield/defect levels can vary over time. Some of these empirical Bayes techniques have been implemented in high-speed electronics manufacturing facilities, where the testing equipment is connected to a central computer and data are taken automatically as product items are being tested. The empirical Bayes methodology takes advantage of this on-line environment by using previous data to choose its own prior. By using recursive equations and maintaining a small database of sufficient statistics, these empirical Bayes techniques provide a very efficient method for monitoring quality.

Some of the key features of this empirical Bayes methodology for process monitoring are:

1. Using the likelihood as the sampling distribution to identify the sampling variability of measurements coming from the current production.
2. Using the empirically estimated Bayes prior to infer the process mean and process variation, or how the process changes over time.

3. Using the posterior distribution to create a distribution estimate for the varying process parameter of interest in the process— how much variability is inherent in this process parameter.
4. By weighting the observations differently, one can create two distributions to see the "long-term" pattern and a "short-term" view of the behavior of the process parameter. Also, embedded in each of the exponentially weighted estimators of the process parameters is an *exponentially weighted moving average* (EWMA) statistic. Both features can be used for process monitoring.
5. Using recursive equations for estimators based on sufficient statistics to make the computation efficient.

Univariate process data monitoring is still the most commonly used SPC technique in real-life applications. In Section 4.2, we first describe some univariate methods for monitoring continuous measurements. However, with enhancing technologies and computer power, more and more frequently we encounter process data of two or more variables that simultaneously impact quality or the yield of the process. In addition, these variables can "interact" in a way that, even when each variable is within its specification limits, the product might not function. Thus, the role of monitoring multivariate process data has become more and more important in quality improvement nowadays. Section 4.3 extends the monitoring techniques for univariate continuous data to techniques for multivariate continuous data.

Because yield and defect data are also frequently observed and of great importance to process control in many industrial applications, empirical Bayes techniques for monitoring binary and polytomous data are presented in Section 4.4 and Section 4.5, respectively. We conclude the chapter with a brief summary and discussion in Section 4.6.

Tremendous amounts of literature are found in both research fields of Bayesian analysis and quality control techniques. In this chapter, references are given to works used or related to developing the techniques presented here as well as the subsequent works in the field. We apologize in advance to those authors whose works might have been inadvertently missed. The literature review pertaining to each technique is included in that section.

4.2 Empirical Bayes Process Monitoring for a Univariate Continuous Variable

Several Bayesian or Bayesian-like analyses, such as Kalman filtering, have been applied to the problem of process control. Kamat [14] developed a Bayesian control procedure for a variable with linear shift tendencies. Hoadley [9] developed a more general Bayesian procedure to monitor and rate quality variables by taking sample audits. Phadke [18] developed a similar procedure using a Kalman filter approach. The following empirical Bayes analysis was developed in Sturm et al. [21]

Let X_t be the continuous quality characteristic of interest, such as voltage or resistance. Since X_t has measurement and sampling error, we assume that X_t is normally distributed with mean μ_t and variance σ^2 at time t. More specifically, the conditional probability density function (p.d.f.) of X_t given μ_t is given by

$$g_1(X_t|\mu_t) = \frac{1}{\sqrt{(2\pi\sigma^2)}} exp\left[\frac{-(X_t - \mu_t)^2}{2\sigma^2}\right].$$

Here σ^2 is usually an unknown constant representing the combined variability of measurement and sampling errors.

For example, suppose in semiconductor fabrication, the SiO_2 layer is targeted to be 720 nm. When the layer is made, its thickness will not always be exactly 720 nm, as there is natural variability (often called sampling variation) exists in the process that forms the layer. Also, measurement errors occur when measuring the thickness by any measuring equipment. Thus, we assume that the thickness values of the same layer at the same place on several plates made at time t are generated by approximately the same process, so that the thickness X_t at time t should have a distribution that is centered at μ_t.

Assume also that the mean quality characteristic μ_t varies over time due to processing variations and environmental changes such as temperature and humidity. This variation (called process variation) can be described by the so-called "prior" distribution in Bayesian analysis. Assume that μ_t is normally distributed with mean μ and variance γ^2. Then the p.d.f. of μ_t is given by

$$g_2(\mu_t) = \frac{1}{\sqrt{(2\pi\gamma^2)}} exp\left[\frac{-(\mu_t - \mu)^2}{2\gamma^2}\right],$$

where μ and γ^2 are unknown constants, often called hyperparameters in Bayesian analysis.

Please note the two sources of variation in this Bayesian model. The first source of variation, which contains measurement errors as well as natural fluctuations of the quality characteristic X_t when the process is centered at μ_t, is represented by the **sampling variance** σ^2. The second source of variation, which is due to the changing behavior of the process parameter μ_t over time, is represented by the **process variance** γ^2.

To estimate the mean μ_t of the quality characteristic X_t at time t using the information contained in X_t, Bayes' theorem is invoked as follows. The p.d.f. of μ_t given X_t is

$$f(\mu_t|X_t) = \frac{g_1(X_t|\mu_t)g_2(\mu_t)}{\int g_1(X_t|\mu_t)g_2(\mu_t)d\mu_t}.$$

In Bayesian terminology, $f(\mu_t|X_t)$ is called the **posterior distribution** of μ_t. For this application, μ_t given X_t has a normal distribution with mean

$$\mu_{pt} = w\mu + (1 - w)X_t \tag{4.1}$$

and variance

$$\sigma_p^2 = \gamma^2 w, \tag{4.2}$$

where $w = \sigma^2/(\gamma^2 + \sigma^2)$, as described in Chapter 1. We can see clearly from Equation (4.1) that the posterior mean is a linear combination of the current observation X_t and the prior mean μ. Also, Equation (4.2) implies that the posterior variance is only a portion of the process variance γ^2.

As the parameters μ, σ^2, and γ^2 are not known, in the empirical Bayes methodology they are estimated empirically from the X_t's (the only observables) and used to estimate the posterior distribution of μ_t, or where the process is at time t. Estimation of these unknown parameters will be discussed next.

4.2.1 Estimation

To be theoretically rigorous, it must be noted that with μ_t varying over time, the X_t's are independent but not identically distributed and the covariance structure between the X_t's and μ_t's is messy. This makes any estimation procedure difficult. By making certain assumptions that approximately hold in many applications, these difficulties can be avoided.

Assume that the process parameter μ_t stays relatively constant over small time intervals, for example, minutes or hours. Then consecutive X_t's are approximately independent and identically distributed (i.i.d.) with the p.d.f. $g_1(X_t|\mu_t)$. Given a set of data of size T, $\{X_1, \ldots, X_T\}$, we can estimate μ, σ^2, and γ^2 as follows.

By the double expectation method, the marginal distribution of X_t has the mean

$$E(X_t) = E_{\mu_t} E_{X_t}(X_t|\mu_t) = \mu.$$

Because each X_t is an unbiased estimate of μ, a reasonable estimator $\hat{\mu}$ of μ is the overall average of the X_t's. That is,

$$\hat{\mu} = \frac{\sum_{t=1}^{T} X_t}{T}. \tag{4.3}$$

Next, we estimate the sampling variance σ^2. By assuming that μ_t is not changing rapidly, adjacent observations — for example, X_t and X_{t-1} — should come from approximately the same distribution. Because $Var(X_t - X_{t-1}) = 2\sigma^2$ for i.i.d. random variables, estimating σ^2 by averaging squared moving ranges is natural:

$$\hat{\sigma}^2 = \frac{\sum_{t=2}^{T}(X_t - X_{t-1})^2/2}{T-1}. \tag{4.4}$$

In estimating the process variance γ^2, we note that the marginal variance of X_t can be obtained by the conditional expectation method as follows:

$$V \equiv Var(X_t) = E_{\mu_t}[Var(X_t|\mu_t)] + Var_{\mu_t}[E(X_t|\mu_t)] = \sigma^2 + \gamma^2.$$

In other words, the total variance in X_t is equal to the **sampling variance** plus the **process variance**. By using $\hat{\mu}$ of Equation (4.3) as the estimator for μ, we can estimate the total variance V by

$$\hat{V} = \frac{\sum_{t=1}^{T}(X_t - \hat{\mu})^2}{T}. \tag{4.5}$$

Let $\hat{\gamma}^2$ be $\hat{V} - \hat{\sigma}^2$ if it is positive; otherwise, let $\hat{\gamma}^2$ be $\kappa\hat{V}$, where κ is a small arbitrary number, say, $0 < \kappa < 0.1$. It is possible that the estimator $\hat{V} - \hat{\sigma}^2$ may be less than zero is possible, especially when the process variance (γ^2) is small relative to the sampling variance (σ^2). If this is the case, selecting a small value for κ will bound the estimate of γ^2 relative to σ^2.

Note that the estimates given in Equations (4.3)–(4.5) would soon be weighted down with "old" data since T gets larger and larger as time goes by. However, if the process changes over time, the current data should have more weight in the estimation of the current process distribution than the data collected far away back in the past. To reflect this, the following exponentially weighted estimators were proposed by Sturm et al. [21] for μ, σ^2, and V. Let the X_t's be ordered in time so that X_T is the most current observation. The weighted version of the estimators are

$$\hat{\mu}_T = \frac{\sum_{t=1}^{T}\lambda^{T-t}X_t}{\sum_{t=1}^{T}\lambda^{T-t}}, \tag{4.6}$$

$$\hat{\sigma}_T^2 = \frac{\sum_{t=2}^{T}\lambda^{T-t}(X_t - X_{t-1})^2/2}{\sum_{t=2}^{T}\lambda^{T-t}}, \tag{4.7}$$

and

$$\hat{V}_T = \frac{\sum_{t=1}^{T}\lambda^{T-t}(X_t - \hat{\mu})^2}{\sum_{t=1}^{T}\lambda^{T-t}}, \tag{4.8}$$

respectively. Here λ^{T-t} is the weight given to time period t, where λ is some positive number less than 1 (usually $0.80 < \lambda < 1.0$). The choice of λ depends on the application. If data taken m observations ago are no longer relevant to the current process, then the λ should be chosen appropriately. To weight away data that is m observations ago, λ can be chosen to be $\epsilon^{1/m}$, where ϵ is a very small number given to weight the observation m periods ago.

For example, to weight away data that are 50, 100, or 200 observations ago, by letting $\epsilon = 0.0001$, we can choose $\lambda = 0.832, 0.912$ or $\lambda = 0.955$, respectively. In our experience, $\lambda = 0.99$ and $\lambda = 0.95$ were used for some applications in factories where the monitoring technique was implemented.

By incorporating the weighting into the empirical Bayes approach, we maintain our distributional structure so that we can partition the variability into the sampling and process variability as well as using the distribution to identify shifts in the process mean. Also, by using the weight λ^{T-t}, the older X_t's gradually have less and less importance in the calculations. When analyzing large amounts of data on-line, these weighted estimators have the additional benefit that old data are automatically and gradually weighted out using the above equations, so that time-expensive reads and purges of databases are not necessary. One needs only update the sufficient statistics with the current observations. For efficient computation, we rewrite the above equations in the recursive form as follows.

Let $\theta_t = 1/\sum_{k=1}^{t} \lambda^{t-k}$ and note that $\theta_1 = 1$. After some algebra, the following EWMA-like recursive equations can be obtained:

$$\hat{\mu}_0 = 0, \quad \hat{\mu}_t = \theta_t X_t + (1 - \theta_t)\hat{\mu}_{t-1}, \quad t = 1, \ldots, T,$$

$$\hat{\sigma}_1^2 = 0, \quad \hat{\sigma}_t^2 = \theta_t (X_t - X_{t-1})^2/2 + (1 - \theta_t)\hat{\sigma}_{t-1}^2, \quad t = 2, \ldots, T,$$

$$\hat{V}_0 = 0, \quad \hat{V}_t = \theta_t (X_t - \hat{\mu}_t)^2 + (1 - \theta_t)(\hat{V}_{t-1} + (\hat{\mu}_t - \hat{\mu}_{t-1})^2), \quad t = 1, \ldots, T.$$

The recursive equations for the non-weighted estimators (4.3)–(4.5) can also be obtained from the above recursive equations by taking $\lambda = 1$. For this special case, $\theta_t = 1/t$.

Finally, by using the exponentially decaying weights, one can see that the estimator for μ is some sort of EWMA, but not exactly a usual EWMA statistic used in SPC. Note the smoothing parameter in an EWMA statistic is a constant, whereas the θ_t here depends on t. Nevertheless, θ_t approaches the constant $1 - \lambda$ as time t goes on.

4.2.2 A Monitoring Scheme

One approach for identifying process changes would be to select two different smoothing parameters for λ, λ_{long} and λ_{short}, to create a "long-term" and a "short-term" posterior distribution for μ_t, respectively. With $\lambda_{short} < \lambda_{long}$, the short-term distribution uses fewer "effective" observations and hence is more volatile than the long-term distribution. Also, by giving more weights to more recent historical data, it also reflects more of the current situation. When the two distributions start to separate, a change in the distribution of μ_t is flagged. This is the approach suggested by Sturm et al. [21], although details were not given there. Specifically, Sturm et al. proposed to flag an out-of-control signal if the short-term posterior mean estimate falls outside of the 10th to 90th percentile interval of the long-term posterior distribution. By using a box-plot to represent the estimated long-term distribution and a symbol "s" to locate the estimated short-term posterior mean, they graphically displayed an easy-to-see status report of the process for on-line operators.

4.2.3 An Example

We illustrate the monitoring scheme of Sturm et al. [21] described above by the following example. Suppose that the SiO_2 layer in a semiconductor fabrication process is supposed to be deposited to a target thickness of 720 nm on a wafer. However, due to a variety of mechanical and composition factors, variation of the deposition thickness exists. In fact, the sampling variability in the past is approximately 90nm^2. Also, due to mechanical and environmental conditions, the mean thickness deposited can change over time to give some process variability. For our example, suppose the process has been stable for some time, so that the short-term distribution is centered over the long-term distribution.

Suppose we monitor 10 observations of thickness of the SiO_2 layer for sequential wafers. For this example, let $\lambda_{short} = 0.95$ and $\lambda_{long} = 0.99$. The effect of each observation is given in Table 4.1, where the box-plot denotes the long-term posterior distribution and the "s" indicates the location of the "short-term" posterior mean.

If the process is under control, the "short-term" posterior mean should be

TABLE 4.1

Short-Term and Long-Term Posterior Distributions

Obs	Short-Term Dist (λ=0.95) of $\mu_t\|X_t$					Long-Term Dist (λ=0.99) of $\mu_t\|X_t$					Box-Plots
	$\hat{\mu}_{pt}$	$\hat{\sigma}_p^2$	$\hat{\mu}$	$\hat{\sigma}^2$	$\hat{\gamma}^2$	$\hat{\mu}_{pt}$	$\hat{\sigma}_p^2$	$\hat{\mu}$	$\hat{\sigma}^2$	$\hat{\gamma}^2$	
714.3	719.1	8.2	719.7	77.8	9.3	719.4	8.1	719.9	80.3	9.1	⊢[s]⊣
729.9	721.1	7.0	720.2	80.0	7.7	721.0	7.9	720.0	80.8	8.7	⊢[s]⊣
719.3	720.1	4.3	720.2	78.8	4.5	720.0	7.4	720.0	80.5	8.1	⊢[s]⊣
710.3	719.0	6.3	720.0	76.9	6.9	719.6	7.5	720.0	79.5	8.3	⊢[s]⊣
717.0	719.4	5.4	719.5	74.1	5.8	719.6	7.5	719.9	79.5	8.3	⊢[s]⊣
730.8	721.0	6.2	720.1	75.2	6.7	721.1	7.7	720.0	79.7	8.5	⊢[s]⊣
741.7	726.4	19.0	721.2	74.4	25.6	723.1	10.8	720.2	79.5	12.4	⊢[s]⊣
755.2	739.0	37.5	722.9	75.2	74.7	728.5	18.2	720.6	79.6	23.5	⊢[ ]s⊣
746.8	737.0	41.6	724.1	73.2	96.3	727.9	21.6	720.8	79.2	29.7	⊢[ ]s⊣
752.4	742.9	45.5	725.5	70.4	128.8	731.5	26.1	721.2	78.5	39.1	⊢[ ]s⊣
752.4	742.9	45.5	725.5	70.4	128.8	731.5	26.1	721.2	78.5	39.1	⊢[ ]s⊣

```
        ├──────────────┼──────────────┤
       680            720            760
```

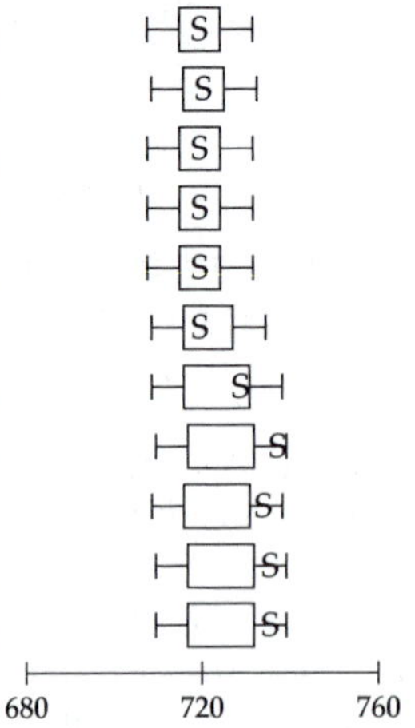

Note: ("s" indicates the location of the "short-term" posterior mean.)

hovering around the center of the "long-term" posterior distribution. However, with these observations, one can see from the box-plots in Table 4.1 that the "short-term" posterior mean drifts outside of the "long-term" posterior distribution. This indicates something is causing the distribution of thickness to change. When the "short-term" posterior mean moves below (above) the 10th (90th) percentile of the "long-term" posterior distribution, a report will be produced to alert the factory personnel.

The above example shows the thickness of SiO_2 deposition is changing to be larger.

4.3 Empirical Bayes Process Monitoring for a Multivariate Continuous Variable

Jain et al. [11] described a Bayesian approach for monitoring multivariate continuous process data. In their paper, they showed that their multivariate control chart procedure was better at identifying out-of-control processes than existing procedures. In this section, we first describe an extension of Jain et al. [11] given in Feltz and Shiau [6] and then we will propose some new monitoring schemes.

The techniques of empirical Bayes process monitoring in the multivariate case mirror that of the univariate case described in Section 4.2. The main difference between the two cases is the availability and estimation of the covariance between variables, for both process variation and sampling variation.

Suppose that we want to monitor p quality characteristics of the process simultaneously. Denote the quality characteristics observed at time t by a $p \times 1$ vector $\mathbf{X}_t$. This is an observation vector with each component being an observation of one of the p quality characteristics. Because our multivariate observation $\mathbf{X}_t$ has sampling error as well as measurement error, we assume that at a given time t, $\mathbf{X}_t$ is normally distributed with mean vector $\boldsymbol{\mu}_t$ and covariance matrix $\boldsymbol{\Sigma}$. Then the p.d.f. of $\mathbf{X}_t$ given $\boldsymbol{\mu}_t$ can be written as

$$g_1(\mathbf{X}_t|\boldsymbol{\mu}_t) = \frac{1}{(2\pi)^{\frac{p}{2}}\,|\boldsymbol{\Sigma}|} exp\left[-\frac{1}{2}(\mathbf{X}_t - \boldsymbol{\mu}_t)'\boldsymbol{\Sigma}^{-1}(\mathbf{X}_t - \boldsymbol{\mu}_t)\right],$$

where $\boldsymbol{\Sigma}$ is an unknown nonnegative-definite matrix. Although we allow the process mean $\boldsymbol{\mu}_t$ to vary over time, we assume that the sampling variability $\boldsymbol{\Sigma}$ stays constant. We further assume that $\boldsymbol{\mu}_t$ is distributed as a multivariate normal with mean vector $\boldsymbol{\mu}$ and covariance matrix $\mathcal{G}$ to model the time-varying process mean.

Similar to the univariate case, the above model allows two sources of variability: (i) the sampling variability, representing the amount of variation present among samples when the quality characteristic $\mathbf{X}_t$ is centered at its mean $\boldsymbol{\mu}_t$; and (ii) process variability, representing the amount of variability due to process (mean) changing over time. In our previous work in the electronics industry, we found that by allowing the process (mean) to have its own variability, we gained information about the process behavior as well as information about how the sample behaves around the process mean. We remark that the above setup generalizes Jain et al. approach, in that the underlying process variability was assumed to be the same as the sampling variability, i.e., $\mathcal{G} = \Sigma$, which puts a very strong restriction on applications.

By treating the distribution of $\boldsymbol{\mu}_t$ as the prior distribution for $\boldsymbol{\mu}_t$, as in the Bayesian approach, we have the well-known result that the posterior distribution of $\boldsymbol{\mu}_t$ is a multivariate normal distribution with mean

$$\boldsymbol{\mu}_{pt} = E(\boldsymbol{\mu}_t|\mathbf{X}_t) = \mathbf{X}_t - \Sigma(\Sigma + \mathcal{G})^{-1}(\mathbf{X}_t - \boldsymbol{\mu}) \tag{4.9}$$

and covariance matrix

$$V_p = Cov(\boldsymbol{\mu}_t|\mathbf{X}_t) = \Sigma - \Sigma(\Sigma + \mathcal{G})^{-1}\Sigma. \tag{4.10}$$

Let $W = \Sigma(\Sigma + \mathcal{G})^{-1}$, a weighting matrix. The posterior mean of Equation (4.9) can be rewritten as

$$\boldsymbol{\mu}_{pt} = \mathbf{W}\boldsymbol{\mu} + (\mathbf{I} - \mathbf{W})\mathbf{X}_t,$$

a form analogous to Equation (4.1) in the univariate case. It is not intuitive to see the effects of $\mathbf{W}$ on the posterior mean for general W's . However, for the special case that $\mathbf{W}$ is a diagonal matrix— which can happen when there is no correlation between process characteristics— we can easily see the effects of the weighting matrix $\mathbf{W}$. In this case, large components of $\mathbf{W}$ indicate that the corresponding process characteristics have a large sampling variation, which will pull their posterior means at time t toward their prior means. Alternately, small components of $\mathbf{W}$ indicate small sampling variability on those quality characteristics and hence pull the corresponding posterior means at time t toward their current observations. When process quality characteristics are correlated, the effects of $\mathbf{W}$ can be interpreted in terms of the principal components. That is, for the principal component directions with large (small) eigenvalues of $\mathbf{W}$, their posterior means are pulled toward their prior means (observations).

4.3.1 Estimation

Similar to the univariate case, the parameters that must be estimated are $\boldsymbol{\mu}$, Σ, and $\mathcal{G}$. The estimation method is also similar. Note that the marginal mean and covariance matrix of $\mathbf{X}_t$ are $\boldsymbol{\mu}$ and $V = \Sigma + \mathcal{G}$, respectively. Assume that T data vectors $\mathbf{X}_1, \ldots, \mathbf{X}_T$ are observed sequentially in time. It is natural to estimate $\boldsymbol{\mu}$ and V by the sample mean

$$\hat{\boldsymbol{\mu}} = \frac{\sum_{t=1}^{T} \mathbf{X}_t}{T} \tag{4.11}$$

and the sample covariance matrix

$$\hat{V} = \frac{\sum_{t=1}^{T}(X_t - \hat{\mu})(X_t - \hat{\mu})'}{T}, \tag{4.12}$$

respectively.

To estimate the sampling variability, we would need multiple independent observations at time t since μ_t is varying over time. However, in our experience replications are seldom available. To circumvent this obstacle, we must assume that the μ_t's are fairly stable. Assume that the process mean at time t remains relatively constant over short time intervals. Thus, consecutive X_t's can be thought of as independent random variables from the same distribution if there is very little time lag between them. Under this assumption, one common estimate of Σ is

$$\hat{\Sigma} = \frac{\sum_{t=2}^{T}(X_t - X_{t-1})(X_t - X_{t-1})'/2}{T-1}. \tag{4.13}$$

To estimate $\mathcal{G}$, because $V = \Sigma + \mathcal{G}$, we simply use the subtraction:

$$\hat{\mathcal{G}} = \hat{V} - \hat{\Sigma}.$$

Similar to the univariate case, if $\hat{\mathcal{G}}$ is not positive definite (which can mean that the process variation is relatively small compared to the sampling variation), let $\hat{\mathcal{G}}$ be $\kappa\hat{V}$, where κ is a small number $(0 < \kappa < 0.1)$, subject to the application.

To incorporate weighting into the estimators of the above equations as in the univariate case, let

$$\hat{\mu}_T = \frac{\sum_{t=1}^{T}\lambda^{T-t}X_t}{\sum_{t=1}^{T}\lambda^{T-t}}, \tag{4.14}$$

$$\hat{\Sigma}_T = \frac{\sum_{t=2}^{T}\lambda^{T-t}(X_t - X_{t-1})(X_t - X_{t-1})'/2}{\sum_{t=2}^{T}\lambda^{T-t}}, \tag{4.15}$$

and

$$\hat{V}_T = \frac{\sum_{t=1}^{T}\lambda^{T-t}(X_t - \hat{\mu})(X_t - \hat{\mu})'}{\sum_{t=1}^{T}\lambda^{T-t}}. \tag{4.16}$$

The choice of λ is as discussed in the univariate case.

To make the analysis more computationally efficient, the above weighted estimators can be written in the form of recursive equations as follows. With the initial values $\hat{\mu}_0 = \mathbf{0}$, $\hat{\Sigma}_1 = \mathbf{0}$, and $\hat{V}_0 = \mathbf{0}$,

$$\hat{\mu}_t = \theta_t X_t + (1-\theta_t)\hat{\mu}_{t-1}, \quad \text{for } t = 1, , \ldots, T, \tag{4.17}$$

$$\hat{\Sigma}_t = \theta_t(X_t - X_{t-1})(X_t - X_{t-1})'/2 + (1-\theta_t)\hat{\Sigma}_{t-1}, \quad \text{for } t = 2, \ldots, T, \tag{4.18}$$

$$\hat{V}_t = \theta_t(X_t - \hat{\mu}_t)(X_t - \hat{\mu}_t)'$$
$$+(1-\theta_t)(\hat{V}_{t-1} + (\hat{\mu}_t - \hat{\mu}_{t-1})(\hat{\mu}_t - \hat{\mu}_{t-1})'), \quad \text{for } t = 1,, \ldots, T, \tag{4.19}$$

where $\theta_t = 1/\sum_{k=1}^{t} \lambda^{t-k}$. The recursive equations for the non-weighted estimators (4.11)–(4.13) can also be obtained from the above recursive equations (4.17)–(4.19) above by taking $\lambda = 1$, which implies $\theta_t = 1/t$.

4.3.2 Monitoring Schemes

In this subsection, we present some possible monitoring schemes. These methods are different from the methods in Jain et al. [11] and Feltz and Shiau [6]. Performance studies of these schemes are underway.

Using marginal distribution of X_t. For online process monitoring (Phase II), assuming that the in-control marginal mean μ and covariance matrix V are known or have been well estimated from some historical data, we can simply use the usual Hotelling T^2 statistic [10] to monitor the process:

$$T_t^2 = (X_t - \mu)'V^{-1}(X_t - \mu), \tag{4.20}$$

where X_t is the current observation vector to be monitored. The control limit is $C_p = \chi^2{}_{p,0.0027}$, the 99.73th percentile of the Chi-square distribution with p degrees of freedom, which corresponds to the regular Shewhart's three-sigma control chart limits. Recall that p is the number of variables being monitored simultaneously.

Note that constructing a control chart for determining whether or not a process is in control based on a set of historical data (Phase I) is different from that for on-line process monitoring (Phase II). In Phase I analysis, μ and V are usually unknown and need to be estimated. It is natural to estimate them by Equation (4.11) and Equation (4.12), respectively. It can be easily shown that the following T^2 statistic,

$$T_t^2 = (X_t - \hat{\mu})'\hat{V}^{-1}(X_t - \hat{\mu}),$$

is distributed as an F distribution with freedom p and $T - p$ for each observation vector X_t in the historical data. Thus, the upper control limit for Phase I application is $p(T+1)/(T-p)F_{0.0027,p,T-p}$, where $F_{0.0027,p,T-p}$ is the 99.73th percentile of the F distribution with degrees of freedom p and $T - p$.

Using the posterior distribution. Given a new observation X_t, compute the posterior mean μ_{pt} and covariance matrix V_p by Equation (4.9) and Equation (4.10). Check if the central 99.73% credible set covers the vector of target values. That is, we flag an out-of-control warning if the statistic

$$T_{pt}^2 = (T - \mu_{pt})'V_p^{-1}(T - \mu_{pt}) \tag{4.21}$$

is greater than $\chi^2_{p,0.0027}$, where T is the target vector. When μ, V, and Σ are unknown, estimate them by the equations (4.11)–(4.13) from a set of in-control data.

Let $w_t = \alpha_s/(\alpha_s + n_t)$. The posterior mean of $\boldsymbol{p}_t$ given $\mathbf{X}_t$ can be re-written as

$$\widetilde{\boldsymbol{p}}_t(\boldsymbol{\alpha}) = w_t\boldsymbol{\alpha}^* + (1 - w_t)\frac{\mathbf{X}_t}{n_t}, \tag{4.32}$$

which indicates that the posterior mean $\widetilde{\boldsymbol{p}}_t(\boldsymbol{\alpha})$ is a weighted average of the prior mean $\boldsymbol{\alpha}^*$ and the observed proportion vector $\mathbf{X}_t/n_t$. A large weight w_t (i.e., more weight on the prior information) pulls the posterior mean toward the prior mean, whereas a small weight w_t (i.e., more weight on current data) pulls the posterior mean toward the observed proportions.

4.5.1 Estimation

Using an empirical Bayes approach, we let data speak for themselves in estimating $\boldsymbol{\alpha}$ as follows. First, the marginal p.m.f. of $\mathbf{X}_t$ is

$$\begin{aligned}
f(\boldsymbol{x}_t\,;\,\boldsymbol{\alpha}) &= \frac{f(\boldsymbol{x}_t, \boldsymbol{p}_t\,;\,\boldsymbol{\alpha})}{f(\boldsymbol{p}_t|\boldsymbol{x}_t\,;\,\boldsymbol{\alpha})} = \frac{f(\boldsymbol{x}_t|\boldsymbol{p}_t)f(\boldsymbol{p}_t\,;\,\boldsymbol{\alpha})}{f(\boldsymbol{p}_t|\boldsymbol{x}_t\,;\,\boldsymbol{\alpha})} \\
&= \frac{n_t!}{x_{0t}!x_{1t}!\cdots x_{kt}!}\,\frac{\Gamma(\alpha_s)}{\Gamma(\alpha_0)\Gamma(\alpha_1)\cdots\Gamma(\alpha_k)} \\
&\quad\times\frac{\Gamma(\alpha_0 + x_{0t})\Gamma(\alpha_1 + x_{1t})\cdots\Gamma(\alpha_k + x_{kt})}{\Gamma(\alpha_s + n_t)} \\
&= \exp\left[\sum_{j=1}^{n_t}\log\left(\frac{j}{\alpha_s + j - 1}\right) - \sum_{i=0}^{k}\sum_{j=1}^{x_{it}}\log\left(\frac{j}{\alpha_i + j - 1}\right)\right]
\end{aligned} \tag{4.33}$$

for $x_{0t}, x_{1t}, \ldots, x_{kt} \in \{0, 1, \ldots, n_t\}$ and $\sum_{i=0}^{k} x_{it} = n_t$. The last equality can be easily derived by noting $\Gamma(x) = (x-1)\Gamma(x-1)$ for $x > 0$. The above marginal distribution of $\mathbf{X}_t$ is called the multivariate Pólya-Eggenberger distribution, or the Dirichlet-compound multinomial distribution with parameters n_t and $\boldsymbol{\alpha}$ (e.g., see page 80 of Johnson et al. [12]). When $k = 1$ (i.e., for the binary data), the marginal distribution of X_{it} is the Pólya distribution with parameters n_t, α_i, and α_s for each $i = 0, 1, \ldots, k$.

The marginal mean of $\mathbf{X}_t/n_t$ can be obtained by double expectation as

$$E(\mathbf{X}_t/n_t) = E[E(\mathbf{X}_t/n_t|\boldsymbol{p}_t)] = E(\boldsymbol{p}_t) = \boldsymbol{\alpha}^* \tag{4.34}$$

and the marginal covariance matrix of $\mathbf{X}_t/n_t$, using conditional expectation, is

$$\begin{aligned}
Cov(\mathbf{X}_t/n_t) &= E[Cov(\mathbf{X}_t/n_t|\boldsymbol{p}_t)] + Cov[E(\mathbf{X}_t/n_t|\boldsymbol{p}_t)] \\
&= (\alpha_s + n_t)(\mathrm{diag}\{\boldsymbol{\alpha}^*\} - \boldsymbol{\alpha}^*\boldsymbol{\alpha}^{*\prime})/[n_t(\alpha_s + 1)].
\end{aligned} \tag{4.35}$$

Let $\boldsymbol{X}_1, \ldots, \boldsymbol{X}_T$ be independent observations such that each $\boldsymbol{X}_t$ has the marginal p.m.f. of Equation (4.33). Since by Equation (4.34), $\boldsymbol{X}_t/n_t$ is an unbiased estimator of $\boldsymbol{\alpha}^*$ for each $t = 1, \ldots, T$, we estimate $\boldsymbol{\alpha}^*$ by the following weighted average:

$$\widehat{\boldsymbol{\alpha}}^* = \frac{\sum_{t=1}^{T} \boldsymbol{X}_t}{\sum_{t=1}^{T} n_t}. \tag{4.36}$$

It is easy to see by Equation (4.34) and Equation (4.35) that

$$E(\widehat{\boldsymbol{\alpha}}^*) = \boldsymbol{\alpha}^* \tag{4.37}$$

$$Cov\left(\widehat{\boldsymbol{\alpha}}^*\right) = \frac{\sum_{t=1}^{T} n_t(\alpha_s + n_t)}{(\alpha_s + 1)\left(\sum_{t=1}^{T} n_t\right)^2} \left(\text{diag}\{\boldsymbol{\alpha}^*\} - \boldsymbol{\alpha}^*\boldsymbol{\alpha}^{*\prime}\right). \tag{4.38}$$

Shiau et al. [19] stated that under regularity conditions, $\widehat{\boldsymbol{\alpha}}^*$ is a strongly consistent estimator of $\boldsymbol{\alpha}^*$ (i.e., $\widehat{\boldsymbol{\alpha}}^*$ converges to $\boldsymbol{\alpha}^*$ almost surely) and

$$\frac{(\alpha_s + 1)^{1/2} \sum_{t=1}^{T} n_t}{\left[\sum_{t=1}^{T} n_t(\alpha_s + n_t)\right]^{1/2}} (\widehat{\boldsymbol{\alpha}}^* - \boldsymbol{\alpha}^*) \xrightarrow{d} N(\boldsymbol{0}, \text{diag}\{\boldsymbol{\alpha}^*\} - \boldsymbol{\alpha}^*\boldsymbol{\alpha}^{*\prime})$$

as $T \to \infty$, where the symbol "$\xrightarrow{d}$" represents "convergence in distribution".

Having estimated $\boldsymbol{\alpha}^*$ by Equation (4.36), we can estimate $\boldsymbol{\alpha}$ by $\widehat{\alpha}_s \widehat{\boldsymbol{\alpha}}^*$ if an estimate $\widehat{\alpha}_s$ of α_s is available. By treating α_s as the only hyperparameter to be estimated in the model, Shiau et al. considered two methods for estimating α_s: the method of moments and the pseudo-maximum likelihood method.

The method of moments estimator (MME) of α_s derived by Shiau et al. [19] is

$$\widehat{\alpha}_{s,MM} = \frac{\sum_{t=1}^{T} n_t \sum_{i=0}^{k} \widehat{\alpha}_i^*(1 - \widehat{\alpha}_i^*) - \sum_{t=1}^{T} n_t \sum_{i=0}^{k} (X_{it}/n_t - \widehat{\alpha}_i^*)^2}{\sum_{t=1}^{T} n_t \sum_{i=0}^{k} (X_{it}/n_t - \widehat{\alpha}_i^*)^2 - T \sum_{i=0}^{k} \widehat{\alpha}_i^*(1 - \widehat{\alpha}_i^*)}, \tag{4.39}$$

which, under regularity conditions, is a weakly consistent estimator of α_s (i.e., $\widehat{\alpha}_{s,MM}$ converges to α_s in probability) and asymptotically normally distributed as $T \to \infty$.

The pseudo-maximum likelihood method was originally introduced by Gong and Samaniego [8]. It is called "pseudo" because only part of the parameters are estimated by the maximum likelihood method. Here, with Equation (4.36), α_s is the only hyperparameter under consideration.

The pseudo-maximum likelihood estimator (PMLE) $\widehat{\alpha}_{s,PML}$ of α_s can be obtained by the Newton-Raphson method as follows: first, choose a good initial value $\widehat{\alpha}_{s,PML}^{(0)}$ of α_s, for example, the method-of-moment estimate $\widehat{\alpha}_{s,MM}$ given in Equation (4.39). Then iterate the following recursive equation

$$\widehat{\alpha}_{s,PML}^{(u+1)} = \widehat{\alpha}_{s,PML}^{(u)} + \frac{s_P\left(\widehat{\alpha}_{s,PML}^{(u)}; \boldsymbol{x}_1, \ldots, \boldsymbol{x}_T\right)}{j_P(\cdot; \boldsymbol{x}_1, \ldots, \boldsymbol{x}_T)}$$

for $u = 0, 1, 2, \ldots$ until convergence, where $s_P(\cdot\,; \boldsymbol{x}_1, \ldots, \boldsymbol{x}_T)$ and $j_P(\cdot\,; \boldsymbol{x}_1, \ldots, \boldsymbol{x}_T)$ are respectively the pseudo-score function and the pseudo-observed information for α_s (given in the Appendix). Under regularity conditions, $\widehat{\alpha}_{s,PML}$ is a weakly consistent estimator of α_s and asymptotically normally distributed as $T \to \infty$ according to Shiau et al. [19]. The derivations of the MME and PMLE of α_s are presented in the Appendix. With $\widehat{\alpha}_s = \widehat{\alpha}_{s,MM}$ or $\widehat{\alpha}_{s,PML}$, plugging $\widehat{\boldsymbol{\alpha}} = \widehat{\alpha}_s \widehat{\boldsymbol{\alpha}}^*$ into the posterior mean of Equation (4.31) or Equation (4.32), we get an empirical Bayes estimator of $\boldsymbol{p}_t$:

$$\widehat{\boldsymbol{p}}_t = \frac{\widehat{\alpha}_s \widehat{\boldsymbol{\alpha}}^* + \mathbf{X}_t}{\widehat{\alpha}_s + n_t} = \frac{\widehat{\alpha}_s}{\widehat{\alpha}_s + n_t} \widehat{\boldsymbol{\alpha}}^* + \frac{n_t}{\widehat{\alpha}_s + n_t} \frac{\mathbf{X}_t}{n_t}.$$

Shiau et al. [19] conducted a comparative study by simulation and reported that the PMLE performs slightly better than the MME in their study.

For the second monitoring scheme proposed in this chapter, we need the exponentially weighted estimators of $\boldsymbol{\alpha}^*$, α_s, and

$$V_{total} \equiv \sum_{t=1}^{T} n_t V_t \Big/ \sum_{t=1}^{T} n_t,$$

where $V_t = \sum_{i=0}^{k} Var(X_{it}/n_t)$ for $t = 1, \ldots, T$. Consider

$$\widehat{\boldsymbol{\alpha}}^* = \frac{\sum_{t=1}^{T} \lambda^{T-t} \mathbf{X}_t}{\sum_{t=1}^{T} \lambda^{T-t} n_t} \tag{4.40}$$

$$\hat{V}_{total} = \frac{\sum_{t=1}^{T} \lambda^{T-t} n_t \sum_{i=0}^{k} \left(X_{it}/n_t - \widehat{\alpha}_i^*\right)^2}{\sum_{t=1}^{T} \lambda^{T-t} n_t} \tag{4.41}$$

$$\widehat{\alpha}_{s,MM} = \frac{\sum_{t=1}^{T} \lambda^{T-t} n_t \sum_{i=0}^{k} \widehat{\alpha}_i^*(1 - \widehat{\alpha}_i^*) - \sum_{t=1}^{T} \lambda^{T-t} n_t \sum_{i=0}^{k} (X_{it}/n_t - \widehat{\alpha}_i^*)^2}{\sum_{t=1}^{T} \lambda^{T-t} n_t \sum_{i=0}^{k} (X_{it}/n_t - \widehat{\alpha}_i^*)^2 - \sum_{t=1}^{T} \lambda^{T-t} \sum_{i=0}^{k} \widehat{\alpha}_i^*(1 - \widehat{\alpha}_i^*)}. \tag{4.42}$$

Similar to the binary case, the recursive equations for (4.40)–(4.42) can be obtained as follows. Let $N_0 = 0$, $\mathbf{S}_0 = \mathbf{0}$, and $S_{i0}^* = 0$ for $i = 0, \ldots, k$, and

$$N_t = \sum_{j=1}^{t} \lambda^{t-j} n_j$$

$$\mathbf{S}_t = \sum_{j=1}^{t} \lambda^{t-j} \mathbf{X}_j$$

$$S_{it}^* = \sum_{j=1}^{t} \lambda^{t-j} \frac{X_{ij}^2}{n_j}$$

for $t = 1, \ldots, T$. It can be easily seen that Equation (4.40) can be computed recursively by

$$\widehat{\alpha}_t^* = \frac{S_t}{N_t} = \frac{X_t + \lambda S_{t-1}}{n_t + \lambda N_{t-1}}, \quad \text{for} \quad t = 1, \ldots, T. \tag{4.43}$$

Also, after some simple algebra, Equation (4.41) can be rewritten as

$$\hat{V}_{total,t} = \sum_{i=0}^{k} \left(\frac{S_{it}^*}{N_t} - (\widehat{\alpha}_{it}^*)^2 \right) = \sum_{i=0}^{k} \left[\frac{X_{it}^2/n_t + \lambda S_{i,t-1}^*}{n_t + \lambda N_{t-1}} - \left(\frac{X_{it} + \lambda S_{i,t-1}}{n_t + \lambda N_{t-1}} \right)^2 \right] \tag{4.44}$$

for $t = 1, \ldots, T$. Let

$$U_{it} = \left(X_{it}^2/n_t + \lambda S_{i,t-1}^*\right)/(n_t + \lambda N_{t-1}).$$

It is interesting to note that the recursive equations for $\widehat{\alpha}_t^*$ in Equation (4.43) and U_{it} in $\hat{V}_{total,t}$ Equation (4.44) have the following EWMA-like forms:

$$\widehat{\alpha}_0^* = \mathbf{0}, \quad \widehat{\alpha}_t^* = \theta_t \frac{X_t}{n_t} + (1 - \theta_t)\widehat{\alpha}_{t-1}^*, \quad \text{for} \quad t = 1, \ldots, T$$

$$U_{i1} = 0, \quad U_{it} = \theta_t \frac{X_{it}^2}{n_t^2} + (1 - \theta_t)U_{it-1}, \quad \text{for} \quad t = 1, \ldots, T$$

where

$$\theta_t = n_t \bigg/ \left(\sum_{j=1}^{t} \lambda^{t-j} n_j \right).$$

As in Section 4.4, let $N_t^* = \sum_{j=1}^{t} \lambda^{t-j} = 1 + \lambda N_{t-1}^*$. In each iteration, having computed N_t, N_t^*, $\hat{V}_{total,t}$, $\widehat{\alpha}_t^*$ recursively, we can first compute $A_t \equiv \sum_{i=0}^{k} \widehat{\alpha}_{it}^*(1 - \widehat{\alpha}_{it}^*)$ and then $\widehat{\alpha}_{s,MM,t}$ by

$$\widehat{\alpha}_{s,MM,t} = \frac{N_t A_t - N_t \hat{V}_{total,t}}{N_t \hat{V}_{total,t} - N_t^* A_t}.$$

4.5.2 Monitoring Schemes

In this subsection, we describe two monitoring schemes for polytomous data. The first one is based on the long-term/short-term posterior distributions, extending the method by Yousry et al. [25] for the univariate case. Recall that the posterior distribution of $\boldsymbol{p}_t$ given $\boldsymbol{X}_t$ is the Dirichlet distribution given in Equation (4.30). However, finding an appropriate out-of-control region for a Dirichlet distribution such that the monitoring scheme has an about-right in-control ARL is difficult.

Two major problems arise here: (i) Dirichlet distributions are multivariate, for which the numerical computation of the boundary-region probability is not an easy task; and (ii) even if (i) can be solved, the correlation among the monitoring statistics due to exponential weighting is still an issue. Also, using a T^2-like statistic for defining the out-of-control region has the disadvantage that users cannot tell which defect types are having problems when an out-of-control signal is flagged. Thus for simplicity and for being more informative, we suggest monitoring each of the components individually. Note that the posterior distribution of p_{it} given data $X_{it} = x_{it}$ is a beta distribution with parameter $\alpha_i + x_{it}$. Again, obtain a long-term and a short-term posterior distribution of $\boldsymbol{p}_t$ with weights λ_{long} and λ_{short}, respectively. Select an appropriate interval of the long-term posterior distribution such that the scheme has an about-right in-control ARL or is based on some criteria according to the real situations. If the short-term posterior mean is in the interval, then the process is considered in control.

Next, we present the monitoring scheme proposed by Shiau et al. [19], a scheme based on the marginal distribution of the sample proportion X_{it}/n_t for $i = 0, 1, \ldots, k$. Recall that $\boldsymbol{X}_t$ has a Dirichlet-compound multinomial distribution with the p.m.f. given in Equation (4.33). Again, finding a reasonable out-of-control set for such a multivariate discrete distribution is difficult. Thus for simplicity and better interpretation, Shiau et al. [19] suggested monitoring each component of the observed proportions $\boldsymbol{X}_t/n_t$. Note that the marginal distribution of X_{it} is a Pólya distribution with parameters n_t, α_i, and α_s for each $i = 0, 1, \ldots, k$ with the p.m.f.

$$f(x_{it}\,;\,\alpha_i, \alpha_s) = \exp\left[\sum_{j=1}^{n_t} \log\left(\frac{j}{\alpha_s + j - 1}\right) - \sum_{j=1}^{x_{it}} \log\left(\frac{j}{\alpha_i + j - 1}\right) - \sum_{j=1}^{n_t - x_{it}} \log\left(\frac{j}{\alpha_s - \alpha_i + j - 1}\right)\right] \quad (4.45)$$

for $x_{it} \in \{0, 1, \ldots, n_t\}$.

Because X_{it}/n_t is a discrete random variable, the conventional false-alarm rate a is almost impossible to attain if deterministic control limits are used. Based on the concept of the randomized test in hypothesis testing, Shiau et al. [19] proposed a randomized-control-limits approach as follows.

To find the lower control limit (LCL) by Equation (4.45), start accumulating the tail probability from 0 until the first l such that $\sum_{x_{it}=0}^{l} f(x_{it}\,;\,\alpha_i, \alpha_s) \geq a/2$. If the equality holds (which is very unlikely), then no need exists for randomization and LCL $= l/n_t$. If the equality does not hold, then the randomized lower control limit RLCL $= l/n_t$. The randomization is done by signaling an out-of-control condition with probability

$$\gamma_{\text{RLCL}}(\alpha_i, \alpha_s) = \frac{a/2 - \sum_{x_{it}=0}^{l-1} f(x_{it}\,;\,\alpha_i, \alpha_s)}{f(l\,;\,\alpha_i, \alpha_s)}$$

when $X_{it} = l$ is observed. A randomized upper control limit (RUCL) can be obtained similarly.

Shiau et al. [19] illustrated the above described monitoring scheme by a numerical example of four different defect types. To demonstrate the effectiveness of the above control scheme, they shifted $\boldsymbol{\alpha}$ from $= (60, 15, 10, 10, 5)'$ to $(55, 15, 10, 10, 10)'$, which means the chance of the fourth defect type to occur has increased whereas the chances of other defect types remain the same. The control charts for individual components of $\mathbf{X}_t/n_t$ successfully demonstrated that only the control charts for X_{0t}/n_t and X_{4t}/n_t flagged out-of-control warnings, reflecting exactly the true situation.

4.6 Summary and Discussion

This chapter gives a framework for analyzing measurements from a manufacturing process in which the variation in the observations comes from two sources: sampling variation and process variation. By allowing the process parameter of interest to vary over time, such as process mean or yield/defect levels, Bayesian models are conveniently borrowed to model this type of process data. Bayesian models, including the Gaussian model with Gaussian prior for univariate/multivariate continuous data, binomial model with beta prior for binary data, and multinomial model with Dirichlet prior for polytomous data are investigated. For estimating the parameters in the model from data, the empirical Bayes approach is used. Some monitoring schemes based on statistics relating to the process parameters of interest are developed. One advantage of adopting this Bayesian framework is that we can create estimates for the distribution of the process parameter at time t, such as process mean (μ_t or $\boldsymbol{\mu}_t$) or yield/defect level (p_t or $\boldsymbol{p}_t$), that can be used in many ways by quality practitioners. Furthermore, in addition to the usual process monitoring, one can use these distribution estimates to compare distributions of the same component across product lines or across different testing equipment. For more examples on the comparison of different distributions, see Sturm et al. [21]

By weighting data exponentially in time, the empirical Bayes estimators inherently become EWMA-like. However, with the usual EWMA methods alone, one does not have a distribution for the varying process parameter, nor does one have estimators for the sampling variation and process variation. One could view the empirical Bayes model with weights as an enhancement of EWMA control charts, which has been shown to be very useful in quality monitoring. For example, see [2, 4, 13, 20, 24].

For being able to implement the monitoring schemes for real-time data analysis of on-line data, efficient computation is necessary. Recursive equations for both weighted and non-weighted estimators are derived for all the statistics needed for implementation.

We remark that some techniques presented in this chapter have been implemented in factories and proved quite useful in monitoring some high-volume, data-intensive manufacturing lines (see [21] and [25]). We also remark that the performance studies of some monitoring schemes proposed in this chapter are currently under study.

The empirical Bayes methodology is a very rich and fruitful methodology for monitoring production data. This chapter covers the methodology and models for univariate and multivariate, discrete (binomial/multinomial), and continuous (Gaussian) data. Other types of data have been considered in the literature. Just recently, Bayarri and Garcia-Donato [1] proposed a sequential empirical Bayes u-control chart for attribute data. They used the Poisson model with a gamma prior to account for the extra variability (also called over-dispersion) that is often present in processes for which the assumption of the usual Poisson model is not appropriate. However, the quality control field has a rich array of types of data, so there is much room for new models handling different types of data. The models can be as varied as the data produced in the manufacturing and service industries [5].

Appendix

Derivations of MME and PMLE

We first derive the method of moments estimator given in Equation (4.39). By Equation (4.35), the marginal variance of X_{it}/n_t can be rewritten as

$$V_{it} \equiv Var(X_{it}/n_t) = \frac{\alpha_s + n_t}{\alpha_s + 1} \frac{\alpha_i^*(1-\alpha_i^*)}{n_t}$$

for $i = 0, 1, \ldots, k$ and $t = 1, \ldots, T$. Thus, let

$$V_{total} \equiv \frac{\sum_{t=1}^T n_t \sum_{i=0}^k V_{it}}{\sum_{t=1}^T n_t} = \frac{T\alpha_s + \sum_{t=1}^T n_t}{\alpha_s + 1} \frac{\sum_{i=0}^k \alpha_i^*(1-\alpha_i^*)}{\sum_{t=1}^T n_t}. \qquad \text{(A.1)}$$

Solving Equation (A.1) for α_s, we have

$$\alpha_s = \frac{\sum_{t=1}^T n_t \sum_{i=0}^k \alpha_i^*(1-\alpha_i^*) - \sum_{t=1}^T n_t \sum_{i=0}^k V_{it}}{\sum_{t=1}^T n_t \sum_{i=0}^k V_{it} - T\sum_{i=0}^k \alpha_i^*(1-\alpha_i^*)}. \qquad \text{(A.2)}$$

By Equation (4.34) and Equation (4.37), V_{total} can be estimated by

$$\hat{V}_{total} = \frac{\sum_{t=1}^T n_t \sum_{i=0}^k \left(X_{it}/n_t - \widehat{\alpha}_i^*\right)^2}{\sum_{t=1}^T n_t}.$$

Estimate $\sum_{t=1}^{T} n_t \sum_{i=0}^{k} V_{it}$ by $\hat{V}_{total} \sum_{t=1}^{T} n_t$. Then plugging all the parameter estimators into Equation (A.2), Equation (4.39) is obtained.

To derive the PMLE, we need the pseudo-likelihood function for α_s. Given $\boldsymbol{X}_t = \boldsymbol{x}_t$ for $t = 1, \ldots, T$, by Equation (4.33), the pseudo-likelihood function for α_s can be defined as

$$L_P(\alpha_s\,; \boldsymbol{x}_1, \ldots, \boldsymbol{x}_T) \equiv \prod_{t=1}^{T} f(\boldsymbol{x}_t\,; \boldsymbol{\alpha})|_{\boldsymbol{\alpha}^*=\widehat{\boldsymbol{\alpha}}^*} = \exp\left\{\sum_{t=1}^{T}\left[\sum_{j=1}^{n_t}\log\left(\frac{j}{\alpha_s+j-1}\right) - \sum_{i=0}^{k}\sum_{j=1}^{x_{it}}\log\left(\frac{j}{\widehat{\alpha}_i^*\alpha_s+j-1}\right)\right]\right\}. \tag{A.3}$$

The corresponding pseudo-log-likelihood function for α_s is then

$$\ell_P(\alpha_s\,; \boldsymbol{x}_1, \ldots, \boldsymbol{x}_T) \equiv \log[L_P(\alpha_s\,; \boldsymbol{x}_1, \ldots, \boldsymbol{x}_T)] = \sum_{t=1}^{T}\left[\sum_{j=1}^{n_t}\log\left(\frac{j}{\alpha_s+j-1}\right) - \sum_{i=0}^{k}\sum_{j=1}^{x_{it}}\log\left(\frac{j}{\widehat{\alpha}_i^*\alpha_s+j-1}\right)\right].$$

The pseudo-score function for α_s is

$$s_P(\alpha_s\,; \boldsymbol{x}_1, \ldots, \boldsymbol{x}_T) \equiv \frac{\partial \ell_P(\alpha_s\,; \boldsymbol{x}_1, \ldots, \boldsymbol{x}_T)}{\partial \alpha_s} = \sum_{t=1}^{T}\left\{\left[\sum_{i=0}^{k}\widehat{\alpha}_i^*\sum_{j=1}^{x_{it}}\frac{1}{\widehat{\alpha}_i^*\alpha_s+j-1}\right] - \sum_{j=1}^{n_t}\frac{1}{\alpha_s+j-1}\right\} \equiv \sum_{t=1}^{T} s_{P,t}(\alpha_s\,; \boldsymbol{x}_1, \ldots, \boldsymbol{x}_T)$$

and the pseudo-observed information for α_s is

$$j_P(\alpha_s\,; \boldsymbol{x}_1, \ldots, \boldsymbol{x}_T) \equiv -\frac{\partial^2 \ell_P(\alpha_s\,; \boldsymbol{x}_1, \ldots, \boldsymbol{x}_T)}{\partial \alpha_s^2} = \sum_{t=1}^{T}\left\{\left[\sum_{i=0}^{k}\widehat{\alpha}_i^{*2}\sum_{j=1}^{x_{it}}\frac{1}{(\widehat{\alpha}_i^*\alpha_s+j-1)^2}\right] - \sum_{j=1}^{n_t}\frac{1}{(\alpha_s+j-1)^2}\right\}.$$

With the above pseudo-score function and pseudo-observed information, the PMLE can be obtained by the Newton-Raphson method as described in Section 4.5.1.

5

A Bayesian Approach to Monitoring the Mean of a Multivariate Normal Process

Frank B. Alt

Robert H. Smith School of Business, University of Maryland

CONTENTS

ABSTRACT A Bayesian multivariate quality control procedure for monitoring the process mean vector is introduced and compared with existing multivariate procedures. Under the Bayesian framework, one blends prior information about the process mean vector with the sample data to estimate the posterior mean vector and use this to decide if the process is in control.

Several multivariate Bayesian procedures are proposed in this paper. The first procedure, denoted as IMBP, performed significantly better for detecting small shifts than Crosier's multivariate CUSUM procedure (CMCUSUM) and Lowry et al's multivariate exponentially weighted moving average (MEWMAE) procedure with exact covariance matrix. The IMBP was revised (RMBP) so that it could be designed without resorting to simulation. The RMBP also performed better than CMCUSUM and MEWMAE for detecting moderate and larger shifts. The RMBP was further revised to overcome the problem of inertia.

Furthermore, it appears that the RMBPs are quite robust to departures from the assumption of multivariate normality. The robustness study also indicated that the MBPs are able to detect increased variability as well as a shift in the process mean vector.

5.1 Introduction

In process control applications, we often need to monitor several correlated quality characteristics simultaneously. Hotelling [10] was among the first to do so when he proposed the T^2 chart to analyze multivariate bombsight data. When the covariance matrix of the process is known, Alt [1] suggested the use of a χ^2 control chart.

Multivariate versions of cumulative sum and exponentially weighted moving average charts were developed later, including those proposed by Woodall and Ncube [19], Pignatiello and Kasunic [13], Pignatiello, Runger and Korpela [15], Pignatiello and Runger [14], and Crosier [6]. One of Crosier's two procedures turned out to be the best for early detection of an out-of-control process.

Lowry at al [12] developed a multivariate exponentially weighted moving average (MEWMA) procedure which, when used with the exact covariance matrix of the MEWMA vector, leads to a fast initial response.

In this paper, we propose a multivariate control chart procedure using a Bayesian approach, together with several variations of the basic Bayesian approach. We use the acronym MBP to denote the multivariate Bayesian procedures. Based on simulation results, we show that the refined MBPs perform better than the existing procedures of Crosier [6] and Lowry et al [12]. With much higher in-control average run length (ARL), the out-of-control ARLs for MBPs are, in general smaller than those for the other procedures.

We also look at different interpretations of a MBP, their design and robustness.

Although the emphasis is on MBPs, the univariate equivalent of the MBP is also considered.

5.2 Existing Multivariate Procedures

In what follows, we will compare the Bayesian procedures with Crosier's better multivariate cusum procedure (MCUSUM), and Lowry et al's multivariate exponentially weighted moving average (MEWMAE) procedure with exact covariance matrix.

At time i, we observe a vector $\mathbf{x}_i$ of p quality characteristics. The vector $\mathbf{x}_i$ may represent a vector of individual observations or a vector of sample means. Assume that the $\mathbf{X}_i$'s are independent multivariate normal random vectors with mean vectors $\boldsymbol{\mu}_i$ for $i = 1, 2, \ldots$. Without loss of generality, it is assumed that the in-control process mean vector is $\boldsymbol{\mu}_0 = (0, 0, \ldots, 0)' = \mathbf{0}$.

In Lowry's MEWMAE procedure, we let $\mathbf{Z}_t = r\mathbf{X}_t + (1 - r)\mathbf{Z}_{t-1}$ where $\mathbf{Z}_0 = 0$ and $0 < r \leq 1$. The MEWMAE procedure gives an out-of-control signal as soon as $T_1^2 > h_2$, where $T_1^2 = \mathbf{Z}_t' \Sigma_{Zt}^{-1} \mathbf{Z}_t$. The value of h_2 is found by simulation.

In Crosier's CMCUSUM procedure, define $C_t = \{(\mathbf{S}_{t-1} + \mathbf{X}_t)' \Sigma^{-1} (\mathbf{S}_{t-1} + \mathbf{X}_t)\}^{1/2}$ where $\mathbf{S}_t = \mathbf{0}$ if $C_t \leq k$. Otherwise, $\mathbf{S}_t = (\mathbf{S}_{t-1} + \mathbf{X}_t)(1 - k/C_t)$ for $t = 1, 2, \ldots$. $\mathbf{S}_0 = \mathbf{0}$, and $k > 0$. This procedure gives an out-of-control signal when $Y_t = (\mathbf{S}_t' \Sigma^{-1} \mathbf{S}_t)^{1/2} > h$. The values of h and k are found by simulation.

In the design of the **CMCUSUM** and **MEWMAE** procedures, it is assumed that we want to detect any shifts from the in-control mean vector as early as possible and that shifts of the same non-centrality measure (λ) are equally important.

5.3 Bayesian Approaches

5.3.1 Univariate Bayesian Procedures

Girshick and Rubin [9] were among the first to adopt a Bayesian perspective for univariate process control. Other early work included the economic design approaches of Bather [2] and Carter [5]. A quasi-Bayesian approach for finding control limits was suggested by Calvin [4]. Calvin's procedure hinges on whether or not one expects the process to be stable. Although Calvin suggested the use of a Shewhart-type control chart, he did not compare his procedure with existing procedures. The problem of determining whether or not the process mean has changed can also be viewed as determining the point in a sequence of random variables at which the underlying distribution changes. The Bayesian approach to a change point was addressed by Smith [16].

Joseph and Bowen [11] developed a univariate Bayesian procedure (referred to as JUBP) and compared it with the univariate CUSUM procedure. They found that the Bayesian technique was inferior to the CUSUM procedure in the case of a full drift (process mean is off by D units from the target mean) in one sampling period, but provided improved average run lengths in the case of incremental drifts of the process mean from the target in 5 and 10 successive sampling intervals of time. The JUBP is considered more extensively in Section 5 of this paper.

A more extensive overview of the univariate Bayesian approach is found in Tagaras and Nenes [18] (see Chapter 6 in this volume).

5.3.2 A Multivariate Bayesian Approach

Assume that the process mean vector for a multivariate random variable is not known but a prior distribution for the process mean vector can be specified. Then, using sample information, one can find the posterior distribution of the mean vector. Details are in Berger [3].

Let $\pi(\boldsymbol{\mu})$ denote a prior probability density function for $\boldsymbol{\mu}$ and let $f(\mathbf{x}|\boldsymbol{\mu})$ be the probability density function of $\mathbf{x}$ given $\boldsymbol{\mu}$, referred to as the sampling distribution. The posterior distribution of $\boldsymbol{\mu}$ given $\mathbf{x}$ is denoted $\pi(\boldsymbol{\mu}|\mathbf{x})$. Note that $\boldsymbol{\mu}$ and $\mathbf{x}$ have joint density

$$h(\mathbf{x}, \boldsymbol{\mu}) = \pi(\boldsymbol{\mu}) f(\mathbf{x}|\boldsymbol{\mu})$$

and that the marginal density of $\mathbf{x}$ is

$$m(\mathbf{x}) = \int h(\mathbf{x}, \boldsymbol{\mu}) d\boldsymbol{\mu}.$$

If $m(\mathbf{x})$ is not equal to 0,

$$\pi(\boldsymbol{\mu}|\mathbf{x}) = h(\mathbf{x}, \boldsymbol{\mu})/m(\mathbf{x}).$$

In order to find the posterior distribution of an unknown parameter vector $\boldsymbol{\mu}$, we need to specify a prior density. For the multivariate Bayesian procedures (MBP), we will assume a conjugate prior.

Assume that the multiple quality characteristics of interest have a multivariate normal distribution with a known covariance matrix $\boldsymbol{\Sigma}$ but unknown mean vector $\boldsymbol{\mu}$. Using each new observation and a conjugate multivariate normal prior, the process mean vector is estimated at that step. An advantage of using a conjugate multivariate normal prior is that the posterior distribution will also be multivariate normal.

Suppose $\mathbf{X} \sim \mathbf{N}_p(\boldsymbol{\mu}, \boldsymbol{\Sigma})$ and $\pi(\boldsymbol{\mu}) \sim \mathbf{N}_p(\boldsymbol{\theta}, \mathbf{A})$, where $\boldsymbol{\theta}$ is a known $(p \times 1)$ vector and $\boldsymbol{\Sigma}$ and $\mathbf{A}$ are known $(p \times p)$ positive definite matrices.

It is well known that $\pi(\boldsymbol{\mu}|\mathbf{x}) \sim \mathbf{N}_p(\boldsymbol{\mu}^{\pi}(\mathbf{x}), \mathbf{V}^{\pi}(\mathbf{x}))$, where the posterior mean is given by

$$\boldsymbol{\mu}^{\pi}(\mathbf{x}) = \mathbf{x} - \boldsymbol{\Sigma}(\boldsymbol{\Sigma} + \mathbf{A})^{-1}(\mathbf{x} - \boldsymbol{\theta})$$

and the posterior covariance matrix by

$$\mathbf{V}^{\pi}(\mathbf{x}) = (\mathbf{A}^{-1} + \boldsymbol{\Sigma}^{-1})^{-1} = \boldsymbol{\Sigma} - \boldsymbol{\Sigma}(\mathbf{A} + \boldsymbol{\Sigma})^{-1}\boldsymbol{\Sigma}.$$

5.3.3 An Initial Multivariate Bayesian Procedure (IMBP)

At time i, we observe a vector $\mathbf{x}_i$ of p quality characteristics. The vector $\mathbf{x}_i$ may represent a vector of individual observations or a vector of sample means. Assume that the $\mathbf{X}_i$'s are independent multivariate normal random vectors with mean vectors $\boldsymbol{\mu}_i$ for $i = 1, 2, \ldots$. Without loss of generality, it is assumed that the in-control process mean vector is $\boldsymbol{\mu}_0 = (0, 0, \ldots, 0)' = \mathbf{0}$.

If we assume that the process is in control at the start (time 0), the prior distribution of the unknown process mean vector can be considered multivariate normal with mean vector $\boldsymbol{\mu}_0$ and covariance matrix $\boldsymbol{\Sigma}$. Then, $\boldsymbol{\theta} = \boldsymbol{\mu}_0$ and $\mathbf{A} = \boldsymbol{\Sigma}$. To estimate $\boldsymbol{\mu}$ at time 1 using the first sample observation, use the posterior distribution of the unknown process mean vector at time 1.

For the IMBP, assume that $\mathbf{X}$ is $\mathbf{N}_p(\boldsymbol{\mu}, \boldsymbol{\Sigma})$ and that the prior distribution, $\pi(\boldsymbol{\mu})$, of $\boldsymbol{\mu}$ is $\mathbf{N}_p(\boldsymbol{\mu}_0, \boldsymbol{\Sigma})$. Once the vector $\mathbf{x}_1$ is observed, the posterior distribution of $\boldsymbol{\mu}$ given $\mathbf{x}_1$, $\pi(\boldsymbol{\mu}|\mathbf{x}_1)$, is $\mathbf{N}_p(\boldsymbol{\mu}^1, \boldsymbol{\Sigma}_1)$ where

$$\begin{aligned}
\boldsymbol{\mu}^1 &= \mathbf{x}_1 - \boldsymbol{\Sigma}(\boldsymbol{\Sigma} + \text{covariance of prior distribution})^{-1}(\mathbf{x}_1 - \boldsymbol{\mu}_0)\\
&= \mathbf{x}_1 - \boldsymbol{\Sigma}(\boldsymbol{\Sigma} + \boldsymbol{\Sigma})^{-1}(\mathbf{x}_1 - \boldsymbol{\mu}_0) = \mathbf{x}_1/2,\\
\boldsymbol{\Sigma}_1 &= \boldsymbol{\Sigma} - \boldsymbol{\Sigma}(\boldsymbol{\Sigma} + \text{covariance of prior distribution})^{-1}\boldsymbol{\Sigma}\\
&= \boldsymbol{\Sigma} - \boldsymbol{\Sigma}(\boldsymbol{\Sigma} + \boldsymbol{\Sigma})^{-1}\boldsymbol{\Sigma} = \boldsymbol{\Sigma}/2
\end{aligned}$$

Thus, $\pi(\boldsymbol{\mu}|\mathbf{x}_1)$ is distributed as $\mathbf{N}_p(\mathbf{x}_1/2, \boldsymbol{\Sigma}/2)$ and $\mathbf{x}_1/2$ is an estimate of the process mean vector at time 1.

Simulation was used to find a decision interval for IMBP. The initial procedure is as follows:

1. At time 1, assume the prior distribution of $\boldsymbol{\mu}$ to be $\mathbf{N}_p(\boldsymbol{\mu}_0, \boldsymbol{\Sigma})$.
2. At time i, observe $\mathbf{x}_i$.
3. For all $i > 1$, find the posterior distribution $\pi(\boldsymbol{\mu}|\mathbf{x}_i)$ using the posterior distribution at time $i-1$ as the prior distribution at time i of $\boldsymbol{\mu}$. For $i = 1$, use the prior as specified in step 1 to compute the posterior distribution. It can be shown that the posterior distribution at time i, $\pi(\boldsymbol{\mu}|\mathbf{x}_i)$, is $\mathbf{N}_p(\boldsymbol{\mu}^i, \boldsymbol{\Sigma}_i)$ where

$$\boldsymbol{\mu}^i = (\mathbf{x}_1 + \mathbf{x}_2 + \ldots + \mathbf{x}_i)/(i+1) \quad \text{and} \quad \boldsymbol{\Sigma}_i = \boldsymbol{\Sigma}/(i+1).$$

4. Calculate $B_i = \boldsymbol{\mu}^{i\prime}(\boldsymbol{\Sigma}_i)^{-1}\boldsymbol{\mu}^i$. If $B_i > h_4$, this is interpreted as a signal that the process is out of control. One searches for an assignable

cause and takes corrective action. Then, set $i = 1$ and go back to step 1. Otherwise, set $i = i + 1$ and go to step 2.

Using simulation, we can find a value of h_4 such that the in-control ARL is at least 200 and yet the out-of-control ARLs are smaller than those for competing multivariate procedures.

5.3.4 Different Interpretations of IMBP

The multivariate Bayesian procedure as developed above can be interpreted in several different ways.

1. IMBP can be considered a MEWMA procedure with the modification that the smoothing constant r is not fixed but is dependent on the observation number.

   Let the MEWMA vector $\mathbf{Z}_0 = \mathbf{0}$, For observation i, let $r_i = 1/(i + 1)$. Then $\mathbf{Z}_i$ equals $(\mathbf{x}_1 + \mathbf{x}_2 + \ldots + \mathbf{x}_i)/(i + 1)$, the posterior mean at time i. Even though the vector $\mathbf{Z}_i$ is the same as the posterior mean at time i, the covariance matrices used in the computation of B_i and the MEWMAE statistic (T_1^2) are different.
2. By modifying the third step of the IMBP, we can show the equivalence of the Bayesian procedure to the MEWMA procedure. For all $i > 1$, let the mean of the prior for $\boldsymbol{\mu}$ be the same as the posterior mean at time $i - 1$ but the covariance matrix of the prior is assumed to be the same as at the start of the procedure. For $i = 1$, use the prior as specified in step 1. It can be shown that when the prior covariance matrix is $\boldsymbol{\Sigma}$ for all i, the posterior mean at step i is equal to the MEWMA statistic, $\mathbf{Z}_i$, with the smoothing constant $r = 1/2$. If the prior covariance matrix is assumed to be $\boldsymbol{\Sigma}/9$, the smoothing constant will be 0.1.
3. The posterior mean vector generated by the initial multivariate Bayesian procedure could also be considered the vector obtained if the cumulative sum vector at time i is shrunk by a factor of $[1 - i/(i + 1)]$.
4. IMBP provides a shrinkage-type estimator of the process mean vector. After an observation is obtained at time i, we compute the mean vector using all i observations and shrink this mean vector by a factor of $[1 - 1/(i + 1)]$ to compute the posterior mean at time i.

5.3.5 Average Run Length Comparisons

Table 5.1 compares the in-control run length distribution of IMBP with those of Crosier's CMCUSUM and Lowry's EWMAE procedures for values of $p = 2$, 5 and 10, respectively. Based on these results, we find that almost 65% to 70% of the run lengths are greater than 200 in the case of IMBP. For the

TABLE 5.1

In-Control Run Length Distribution — Initial Procedure

	p = 2			p = 5			p = 10		
Run Length	CMCU-SUM	MEWMAE	IMBP	CMCU-SUM	MEWMAE	IMBP	CMCU-SUM	MEWMAE	IMBP
<10	118	471	392	16	474	379	1	429	415
10–19	323	318	322	198	311	357	51	314	472
20–29	321	296	181	342	305	223	211	313	283
30–39	317	287	121	308	237	174	330	294	201
40–49	280	278	110	321	248	131	328	282	162
50–99	1280	1142	336	1298	1176	413	1555	1187	476
100–149	957	925	204	998	882	238	1154	932	282
150–199	763	699	129	823	717	136	840	713	182
200–249	609	559	106	642	594	104	656	560	159
>249	2032	2025	5099	2054	2026	4845	1874	1976	4368

other procedures, at most 37% to 40% of the run lengths are greater than 200. The median run length for all procedures, other than IMBP, is significantly less than 200; the median run length for the IMBP exceeds 200. For both the IMBP and the MEWMAE procedures, the frequency of run lengths less than 10 is comparatively much higher than the corresponding frequencies for the CMCUSUM procedure.

These comparisons indicate that while the ARL is frequently used as the basis for comparison, it is not an ideal performance measure and the distribution of run length can be much more revealing.

Table 5.2 shows ARL comparisons of Crosier's CUSUM, Lowry's EWMA with exact covariance matrix and the IMBP for $p = 2$, 5 and 10, respectively. Crosier and Lowry had shown that the performance of their procedures depended on the process mean vector and covariance matrix only through the non-centrality parameter. In the Appendix of this paper, it is shown that the IMBP's performance depends on the process mean vector and covariance matrix only through the value of the non-centrality parameter.

In Table 5.2, the comparison of the IMBP with Crosier's procedure (CMCUSUM), and the MEWMA procedure with exact covariance matrix (MEWMAE) is based on 7,000 independent run lengths and have been computed by truncating the run length at 500. The in-control ARLs would have been higher if the run length truncation was chosen at values higher than 500. We also assume that we want to detect any shifts from the in-control mean vector as early as possible and that shifts of equal non-centrality measure are equally important.

Based on these simulation results, IMBP is found to perform significantly better than MEWMAE for detecting small shifts, that is, shifts of non-centrality parameter $\lambda \leq 1$. Furthermore, for $\lambda \leq 1$, the performance of the IMBP increases as p increases. The performance of the MEWMAE procedure, however, is the best for detecting large shifts.

TABLE 5.2
ARL Comparisons — Initial Procedure

	p = 2			p = 5			p = 10		
λ	CMCUSUM $h = 5.5$	MEWMAE $h_2 = 8.79$	IMBP $h_5 = 6.91$	CMCUSUM $h = 9.46$	MEWMAE $H_2 = 14.74$	IMBP $h_4 = 11.9$	CMCUSUM $h = 14.9$	MEWMAE $h_2 = 22.91$	IMBP $h_4 = 18.2$
0	200.9 (2.28)	199.3 (2.34)	374.5 (2.36)	207.7 (2.27)	201.8 (2.45)	357.2 (2.44)	195.3 (2.05)	198.3 (2.43)	326.5 (2.53)
.5	29.6 (.266)	25.2 (.247)	23.8 (.214)	35.4 (.265)	34.1 (.348)	29.4 (.253)	42.8 (.256)	44.7 (.468)	33.5 (.296)
1.0	9.9 (.057)	7.8 (.061)	7.7 (.054)	13.8 (.062)	10.2 (.080)	9.8 (.067)	18.7 (.067)	12.7 (.098)	11.3 (.074)
1.5	5.8 (.025)	4.0 (.028)	4.3 (.025)	8.4 (.029)	5.2 (.035)	5.4 (.030)	11.9 (.033)	6.3 (.042)	6.2 (.034)
2.0	4.1 (.015)	2.6 (.016)	2.9 (.015)	6.1 (.017)	3.2 (.016)	3.6 (.017)	8.8 (.020)	4.0 (.024)	4.3 (.019)
2.5	3.2 (.010)	1.9 (.011)	2.3 (.010)	4.9 (.012)	2.3 (.013)	2.8 (.011)	5.9 (.011)	2.8 (.016)	3.2 (.013)
3.0	2.7 (.008)	1.5 (.008)	1.8 (.008)	4.0 (.009)	1.8 (.010)	2.3 (.008)	4.06 (.04)	2.1 (.011)	2.6 (.009)

Note: Numbers in parentheses ( ) are the standard errors of the run lengths.

Although it is not shown, it should be noted that the MEWMAE procedure performs better than the Shewhart-based χ^2 procedure for $\lambda = 3$. This is mainly due to the fact that, even though the MEWMAE statistic is exactly the same as the χ^2 statistic for the first observation, the decision interval for the MEWMAE procedure is smaller than the decision interval for the Shewhart-based χ^2 procedure.

IMBP is better than Crosier's procedure for detecting all magnitudes of shifts. We also notice that the standard errors of out-of-control run lengths associated with IMBP are smaller than the corresponding standard errors associated with the MEWMAE procedure, with the difference being most notable as p increases.

Based on all of these results, it is seen that IMBP is better than competing multivariate quality control procedures as it detects small shifts in the process mean vector faster than other procedures and also results in a significantly higher in-control median run length.

5.3.6 A Revised Multivariate Bayesian Procedure

In this section, the IMBP will be revised by varying the prior distribution so that the procedure can be designed without using simulation. The revised procedure (RMBP) will then be compared with the MEWMAE and CMCUSUM procedures.

Suppose $\mathbf{X} \sim \mathbf{N}_p(\boldsymbol{\mu}, \boldsymbol{\Sigma})$ and $\pi(\boldsymbol{\mu}) \sim \mathbf{N}_p(\boldsymbol{\theta}, \mathbf{A})$, where $\boldsymbol{\theta}$ is a known $(p \times 1)$ vector and $\boldsymbol{\Sigma}$ and $\mathbf{A}$ are known $(p \times p)$ covariance matrices.

Berger [3] suggested that another way of specifying the parameters of a multivariate normal prior is to consider the ellipsoid

$$\{\boldsymbol{\mu}:(\boldsymbol{\mu} - \boldsymbol{\theta})'\mathbf{A}^{-1}(\boldsymbol{\mu} - \boldsymbol{\theta}) \leq p - .6\}$$

which has approximately a 50% chance of containing $\boldsymbol{\mu}$. He noted that $p - .6$ was approximately equal to $\chi^2_{p,.50}$. For example, for $p = 2$, $\chi^2_{2,.50} = 1.386$ while for $p = 3$, $\chi^2_{3,.50} = 2.366$.

Taking the covariance matrix $\mathbf{A}$ of the prior as $R\boldsymbol{\Sigma}$ where $R = (\chi^2_{p,(1-\alpha)})/(p - .6)$, the ellipsoid becomes

$$\left\{\boldsymbol{\mu}:(\boldsymbol{\mu} - \boldsymbol{\theta})'\boldsymbol{\Sigma}^{-1}(\boldsymbol{\mu} - \boldsymbol{\theta}) \leq \chi^2_{p,(1-\alpha)}\right\},$$

which is the $100(1 - \alpha)\%$ highest posterior density (HPD) credible set for the process mean vector based on an estimator $\boldsymbol{\mu}$.

A credible set to a Bayesian corresponds to a traditional confidence set while a HPD credible set corresponds to the method used to obtain a likelihood set. According to Berger [3], credible sets do not have a clear decision theoretic role and only provide an easily reportable crude summary of the posterior distribution. The credible sets are also not necessarily invariant under transformations.

For the revised procedure, we set $R = (\chi^2_{p,(1-\alpha)})/(p - .6)$. When $R = 1$, this implies that the prior covariance matrix is equal to the covariance matrix of the observation vector, whereas $R > 1$ implies greater dispersion associated with the prior process mean vector.

Multiplying Σ by R preserves the correlations between the variables. In other words, the correlation matrices for Σ and $R\Sigma$ are exactly the same.

Assuming that $\pi(\boldsymbol{\mu})$ is $\mathbf{N}_p(\boldsymbol{\mu}_0, R\Sigma)$, the revised posterior distribution, $\pi(\boldsymbol{\mu}|\mathbf{x}_i)$, at step i is $\mathbf{N}_p(\boldsymbol{\mu}^i, \Sigma_i)$, where

$$\boldsymbol{\mu}^i = R(\mathbf{x}_1 + \mathbf{x}_2 + \ldots + \mathbf{x}_i)/(1 + iR),$$

and

$$\Sigma_i = R\Sigma/(1 + iR).$$

Using the revised posterior with $h_5 = \chi^2_{p,(1-\alpha)}$, steps 1 to 4 previously outlined in Section 3.3 for the initial procedure can now be used to implement the revised procedure. The revised procedure with $R = \chi^2_{p,.99}/\chi^2_{p,.5}$ and $h_5 = \chi^2_{p,.99}$ is quite comparable, with respect to the out-of-control ARLs, to the CMCUSUM procedure with an in-control ARL of 200. The in-control ARL for the RMBP is at least twice as large as the in-control ARLs of the CMCUSUM procedure.

5.3.7 Average Run Length Comparisons for the RMBP

Table 5.3 presents the ARL comparisons of the RMBP for $p = 2$, 5 and 10, respectively, with the CMCUSUM and MEWMAE procedures. The in-control ARLs given in these tables have been achieved when the run length was truncated at 500. The in-control ARLs would have been higher if the run length was truncated at values higher than 500.

The results in these tables indicate that RMBP performs significantly better than the CMCUSUM procedure for detecting shifts of non-centrality parameter $\lambda > .5$. The performance of the MEWMAE procedure, however, is the best for detecting shifts of all magnitudes.

Table 5.4 gives the in-control run length distribution for the RMBP procedure for $p = 2$, 5 and 10. The results of Table 5.4 indicate that in more than 86% of the 7,000 simulations, the in-control run length is at least 249. Even the frequency with which this procedure gives run lengths less than 10 is significantly smaller than the corresponding frequencies of both the IMBP and MEWMAE procedures.

5.3.8 Interpretation of Out-of-Control Signals

The interpretation of out-of-control signals from multivariate control charts may be problematic. An advantage of the MBPs is that the process mean vector is estimated at time i. The estimate of the process mean vector does give an indication of the direction of the shift. When MBPs signal an out-of-control process, we suggest the use of a univariate Bayesian procedure for each of the variables to identify the quality characteristic(s) responsible for the out-of-control signal.

TABLE 5.3

ARL Comparisons — Revised Procedure

	p = 2, R = 9.21/1.4			p = 5, R = 15.09/4.4			p = 10, R = 23.21/9.4		
	CMCUSUM	**MEWMAE**	**RMBP**	**CMCUSUM**	**MEWMAE**	**RMBP**	**CMCUSUM**	**MEWMAE**	**RMBP**
λ	$h = 5.5$	$h_2 = 8.79$	$h_5 = 9.21$	$h = 9.46$	$h_2 = 14.74$	$h_5 = 15.09$	$h = 14.9$	$h_2 = 22.91$	$h_5 = 23.21$
0	200.9 (2.28)	199.3 (2.41)	437.7 (1.83)	207.7 (2.27)	201.8 (2.45)	440.7 (1.77)	195.3 (2.05)	198.3 (2.43)	440.9 (1.76)
.5	29.6 (.266)	25.2 (.247)	30.7 (.260)	35.4 (.265)	34.1 (.348)	40.2 (.322)	42.8 (.256)	44.7 (.468)	50.8 (.381)
1.0	9.9 (.057)	7.8 (.061)	8.8 (.065)	13.8 (.062)	10.2 (.080)	11.3 (.074)	18.7 (.067)	12.7 (.098)	14.8 (.094)
1.5	5.8 (.025)	4.0 (.028)	4.5 (.030)	8.4 (.029)	5.2 (.035)	5.9 (.036)	11.9 (.033)	6.3 (.042)	7.5 (.041)
2.0	4.1 (.015)	2.6 (.016)	2.9 (.017)	6.1 (.017)	3.2 (.016)	3.8 (.020)	8.8 (.020)	4.0 (.024)	4.8 (.023)
2.5	3.2 (.010)	1.9 (.011)	2.1 (.011)	4.9 (.012)	2.3 (.013)	2.7 (.013)	7.0 (.014)	2.8 (.016)	3.4 (.015)
3.0	2.7 (.008)	1.5 (.008)	1.6 (.008)	4.0 (.009)	1.8 (.010)	2.1 (.010)	5.9 (.011)	2.1 (.011)	2.7 (.011)

Note: Numbers in parentheses ( ) are the standard errors of the run lengths.

TABLE 5.4

In-Control Run Length Distribution for the RMBP

Run Length	p: 2	5	10
< 10	269	199	180
10–19	137	124	122
20–29	78	92	94
30–39	57	59	49
40–49	40	33	50
50–99	149	146	164
100–149	91	119	117
150–199	69	64	78
200–249	49	67	51
> 249	6061	6097	6095

5.3.9 Study of Inertia

Several researchers (e.g., Yashchin [20]; Crosier, [6]; Lowry et al., [12]) have suggested that both univariate and multivariate CUSUM and EWMA procedures could suffer from the problem of inertia. It has been suggested that the use of a Shewhart-type rule along with the CUSUM or the EWMA procedures could partially alleviate this problem. Lowry et al. [12] , however, point out that this solution involves a trade off between protection from inertia and faster detection of small shifts in the mean vector.

The same type of inertia problems can occur with the MBP procedures. Suppose the process has been *in control* and then shifts immediately prior to observation i. Then the posterior mean μ^i has accumulated all but the last observation from an in-control process. Since the divisor is $(i + 1)$, there is no discounting of the $(i - 1)$ observations from an in-control process. In an EWMA scheme and CUSUM scheme, previous observations get discounted.

Table 5.5 give the ARL comparisons of the IMBP, RMBP, MEWMAE and CMCUSUM procedures for the *bivariate case* under the assumption that the process is in control for the first $n(n = 20$ and $50)$ sampling epochs and then goes out of control. In this case, the MEWMAE and MCUSUM procedures perform better than the MBPs. The differences in the in-control run length distributions of the three procedures are responsible for this result. The issue of what sampling epochs to explore has been discussed by Sparks [17]. He states "Generally, the process may fluctuate from in-control to out-of-control situations. It is impossible to simulate all practical examples."

To alleviate the inertia problem, the RMBP procedure will be further refined in the next section so that it has in-control ARL equal to the in-control ARLs of the other procedures.

Based on simulation results, the RMBP is found to be better than the CMCUSUM procedure for detecting shifts in the mean vector with a value of the non-centrality parameter of at least 1 for $p = 2, 5$ and 10. The RMBP not

TABLE 5.5

ARL Comparisons When the Process is in Control

	First 20 Sampling Epochs				First 50 Sampling Epochs			
	CMCUSUM	MEWMAE	IMBP	RMBP	CMCUSUM	MEWMAE	IMBP	RMBP
λ	$h = 5.5$	$h_2 = 8.79$	$h_4 = 6.91$	$h_5 = 9.21$	$h = 5.5$	$h_2 = 8.79$	$h_4 = 6.91$	$h_5 = 9.21$
0.0	200.90	199.30	374.55	437.72	200.00	199.30	374.55	437.72
0.5	45.69	42.82	52.55	63.74	67.92	63.87	86.29	103.60
1.0	28.15	27.12	32.75	36.69	52.91	50.59	62.70	70.74
1.5	24.64	23.78	27.62	29.96	49.92	47.92	56.39	61.97
2.0	23.23	22.36	25.11	26.92	48.55	46.57	53.51	57.89
2.5	22.45	21.54	23.90	25.32	47.83	45.81	51.66	55.68
3.0	22.94	21.22	22.96	24.23	47.47	45.31	50.58	54.33

Note: ARLs are computed for $p = 2$.

only has better in-control ARLs for the values of p considered but also has better out-of-control ARLs. Although one does not need to use simulation to start the procedure, one does have to resort to simulation in order to find the out-of-control ARLs for different values of p and different values of λ.

Another advantage of the MBPs is that the process mean vector is estimated at time i, and the estimate could give an indication of the direction of the shift.

5.4 Refinement of RMBP to Alleviate Inertia

In Section 3.9, it was pointed out that the RMBP suffers from the problem of inertia. To alleviate this problem and to ensure a fair comparison of the RMBP with other procedures, the parameters of RMBP should be chosen to obtain a fixed in-control ARL of 200. One approach for handling this problem is presented here. In this approach, the statistic B_i is based on the k latest observations and results in RMBPs with fixed in-control run lengths.

5.4.1 Moving Window of the k Latest Observations

The approach used is to base the statistic B_i on the k latest observations. In this case, the statistic B_i is computed as follows:

$$B_i = \{R/(1 + iR)\}(\mathbf{x}_1 + \mathbf{x}_2 + \ldots + \mathbf{x}_i)'\Sigma^{-1}(\mathbf{x}_1 + \mathbf{x}_2 + \ldots + \mathbf{x}_i) \quad \text{for} \quad i < k$$

and,

$$B_i = \{R/(1 + kR)\}(\mathbf{x}_{i-k+1} + \mathbf{x}_{i-k+2} + \ldots + \mathbf{x}_i)'\Sigma^{-1}(\mathbf{x}_{i-k+1} + \mathbf{x}_{i-k+2} + \ldots + \mathbf{x}_i) \quad \text{for} \quad i \geq k$$

An out-of-control signal occurs at time i if B_i is greater than h_5.

The parameters to be determined are k, R, and h_5. The values of k used were 19, 29 and 39. Since at time 1, the statistic B_1 equals $\{R/(1+R)\}$ times the χ^2 statistic, the value of h_5 for a given R is taken as $\{R/(1+R)\}\chi^2_{p,(1-\alpha)}$. Combinations of $p = 2, 3, 5$ and 10 and $k = 19, 29$ and 39, were used.

For a given combination of (p, k), the value of R that results in an in-control run length between 199.5 to 200.5 is found based on simulation experiments of 7,000 independent runs each. In the following section, we study the performance of these revised procedures.

5.4.2 ARL Comparisons using Moving Window Bayesian Procedures

Since the MEWMAE procedure is better than the CMCUSUM procedure, the performance of the revised Bayesian procedures is compared with the performance of the MEWMAE procedure only. The revised Bayesian procedures with $k = 19, 29$ and 39 will be referred to as the RMBP1, RMBP2, and RMBP3, respectively.

Tables 5.6 and 5.7 give the ARL comparisons of the RMBP1, RMBP2, and RMBP3 procedures with the MEWMAE procedures for $p = 2, 3, 5$ and 10. The ARLs are based on 7,000 run lengths each. Comparisons of the procedures are based on 99% pairwise confidence intervals for each ARL.

Irrespective of the value of p, the performance of RMBP2 is significantly better than the MEWMAE procedure for detecting small shifts, i.e., shifts of non-centrality parameter $\lambda \leq 1.5$. The RMBP3 procedure performs better than the MEWMAE for detecting shifts of non-centrality parameter $\lambda \leq 2.0$ for all values of $p > 2$. For $p = 10$, RMBP3 is significantly better than MEWMAE for detecting shifts as large as $\lambda = 2.5$. The performance of the MEWMAE procedure, however, is better than the performance of at least one of the RMBPs for detecting large shifts, i.e., shifts of non-centrality parameter $\lambda \geq 2.0$.

Although the comparative in-control run length distributions of the four procedures were studied for $p = 2, 3, 4, 5$ and 10, the results are presented in Table 5.8 for $p = 2$ only. These comparisons indicate that the in-control run length distributions of the four procedures are quite similar. In other words, the better performance of RMBP3 is not at the expense of a worse in-control run length distribution.

However, the probability of false alarm associated with the first observation is at least twice as large for the MEWMAE procedure as the corresponding probability for RMBPs. For example, in the case of RMBPs for $p = 2$, h_5 is chosen so that the probability of false alarm associated with the first observation is .005. The decision interval for the MEWMAE procedure is $[0, 8.79]$ for $p = 2$ and the corresponding probability of false alarm is 0.012.

In this section an approach was introduced for refining the multivariate Bayesian procedure by using a moving window. Through simulation it is shown that the RMBPs based on the latest $k = 29$ and 39 observations perform better than the MEWMAE procedure for detecting small to moderate shifts

TABLE 5.6

ARL Comparisons

	p = 2				p = 3			
	MEWMAE	**RMBP1**	**RMBP2**	**RMBP3**	**MEWMAE**	**RMBP1**	**RMBP2**	**RMBP3**
	$h_2 = 8.79$	$h_5 = 7.773$	$h_5 = 7.043$	$h_5 = 6.523$	$h_2 = 10.97$	$h_5 = 9.871$	$h_5 = 9.086$	$h_5 = 8.502$
λ	$r = .1$	$R = 2.75$	$R = 1.98$	$R = 1.6$	$r = .1$	$R = 3.325$	$R = 2.42$	$R = 1.96$
0.0	199.3 (2.41)	200.0 (2.34)	199.7 (2.42)	199.7 (2.47)	197.7 (2.44)	200.4 (2.42)	200.02 (2.43)	200.4 (2.51)
0.5	25.2 (.247)	26.0 (.258)	22.9 (.221)*	21.2 (.202)*	29.2 (.287)	29.8 (.295)	26.1 (.248)*	24.0 (.226)*
1.0	7.8 (.061)	7.8 (.059)	7.3 (.055)*	7.0 (.053)*	8.7 (.068)	8.7 (.066)	8.1 (.061)*	7.7 (.058)*
1.5	4.0 (.028)†	4.1 (.026)	3.9 (.025)*	3.8 (.024)*	4.5 (.031)	4.6 (.029)	4.3 (.028)*	4.2 (.027)*
2.0	2.6 (.016)†	2.7 (.015)	2.6 (.015)	2.6 (.014)	2.9 (.018)†	3.0 (.017)	2.9 (.016)	2.8 (.015)*
2.5	1.9 (.011)†	2.0 (.010)	2.0 (.010)	2.0 (.018)	2.1 (.012)†	2.2 (.011)	2.2 (.011)	2.1 (.011)
3.0	1.5 (.008)†	1.6 (.008)	1.6 (.008)	1.6 (.008)	1.6 (.008)+	1.8 (.008)	1.7 (.008)	1.7 (.008)

Note: Numbers in ( ) are standard errors of average run lengths.

* Indicates that RMBP is significantly better than MEWMAE at $\alpha = .01$

† Indicates that MEWMAE is significantly better than at least one RMBP at $\alpha = .01$

TABLE 5.7

ARL Comparisons

	p = 5				p = 10			
λ	**MEWMAE** $h_2 = 14.74$ $r = .1$	**RMBP1** $h_5 = 13.498$ $R = 4.15$	**RMBP2** $h_5 = 12.624$ $R = 3.06$	**RMBP3** $h_5 = 11.99$ $R = 2.52$	**MEWMAE** $h_2 = 22.91$ $r = .1$	**RMBP1** $h = 21.526$ $R = 5.875$	**RMBP2** $h = 20.482$ $R = 4.35$	**RMBP3** $h = 19.64$ $R = 3.54$
0.0	201.8 (2.45)	199.9 (2.34)	199.7 (2.42)	200.1 (2.44)	198.3 (2.43)	200.2 (2.40)	200.4 (2.44)	200.2 (2.55)
0.5	34.1 (.348)†	35.4 (.360)	30.2 (.295)*	27.7 (.261)*	44.7 (.468)†	47.1 (.494)	39.2 (.388)*	35.5 (.338)*
1.0	10.2 (.080)	10.1 (.077)	9.5 (.071)*	9.0 (.068)*	12.7 (.098)	12.7 (.096)	11.7 (.086)*	11.0 (.081)*
1.5	5.2 (.035)	5.2 (.033)	4.9 (.031)*	4.7 (.030)*	6.3 (.042)	6.3 (.040)	6.0 (.038)*	5.7 (0.36)*
2.0	3.2 (.016)†	3.3 (.019)	3.2 (.018)	3.1 (.017)*	4.0 (.024)	4.0 (.022)	3.9 (.022)*	3.8 (.021)*
2.5	2.3 (.013)†	2.4 (.012)	2.4 (.011)	2.4 (.012)	2.8 (.016)	2.9 (.015)	2.8 (.014)	2.7 (.014)*
3.0	1.8 (.010)†	1.9 (.009)	1.9 (.009)	1.9 (.009)	2.1 (.011)†	2.2 (.011)	2.2 (.010)	2.2 (.010)

Notes: Numbers in ( ) are standard errors of average run lengths.

* Indicates that RMBP is significantly better than MEWMAE at $\alpha = .01$

† Indicates that MEWMAE is significantly better than at least one RMBP at $\alpha = .01$

TABLE 5.8

In-Control Run Length Distribution Comparisons

Procedure	MEWMAE $h_2 = 8.79$ $r = .1$	RMBP1 $h_5 = 7.773$ $R = 2.75$	RMBP2 $h_5 = 7.043$ $R = 1.98$	RMBP3 $h_5 = 6.523$ $R = 1.6$
Mean	199.32	200.04	199.74	199.66
Median	136.50	142.00	141.00	137.00
SEMEAN	2.41	2.34	2.42	2.47
Q1	54.00	59.00	56.00	53.00
Q3	279.00	277.00	278.00	280.00
MAX	1543.00	1837.00	2572.00	1956.00

Note: Run lengths are computed for $p = 2$.

in the process mean vector. The MEWMAE procedure, however, performs better than RMBPs for detecting large shifts.

It is also pointed out that when the process is in control, the probability of a false alarm associated with the first observation in case of MEWMAE is at least twice as large as the corresponding probability for the RMBPs. This, in turn, leads to a faster detection of out-of-control processes. When the process is in control for $n = 10, 20, 50$ and 100 sampling periods and then goes out of control, RMBP3 outperforms all the other procedures for detecting small shifts.

Another observation which has practical significance is that the values of h_5 are approximately equal to $\chi^2_{p,.98}$, $\chi^2_{p,.97}$ and $\chi^2_{p,.96}$ for $k = 19, 29$ and 39, respectively. Given the approximate value of h_5 for a given p and k, the value of R to be used is given by $h_5/(\chi^2_{p,.995} - h_5)$. Thus, simulation is not needed.

5.4.3 Study of Inertia

Although the ARLs were compared for MEWMAE, RMBP1, RMBP2, and RMBP3 for $p = 2$ under the assumption that the process is in control for the first n ($n = 10, 20, 50$ and 100) sampling epochs and then suddenly goes out of control, we present in Table 5.9 the results only for $n = 20$ and 50.

A comparison of the results in Table 5.9 with those in Table 5.5 shows that the RMBP1, RMBP2, and RMBP3 procedures do not suffer from the problem of inertia to the same extent as the original RMBP.

There is no statistically significant difference between the RMBP2, RMBP3 and MEWMAE procedures for detecting small shifts, i.e., shifts of non-centrality parameter $\lambda \leq 0.5$. The performance of the MEWMAE procedure appears better than that of RMBPs for moderate to large shifts. However, one has to keep in mind that the large probability of false alarms when the process is in-control also results in faster out-of-control signals.

TABLE 5.9

ARL Comparisons

	First 20 Sampling Epochs				First 50 Sampling Epochs			
λ	**MEWMAE** $h_2 = 8.79$	**RMBP1** $R = 2.75$	**RMBP2** $R = 1.98$	**RMBP3** $R = 1.6$	**MEWMAE** $h_2 = 8.79$	**RMBP1** $R = 2.75$	**RMBP2** $R = 1.98$	**RMBP3** $R = 1.6$
0.0	200.00	200.00	200.00	200.00	200.00	200.00	200.00	200.00
0.5	42.82 (.275)	43.87 (.273)	42.83 (.239)	43.33 (.237)	63.87 (.362)	65.69 (.354)	64.56 (.354)	64.34 (.350)
1.0	27.12 (.102)	29.27 (.098)	30.44 (.116)	30.76 (.130)	50.59 (.219)	53.17 (.212)	53.49 (.237)	54.27 (.255)
1.5	23.78 (.078)	26.12 (.079)	26.77 (.094)	26.33 (.101)	47.92 (.194)	50.69 (.191)	50.51 (.216)	50.73 (.230)
2.0	22.36 (.068)	24.60 (.068)	24.66 (.083)	24.08 (.087)	46.57 (.187)	49.23 (.181)	48.96 (.202)	48.69 (.219)
2.5	21.54 (.064)	23.48 (.065)	23.52 (.075)	22.82 (.080)	45.81 (.185)	48.29 (.177)	48.22 (.193)	47.67 (.210)
3.0	21.22 (.059)	22.75 (.062)	22.61 (.071)	22.05 (.075)	45.31 (.183)	47.58 (.175)	47.31 (.192)	45.96 (.206)

Note: ARLs are computed for $p = 2$. Numbers in ( ) indicate the standard errors of average run lengths.

TABLE 5.13

ARL Comparisons of RMBPs

	Observations Drawn from					
	Multivariate Normal			Pearson Type VII*		
λ	MBP1	MBP2	MBP3	MBP1	MBP2	MBP3
0.0	200.0 (2.34)	199.7 (2.42)	199.7 (2.47)	230.3 (2.84)	251.2 (3.07)	259.0 (3.19)
0.5	26.0 (.258)	22.9 (.221)	21.2 (.202)	29.7 (.290)	24.7 (.223)	22.2 (.193)
1.0	7.8 (.059)	7.3 (.055)	7.0 (.053)	8.1 (.055)	7.5 (.050)	7.1 (.048)
1.5	4.1 (.026)	3.9 (.025)	3.8 (.024)	4.2 (.024)	4.0 (.022)	3.9 (.021)
2.0	2.7 (.015)	2.6 (.015)	2.6 (.014)	2.7 (.013)	2.65 (.012)	2.6 (.012)
2.5	2.0 (.010)	2.0 (.010)	2.0 (.010)	2.1 (.009)	2.0 (.008)	2.0 (.008)
3.0	1.6 (.008)	1.6 (.008)	1.6 (.008)	1.7 (.007)	1.65 (.007)	1.65 (.007)

Note: ARLs are computed for $p = 2$. Numbers in ( ) are standard errors of average run lengths.
$^*m = 2.5$

χ_p^2 distribution. The Cov(**X**) for a Pearson Type VII distribution equals Σ for $m = 2.5$ and $p = 2$.

Random vectors having a Pearson Type VII multivariate distribution can be generated using Cambani's approach.

5.6.3 Average Run Length Comparisons When Cov(X) = Σ

Since the MBPs have been designed under the assumption of a known covariance matrix Σ, the parameter m of the Pearson Type VII distribution was chosen to ensure that Cov(**X**) equals Σ. For a given p, $m = (p+3)/2$ has been used in the random variate generation to ensure that feature.

For $p = 2$, 5 and 10, Tables 5.13 through 5.15 give the ARL comparisons of the RMBP1, RMBP2, and RMBP3 when the observations are drawn from

TABLE 5.14

ARL Comparisons of RMBPs

	Observations Drawn from					
	Multivariate Normal			Pearson Type VII*		
λ	MBP1	MBP2	MBP3	MBP1	MBP2	MBP3
0.0	199.9 (2.34)	199.7 (2.42)	200.1 (2.44)	175.9 (2.24)	206.8 (2.69)	226.4 (2.97)
0.5	35.4 (.360)	30.2 (.295)	27.7 (.261)	45.8 (.511)	36.0 (.358)	31.6 (.300)
1.0	10.1 (.077)	9.5 (.071)	9.0 (.068)	10.8 (.075)	10.0 (.067)	9.6 (.064)
1.5	5.2 (.033)	4.9 (.031)	4.7 (.030)	5.5 (.031)	5.3 (.029)	5.1 (.028)
2.0	3.3 (.019)	3.2 (.018)	3.1 (.017)	3.5 (.017)	3.4 (.016)	3.3 (0.16)
2.5	2.4 (.012)	2.4 (.011)	2.4 (.012)	2.6 (.011)	2.5 (.011)	2.5 (.010)
3.0	1.9 (.009)	1.9 (.009)	1.9 (.009)	2.0 (.008)	2.0 (.007)	2.0 (.007)

Note: Comparisons for $p = 5$. Numbers in ( ) are standard errors of average run lengths.
$^*m = 4$.

TABLE 5.15

ARL Comparisons of RMBPs

	Observations Drawn from					
	Multivariate Normal			Pearson Type VII*		
λ	MBP1	MBP2	MBP3	MBP1	MBP2	MBP3
0.0	200.2 (2.40)	200.4 (2.44)	200.2 (2.55)	141.3 (1.88)	175.7 (2.39)	200.0 (2.74)
0.5	47.1 (.494)	39.2 (.388)	35.5 (.338)	60.5 (.737)	49.2 (.566)	41.8 (.451)
1.0	12.7 (.096)	11.7 (.086)	11.0 (.081)	14.3 (.117)	12.9 (.092)	12.21 (.087)
1.5	6.3 (.040)	6.0 (.038)	5.7 (.036)	7.0 (.041)	6.7 (.039)	6.4 (.038)
2.0	4.0 (.022)	3.9 (.022)	3.8 (.021)	4.4 (.023)	4.3 (.022)	4.2 (0.21)
2.5	2.9 (.015)	2.8 (.014)	2.7 (.014)	3.2 (.015)	3.1 (.014)	3.0 (.014)
3.0	2.2 (.011)	2.2 (.010)	2.2 (.010)	2.5 (.010)	2.4 (.010)	2.4 (.010)

Note: Comparisons for $p = 10$. Numbers in ( ) are standard errors of average run lengths.
$^*m = 6.5$

a multivariate normal distribution vs. when they are drawn from a Pearson Type VII multivariate distribution (RMBPs in the Tables have been referred as MBPs). The results in these tables are based on simulations of 7,000 independent run lengths each.

The results show that the RMBPs are quite robust to departures from the multivariate normality distributional assumption used in this paper. For a Pearson Type VII multivariate process, the in-control ARLs for the three RMBPs ranged from 141 to 259. In general, the in-control ARL for RMBP3 is at least 200 for all values of p that were considered. For $p = 2$ and 5, the in-control ARL is at least 200 for RMBP2. The in-control ARL for RMBP1, however, is at least 200 only for $p = 2$ and 3. The out-of-control ARLs for detecting small shifts ($\lambda \leq 1$) are comparatively higher for the Pearson Type VII distribution than the corresponding out-of-control ARLs for a multivariate normal process. There are two reasons for these differences. Firstly, the Pearson Type VII distribution is a heavy-tailed distribution; and secondly, the distributions of the statistics being computed are different.

5.6.4 Average Run Length Comparisons When Cov(X) $\neq \Sigma$ and $p = 2$

The results in Table 5.16 are based on two cases. First, the random vectors **X** have been drawn from a Pearson Type VII bivariate distribution with parameter $m = 2.2$, in-control mean vector = **0**, and Cov (**X**) = 2.5**I**. In the second case, $m = 3.2$, the in-control mean vector = **0**, and Cov(**x**) = $\Sigma/2.4$. The results indicate that in-control ARLs, when Cov(**X**) = $\Sigma/2.4$, are much larger than the in-control ARLs when Cov(**X**) = 2.5Σ. This clearly indicates that the MBPs are able to detect not only the process mean shift but also increased variability.

TABLE 5.16

ARL Comparisons of RMBPs

	Cov(x)=2.51						Cov(x)=I/2.4					
	Observations Drawn from						Observations Drawn from					
	Multivariate Normal			Pearson Type VII*			Multivariate Normal			Pearson Type VII†		
λ	MBP1	MBP2	MBP3	MBP1	MBP2	MBP3	MBP1	MBP2	MBP3	MBP1	MBP2	MBP3
0.0	200.0	199.7	199.7	68.5	73.8	76.5	200.0	199.7	199.7	1300	1314	1317
	(2.34)	(2.42)	(2.47)	(.861)	(.955)	(1.02)	(2.34)	(2.42)	(2.47)	(4.88)	(4.80)	(4.77)
0.5	26.0	22.9	21.2	21.8	19.9	18.5	26.0	22.9	21.2	45.5	30.0	25.8
	(.258)	(.221)	(.202)	(.230)	(.208)	(.196)	(.258)	(.221)	(.202)	(.425)	(.226)	(.164)
1.0	7.8	7.3	7.0	7.6	7.0	6.7	7.8	7.3	7.0	8.5	7.9	7.5
	(.059)	(.055)	(.053)	(.062)	(.057)	(.053)	(.059)	(.055)	(.053)	(.041)	(.038)	(.036)
1.5	4.1	3.9	3.8	4.1	3.9	3.8	4.1	3.9	3.8	4.3	4.1	3.9
	(.026)	(.025)	(.024)	(.026)	(.025)	(.024)	(.026)	(.025)	(.024)	(.018)	(.017)	(.016)
2.0	2.7	2.6	2.6	2.7	2.6	2.6	2.7	2.6	2.6	2.75	2.7	2.6
	(.015)	(.015)	(.014)	(.015)	(.014)	(.014)	(.015)	(.015)	(.014)	(.010)	(.010)	(.009)
2.5	2.0	2.0	2.0	2.0	2.0	2.0	2.0	2.0	2.0	2.1	2.0	2.0
	(.010)	(.010)	(.010)	(.010)	(.010)	(.009)	(.010)	(.010)	(.010)	(.007)	(.006)	(.006)
3.0	1.6	1.6	1.6	1.65	1.64	1.63	1.6	1.6	1.6	1.7	1.68	1.67
	(.008)	(.008)	(.008)	(.008)	(.007)	(.007)	(.008)	(.008)	(.008)	(.006)	(.006)	(.006)

Note: ARLs are computed for $p = 2$. Numbers in parentheses are standard errors of average run lengths.

*m=2.2.

†m=3.2.

5.7 Summary and Conclusions

The main objective of this paper was to develop a simple and effective multivariate process control procedure using a Bayesian approach. An initial multivariate Bayesian procedure (IMBP) was proposed and, based on simulation results, it was shown to be better than the existing multivariate procedures. With a much higher in-control ARL than other procedures, the IMBP had much smaller out-of-control ARLs than those of the other procedures. Thus, IMBP performed better than other multivariate procedures in terms of both in-control and out-of-control ARLs.

The IMBP was revised so that it could be designed without resorting to simulation. The revised procedure (RMBP) was better than the CMCUSUM and MEWMAE procedures for detecting shifts in the mean vector when the non-centrality parameter was at least 1.0.

The CMCUSUM, MEWMAE and MBP suffer from inertia in reacting to shifts in the mean. Initial simulations to study the inertia problem indicated that IMBP and RMBP both performed worse than the CMCUSUM and MEWMAE procedures in this respect. The main reason for this poor performance was that the in-control ARLs of the MBPs were much higher than the in-control ARLs of the other procedures. To overcome this, two different alternatives were considered to make the MBPs comparable to the other procedures in terms of the in-control ARL.

The first alternative was to restart the procedure after a fixed number of observations had been obtained. This alternative did result in procedures with fixed in-control ARLs and partially alleviated the inertia problem. However, the second alternative in which the statistic B_i was based on the k latest observations resulted in RMBPs (for $k = 29$ and 39) that performed better than existing procedures for detecting small shifts in the process mean vector. Also, when the process is in-control for 10, 20, 50 and 100 sampling epochs and then goes out-of-control, the RMBP based on the 39 latest observations outperformed other procedures in detecting small shifts.

As a special case of the RMBPs, the univariate Bayesian procedures based on the k latest observations performed much better than the univariate CUSUM and EWMA procedures.

Appendix

We show that the IMBP's performance depends on the values of the process mean vector and the covariance matrix only through the value of the non-centrality parameter λ. By using Crosier's [6] approach, the desired result can be easily proven by showing that the IMBP's statistic, B_i, is invariant to any full-rank transformation of the data.

by Tagaras [16], the two-sided case may be formulated with no additional conceptual difficulty over the one-sided case. However, the issues of computational requirements and accuracy become much more important because of the expansion of the state space. As a matter of fact no specific numerical results had been reported for two-sided Bayesian charts, until a very recent working paper by Celano et al. [5] that investigated the possibility of using a hybrid genetic – dynamic programming (DP) algorithm to overcome the computational difficulties and arrive at a near-optimal economic design of the two-sided Bayesian $\overline{X}$ chart.

This chapter investigates the computational efficiency and accuracy of the "pure" DP formulation, i.e., without combining it to genetic or similar-type algorithms, as applied to the design and operation of the two-sided Bayesian $\overline{X}$ control chart. Like most previous related work we concentrate on monitoring production runs of finite duration and we adopt an economic optimization perspective. The objective of the investigation is to provide answers to the following interrelated practical questions:

- Given the capabilities of modern computers, how difficult is it today to design and operate a two-sided $\overline{X}$ control chart using the principles of Bayesian process control?
- How should we discretize the state variables (out-of-control probabilities) and how should we choose the allowable values of the adaptive decision variables (sampling intervals and sample sizes) so as to achieve a good balance between economic effectiveness and accuracy on one hand and computational efficiency on the other?

The next section contains a brief description of the problem setting and the DP formulation. Sections 3 and 4 present and discuss the results of the numerical investigation for eight different sets of cost and process parameters, analyzed under a variety of discretizations of the state space (Section 3) and decision space (Section 4). The results of all DP computations are validated by extensive simulations. The last section summarizes the main findings and conclusions of this research, which constitute the answers to the above two questions. Fortunately enough, these answers are very encouraging. In short, we conclude that with a suitable discretization of the state variables and the proper choice of decision variables the two-sided Bayesian $\overline{X}$ control charts can be rather easily optimized and used for economically effective monitoring of production processes.

6.2 Problem Description and Formulation

We consider a production process that is set up for a production run of a specific net operating time T. The critical quality characteristic that is to be monitored is a normally distributed random variable X, with target value μ_0

negligible time. The time to sample, measure and plot the measurements is assumed to be insignificant. If no sampling and monitoring takes place throughout the production run ($I = 0$), the total quality-related cost is the expected cost due to possible out-of-control operation, which remains undetected until the end of the run:

$$TC_0 = M\left(T - \frac{1 - e^{-\lambda T}}{\lambda}\right). \tag{6.6}$$

6.2.2 Monitoring with a Bayesian Chart

A more powerful alternative to monitoring the mean of X in the above setting is to employ a two-sided Bayesian $\overline{X}$-chart, which will incorporate all available relevant information through the continuous update of our knowledge about the actual state of the process. The knowledge about the unobservable (except for times of intervention) process state is sufficiently expressed by the probabilities p_0, p_1 and p_2 that the process is actually in state 0, or state 1 or state 2 respectively. Since $p_0 + p_1 + p_2 = 1$, it obviously suffices to consider only the pair (p_1, p_2). Depending on the values of p_1 and p_2 immediately after a sample, a decision a is made whether to continue production ($a = 0$) or stop and investigate if an assignable cause has occurred ($a = 1$). In addition, the length h of the next sampling interval and the size n of the next sample are also determined if the Bayesian chart is implemented in its fully adaptive version. Alternatively, h and/or n may be fixed and constant for the entire duration of the production run.

The knowledge updating mechanism is the application of Bayes' formula, every time a sample is taken and the sample mean $\overline{X}$ is recorded, to transform the values of p_1 and p_2 taking into account $\overline{X}$ and the average shift rates λ_1 and λ_2. Specifically, let $f_j(\overline{X})$ be the normal density function of $\overline{X}$ when the actual process mean is μ_j, $j = 0,1,2$. Also, let p_1 and p_2 be the probabilities that the process is in state 1 and in state 2, as they have been computed after the previous sample. Note that if $a = 0$ the probabilities p_1 and p_2 retain their values after the decision a is made but if $a = 1$ the probabilities p_1 and p_2 vanish since the investigation of the process in case of a signal is assumed to accurately reveal the true state of the process and the process restoration, if needed, brings the process back to the in-control state 0 with certainty. Suppose now than after a sampling interval of h time units a next sample of size n is taken. Given the values of p_1 and p_2 immediately after the previous sample and the decision a, n, and h, the density function of the new sample mean $f(\overline{X}|p_1, p_2, a, n, h)$ is written as follows for $a = 0$ and for $a = 1$:

$$\begin{aligned} f(\overline{X}|p_1, p_2, 0, n, h) &= [(1-\gamma)(1-p_1-p_2)]f_0(\overline{X}) \\ &\quad + [\gamma_1(1-p_1-p_2)+p_1]f_1(\overline{X}) + [\gamma_2(1-p_1-p_2)+p_2]f_2(\overline{X}), \\ f(\overline{X}|p_1, p_2, 1, n, h) &= (1-\gamma)f_0(\overline{X}) + \gamma_1 f_1(\overline{X}) + \gamma_2 f_2(\overline{X}) \end{aligned} \tag{6.7}$$

optimal sampling parameters n and h as a function of p_1 and p_2, converges to a specific policy that remains the same for all stages beyond some stage N_c, where N_c is usually not very large. This "steady-state" policy is apparently the optimal monitoring and control policy for continuously operating processes with the same parameters λ_1, λ_2, δ, b, c, L_0, L_1, M. In other words, the DP procedure described in this section for obtaining the optimal control policy in short runs can also be used to derive the optimal control policy in the infinite horizon case.

6.3 The Effect of State-Space Discretization on Computational Efficiency and Accuracy

In order to perform the computations required to obtain the optimal monitoring and control policy via the two-sided Bayesian $\overline{X}$-chart DP formulation of the previous section it is necessary to determine first how the two-dimensional state variable (p_1, p_2) will be discretized. Tagaras [16] explains in detail how any specific discretization of the (p_1, p_2) variable can be used to implement the computational procedure.

There is a practically infinite number of ways to discretize the two-dimensional state space defined by

$$\begin{gathered} p_1 \geq 0 \\ p_2 \geq 0 \\ p_1 + p_2 < 1 \\ p_1 p_2 \neq 0 \text{ unless } p_1 = p_2 = 0 \end{gathered}$$

so that it is expressed by means of a grid of points/states, where each point (p_1, p_2), with the exception of (0,0), represents all values in its neighborhood $(p_1 \pm \Delta_1, p_2 \pm \Delta_2)$. Note that Δ_1 and Δ_2 need not be constant for all (p_1, p_2) values. As a matter of fact, extensive computational experience has led to the conclusion that it is preferable to have a finer discretization at low values of p_1 and p_2 (small Δ_1, Δ_2) rather than at high values. For example, Tagaras [16] suggests that a good compromise between accuracy and computational requirements is obtained in the one-sided $\overline{X}$-chart case if the (0, 1) interval for the out-of-control probability p is divided into 100 unequal intervals (states) as follows: 10 points/states of length 0.002 each, covering the interval (0, 0.02); 20 states of length 0.004 for the interval [0.02, 0.10); 50 states of length 0.01 for [0.10, 0.60); and 20 states of length 0.02 for [0.60, 1.00). Adding the state $p = 0$, the total number of states is 101. This discretization has been used in later studies as well [3, 4].

For each of the cases of Table 6.2 we obtained the optimal control policy using fully adaptive two-sided Bayesian $\overline{X}$-charts with six choices for the sample size n and six choices for the sampling interval h, as follows [17]:

- Depending on the optimal sample size n_S of the respective Shewhart chart, the set of allowable values of n is $S_n = \{1, 2, 3, 4, 5, 6\}$ for $n_S = 5$ (Cases 3, 4, 7, 8) and $S_n = \{1, 3, 5, 7, 9, 11\}$ for $n_S = 11$ (Cases 1, 2, 5, 6).
- Depending on the optimal sampling interval $h_S = T/(I_S+1)$ of the respective Shewhart chart, the set of allowable values of h is $S_h = \{0.25h_S, 0.5h_S, 0.75h_S, h_S, 1.5h_S, 2h_S\}$, i.e., $h_{min} = 0.5$ (30 min) in cases 1, 3, 5, 7 (implying 16 stages in the DP formulation) and $h_{min} = 0.167$ (10 min) in cases 2, 4, 6, 8 (48 stages in the DP formulation).

The expected total costs of the optimal policies using each one of the discretizations D1, D2, D3 are shown in Table 6.3 as $C_{16}(0,0)$ for cases 1, 3, 5, 7 and $C_{48}(0,0)$ for cases 2, 4, 6, 8. The values $C_N(0,0)$ are clearly approximations of the true cost figures, since their computation is based on an approximate (discrete) mapping of the continuous state space. In order to estimate the accuracy of these approximations we simulated every one of the 8 cases one million times (production runs), applying each one of the optimal policies derived under D1, D2, D3 with the exact (continuous) values of p_1 and p_2; these values were computed from (8) and (9) using the simulated sample means $\bar{x}$. The simulation point estimates of the expected costs of the optimal policies under discretization D1, D2, D3 will be denoted as C_{D1}, C_{D2}, C_{D3}. Table 6.3 shows the 99% confidence intervals of the expected costs of the optimal policies, including the corresponding point estimates C_{Di} ($i = 1$, 2, 3). The 99% confidence intervals are very narrow; their length is always smaller than 1% of the respective point estimate, indicating that the simulation estimates are quite accurate.

Given that the simulation estimates of the total expected costs are very close to the true values, the main conclusion following from the careful examination of the numbers in Table 6.3 is that the finest of the three discretizations, D3, provides the most accurate results, as expected, which differ from the simulation point estimates by at most 1.6% in the 8 cases examined. At the other end, the expected cost values computed with the coarse discretization D1 differ by as much as 8.2% from the respective simulation estimates. Discretization D2 comes in-between with the largest observed difference being 4%. Naturally, the higher accuracy of D3 comes at the expense of higher computational requirements. Specifically, the computation time required to determine the optimal policy with dynamic programming using D1 is in the range 3 – 15 minutes on a Pentium IV-3.4GHz computer, while it takes 1 – 5 hours to obtain the optimal policy using D3. Discretization D2 is, computationally, about 5 times more demanding than D1 and about 4 times less demanding than D3. It is important to note that it is not necessary to perform the DP computations for all discrete states and all $6 \times 6 = 36$ allowable combinations of n and h at each stage; the computational

TABLE 6.3

Expected Cost of Optimal Policy

Case		Discretizations		
		D1	D2	D3
1	$C_{16}(0,0)$	185.39	184.43	184.15
	99% CI	184.28 ± 0.68	184.09 ± 0.67	184.25 ± 0.67
2	$C_{48}(0,0)$	342.04	334.33	331.68
	99% CI	333.69 ± 1.12	331.12 ± 1.09	330.68 ± 1.07
3	$C_{16}(0,0)$	143.59	142.60	142.30
	99% CI	142.46 ± 0.52	142.25 ± 0.52	142.14 ± 0.52
4	$C_{48}(0,0)$	350.10	340.56	337.17
	99% CI	338.35 ± 1.14	336.33 ± 1.10	335.80 ± 1.09
5	$C_{16}(0,0)$	92.27	90.80	90.14
	99% CI	90.04 ± 0.37	89.82 ± 0.36	89.79 ± 0.35
6	$C_{48}(0,0)$	206.49	196.06	191.01
	99% CI	191.45 ± 0.70	188.71 ± 0.67	187.91 ± 0.65
7	$C_{16}(0,0)$	79.05	77.67	77.05
	99% CI	76.80 ± 0.32	76.58 ± 0.32	76.53 ± 0.31
8	$C_{48}(0,0)$	159.00	150.02	146.06
	99% CI	147.01 ± 0.57	144.22 ± 0.53	143.84 ± 0.52

Note: Expected cost, $C_N(0,0)$, is computed by dynamic programming and estimated by simulation, 99% CI (limits of the 99% confidence interval; the middle of each confidence interval is the respective point estimate C_{Di}).

requirements are significantly reduced by exploiting several properties of the optimal policy [12]. The computation times reported above refer to the implementation of the DP procedure with the reduced requirements.

6.3.2 Quality of Bayesian Control Policies Under Different State-Space Discretizations

From Table 6.3 and the preceding discussion the trade-off between accuracy and computational requirements is obvious. However, the accurate and efficient estimation of the expected cost of a particular policy is of secondary importance, because simulation is a very effective tool than can serve this purpose. The really important question is whether, how and to what extent it is possible to determine a high-quality (near-optimal) Bayesian $\overline{X}$-chart policy given the capabilities of today's computers. To be more specific, the practically interesting trade-off is between the computational requirements for determining the optimal policy using D1, D2 or D3 and the quality of the resulting policies. The former (computational requirements) has already been discussed above. The latter can be evaluated by comparing the "true" (simulated) costs C_{D1}, C_{D2}, C_{D3} of the policies obtained using D1, D2, D3. Although these costs are contained in Table 6.3, their direct comparison is facilitated by Table 6.4, which also contains the percentage differences ΔC_{D32}

TABLE 6.4

Effect of State-Space Discretization on the Quality of the Optimal Control Policy and Cost Penalties for Using Shewhart-Type Rather than Bayesian Charts

		Discretization Cost			**Percentage Differences**		
Case	TC_S	C_{D1}	C_{D2}	C_{D3}	$\Delta C_{D32}(\%)$	$\Delta C_{D31}(\%)$	$\Delta C_{SD3}(\%)$
1	193.03	184.28	184.09	184.25	−0.09	0.02	4.77
2	350.53	333.69	331.12	330.68	0.13	0.90	6.00
3	148.92	142.46	142.25	142.14	0.08	0.22	4.77
4	356.28	338.35	336.33	335.80	0.16	0.75	6.10
5	96.92	90.04	89.82	89.79	0.04	0.28	7.94
6	207.95	191.45	188.71	187.91	0.42	1.85	10.66
7	82.96	76.80	76.58	76.53	0.07	0.35	8.40
8	157.77	147.01	144.22	143.84	0.27	2.16	9.68

Note: TCs = total expected cost.

and ΔC_{D31} that are computed as follows:

$$\Delta C_{D32} = \frac{C_{D2} - C_{D3}}{C_{D2}} \times 100\%$$

$$\Delta C_{D31} = \frac{C_{D1} - C_{D3}}{C_{D1}} \times 100\%$$

ΔC_{D32} (ΔC_{D31}) measures the reduction in expected cost that is achieved when we move from the policy obtained using the discretization D2 (D1) to the generally better optimal policy determined using the finest discretization D3. To put these numbers into proper perspective, Table 6.4 also contains the percentage difference

$$\Delta C_{SD3} = \frac{TC_S - C_{D3}}{C_{D3}} \times 100\%$$

that is the cost penalty from using the optimal Shewhart-type static $\overline{X}$-chart rather than the optimal adaptive Bayesian $\overline{X}$-chart (under D3).

The main conclusion from Table 6.4 is that it is not necessary to employ a very fine and computationally demanding state-space discretization, like D3, to obtain a near-optimal policy. More specifically, discretization D2, which is computationally about 4 times less demanding than D3, yields policies that are practically as good as those obtained with D3. The percentage differences ΔC_{D32} are lower than 0.42% in all 8 cases. In fact, direct comparison of the 99% confidence intervals in Table 6.3 reveals that the difference in the expected costs of the policies determined with D2 and D3 is not significant at the 1% level in any of the 8 test cases. Even the policies obtained with the coarse discretization D1 are not much inferior to those of D3; the average ΔC_{D31} difference is 0.82% while the highest recorded ΔC_{D31} is 2.16% in Case 8. Note that ΔC_{SD3} is 9.68% in Case 8, implying that the improvement from using the Bayesian adaptive $\overline{X}$-chart with D1 instead of the optimal static Shewhart chart would be substantial. More generally, the ΔC_{D31} differences are an order of magnitude smaller than the respective ΔC_{SD3}.

In short, it is possible to obtain in only a few minutes a Bayesian adaptive $\overline{X}$-chart monitoring and control policy that is considerably more economical than that of the standard Shewhart chart. The anticipated cost reduction is around 5% to 10%. In cases of expensive processes where the economic stakes are really high and even a small percentage cost improvement is important we recommend to employ the D2 discretization for deriving and implementing the two-sided Bayesian $\overline{X}$control scheme. The control policy can be obtained in a reasonable time and will be very close to the true optimum.

6.4 The Effect of Allowable Choices for the Sample Size and Sampling Interval

All the numerical examples and results in the previous section concern the fully adaptive Bayesian scheme with 36 specific combinations of allowable sample sizes n and sampling intervals h. Previous research that was focused on one-sided $\overline{X}$-charts [16, 17] has established that the economic effectiveness of the Bayesian scheme strongly depends on the number of feasible (n, h) pairs and on the choice of the exact elements of the sets S_n and S_h, which contain the allowable values. Specifically, with respect to S_h it has been found that the best results are obtained by using as short an h_{min} as possible, while maintaining a relatively long h. As far as S_n is concerned, the previous investigations have concluded that it is advisable to contain mostly small sample sizes, especially if the sampling intervals can be short.

We have conducted a large number of numerical experiments with two-sided Bayesian $\overline{X}$-charts in addition to the test cases of Tables 6.2 and 6.3 and we have arrived at the same as above conclusions and guidelines for the allowable (n, h) pairs, as reflected by our choice of the sets S_n and S_h in Section 6.3. Among other experimentations, we used the cases of Table 6.2 as a vehicle for the investigation of the effectiveness of fully adaptive Bayesian $\overline{X}$-charts with only two allowable sample sizes, namely $n_1 = 1$ and $n_2 = 6$ (if $n_S = 5$) or $n_1 = 1$ and $n_2 = 11$ (if $n_S = 11$), and/or only two allowable sampling intervals, namely $h_1 = 0.25h_S$ and $h_2 = 2h_S$. More specifically, for each of the 8 test cases of Table 6.2 we determined the optimal Bayesian monitoring schemes, using D1, D2 and D3, with three additional sets of allowable (n, h) pairs:

(a) 4 (n, h) pairs: all combinations of $n_1 = 1$ and $n_2 = 6$ or 11 with $h_1 = 0.25h_S$ and $h_2 = 2h_S$;

(b) 12 (n, h) pairs: all combinations of $n_1 = 1$ and $n_2 = 6$ or 11 with the 6 values of S_h of Section 6.3;

(c) 12 (n, h) pairs: all combinations of $h_1 = 0.25h_S$ and $h_2 = 2h_S$ with the 6 values of S_n of Section 6.3.

Although it is not worthwhile providing the detailed results here, it is interesting to highlight some interesting findings:

- The ΔC_{D32} differences are consistently lower than 0.5% with the exception of one case where $\Delta C_{D32} = 0.7\%$. This result provides additional support to the claim that the D2 discretization yields Bayesian schemes that are practically as good as those obtained with the much more computationally demanding discretization D3.
- The average value of the percentage difference ΔC_{D31} over all 32 cases (8 test cases solved with 4 different sets of allowable (n, h) combinations) is only 0.52% while the maximum observed ΔC_{D31} difference is 2.2%. Thus, it is confirmed that D1 constitutes a very efficient and effective discretization alternative for most practical cases.
- Although the adaptive Bayesian charts typically outperform the static Shewhart chart, in Cases 3 and 4 the expected cost TC_S of the optimal Shewhart chart is lower than the simulated expected cost C_{D3} of the Bayesian scheme with only 4 allowable (n, h) pairs. The obvious implication is that the allowable (n, h) combinations of the Bayesian scheme should be chosen with great care, otherwise the use of this potentially more powerful control mechanism may result in worse performance than the use of a much simpler tool.

Before closing this section it is necessary to reiterate and emphasize that the discretization of the decision space (i.e., the elements of the sets S_n and S_h) that was adopted here and used for the purposes of the numerical investigation is based on extensive computational experience but it is certainly just one of infinite such possibilities and in many cases it will not be quite as effective as other alternatives. If maximal effectiveness of the Bayesian $\overline{X}$-chart is desired, the only way to come close to achieving it is through careful and extensive numerical experimentation that will result in a customized decision-space discretization. It is comforting to know, though, that it suffices to use relatively few carefully selected allowable values of the adaptive chart parameters n and h to reap most of the benefits that the Bayesian approach can offer [17].

6.5 Summary and Conclusions

Bayesian process control has been developing for many years now, with some success in the theoretical domain but with much less success in the application field. Its limited progress in terms of acceptance by other researchers and practitioners can be attributed to its relative complexity and the excessive computations required to determine the optimal statistical monitoring and control policy. Because of these difficulties the study of Bayesian control charts has so far concentrated mostly on one-sided $\overline{X}$-charts and on control charts for attributes (percent nonconforming, nonconformities) that are typically implemented as one-sided charts. It is encouraging that an industrial application of a Bayesian one-sided $\overline{X}$-chart for monitoring the quality of tiles at a particular stage of the production process is now appearing for the first time in the academic literature [13].

The computational problems associated with the design of Bayesian control charts are exacerbated when the chart is two-sided, as is usually the case with $\overline{X}$-charts, because of the expansion of the state space from one dimension to two. In order to perform the dynamic programming computations the continuous state space (out-of-control probabilities) must be discretized and represented as a grid in two dimensions, dense enough to result in accurate approximations. The implementation of the dynamic programming approach with a state-space discretization of satisfactory density and accuracy was not possible until a few years ago. However, thanks to the tremendous advances in computer technology, the situation is very different today. The research presented in this chapter, mostly computational in flavour, has proved that it is now feasible to obtain a near-optimal design of a fully adaptive two-sided Bayesian $\overline{X}$-chart in reasonable computer time.

In addition, our findings reaffirm that the choice of the allowable values for the sample size and the sampling interval of the adaptive chart is of paramount importance to the economic effectiveness of the procedure. Even though some general guidelines are available and can be used as a starting point, the economic performance of the Bayesian process control scheme may be optimized only if this choice is made through careful experimentation on a case-by-case basis.

Incidentally, the numerical results presented in this chapter confirm earlier evidence that a properly designed adaptive Bayesian process control scheme can be economically much more effective than the static Shewhart $\overline{X}$-chart. However, a very natural and interesting question is whether and to what extent the economic advantage of the Bayesian chart is due to the additional degrees of freedom in selecting sampling intervals and sample sizes, i.e., to its adaptive nature itself. A well documented answer to this question may only be provided by means of a future systematic performance comparison between adaptive Bayesian and adaptive Shewhart-type control charts.

Finally, it must be noted that despite the focus of the exposition on monitoring short production runs, which is reflected in the title of the chapter as well, the general approach and the main findings and conclusions of this chapter carry over to the cases of long production runs and continuously operating production processes.

create a prior for the next that considers possible changes of state in the interim, and (2) using Bayes' theorem to combine the current prior and observation to produce an updated posterior. These tools provide easy derivations of posteriors, especially for normal approximations. Another application involves mixed effects models outside the normal linear framework. This chapter includes derivations of Bayesian exponentially weighted moving averages (EWMAs) for exponential family/exponential dispersion models including gamma-Poison, beta-binomial and Dirichlet-multinomial. Procedures are also outlined for using these techniques to model changes in the Poisson defect rate between manufacturing equipment and products in the production of integrated circuits. Pathologies that occur with violations of standard assumptions are illustrated with an exponential-uniform model.

7.1 Introduction

Many tools are available for deriving and easily understanding sums of random variables. This chapter presents two comparable (dual) properties of Bayes' theorem. These results concern the "score" and the "information", where the score = the first derivative of the log(likelihood) [3], extended here to include log(prior) and log(posterior); differentiation is with respect to parameter(s) of the distribution of the observations, which are therefore the random variables of the prior and posterior. Similarly, the "observed information" = the negative of the second derivatives. With these definitions, (a) the posterior score is the prior score plus the score from the data, and (b) the posterior observed information is the prior information plus the information from the data. Previous Bayesian analyses have used this mathematics (e.g., [6], [7]) but without recognizing it as having sufficient general utility to merit a name like "Bayes' Rule of Information".

These tools provide relatively easy derivations of extended Kalman filtering and of Laplace approximations for mixed models outside the normal linear case (e.g., [20], which includes software for S-Plus and R). The adequacy of these approximations can then be evaluated using techniques like importance sampling with Monte Carlo integration (including, e.g., importance weighted marginal posterior density estimation within Markov Chain Monte Carlo [5], see Chapter 2 in this book) or in low dimensions adaptive Hermite quadrature [8], [26]. The error in the simple approximation can then be used to decide if the additional accuracy provided by the more sophisticated methods is worth the extra expense.

We therefore focus on the power and simplicity obtainable from "keeping score with Bayes' theorem" and accumulating observed information from prior to posterior. In Sections 7.2 and 7.3, we derive the properties of interest by factoring the joint distribution of observations $\mathbf{y}$ and parameters μ in two ways: (predictive) × (posterior) = (observation) × (prior):

$$\begin{array}{ccccccccc} p(y,\mu) = & p(y) & \times & p(\mu|y) & = & p(y|u) & \times & p(\mu) \\ \text{(joint)} & = \text{(predictive)} & \times & \text{(posterior)} & = & \text{(observation)} & \times & \text{(prior),} \end{array} \tag{7.1}$$

where $p(\cdot)$ = probability density of observations or parameters as indicated. In Kalman filtering applications, we want to track the evolution of unknown, latent parameters μ over time through their influence on the observations. The predictive distribution does not appear in the score and information equations, but can be useful for evaluating if it is plausible to assume that y came from this model; if y seems inconsistent with that model, the posterior computation might be skipped and other action taken [43].

Beta-binomial, gamma-Poisson, and other conjugate exponential family applications appear in Section 7.2. In Section 7.4 (and the appendix), we keep score with Bayes' theorem and apply Bayes' rule of information with approximating normal priors and posteriors. The results are specialized further to normal observations including linear regression in Section 7.5. Section 7.6 reviews the connection between Bayes' and central limit theorems. Specialized systems for monitoring the production of integrated circuits are outlined in Section 7.7. The relationships between alternative definitions of information in statistics are reviewed in Section 7.8, and concluding remarks appear in Section 7.9.

7.2 Factoring Joint Probability and Keeping Score

Taking logarithms of (7.1), letting $l\,(\,.\,) = \log\,[p\,(\,.\,)]$ = the logarithm of the corresponding probability density, we get the following:

$$\begin{array}{ccccccc} l(y) & + & l(\mu|y) & = & l(y|\mu) & + & l(\mu) \\ \text{(predictive)} & + & \text{(posterior)} & = & \text{(observation)} & + & \text{(prior).} \end{array}$$

R. A. Fisher described the first derivative of the log(density) as the "efficient score" [3], [25, p. 470]. In this sense, the "score" from n independent observations is the sum of the scores from the individual observations, and with regular likelihood, prior and posterior, the likelihood is maximized or the posterior mode is located where the applicable score (i.e., the first derivative of the log density) "balances" at 0.

In particular, the posterior score is the prior score plus the score from the data:

$$\frac{\partial l(\mu|y)}{\partial \mu} = \frac{\partial l(y|\mu)}{\partial \mu} + \frac{\partial l(\mu)}{\partial \mu}, \tag{7.2}$$

Expression (7.2) is a powerful tool for computing Bayesian posteriors. This is particularly valuable when a normal distribution is an adequate approximation for both prior and posterior or when a normal distribution is used as a kernel for adaptive Hermite quadrature or for importance sampling in Monte Carlo. As a mnemonic device to make it easier to remember, it describes how to keep score with Bayes' theorem.

Before taking the second derivative, we illustrate the use of (7.2) in examples.

Example 1: *Gamma-Poisson.*
Consider the gamma-Poisson conjugate pair. In this case, the gamma prior $p(\lambda) = \beta^{\alpha}\lambda^{\alpha-1}e^{\lambda\beta}/\Gamma(\alpha)$, so the prior score for λ is $\partial l(\lambda)/\partial\lambda = \{[(\alpha-1)/\lambda]-\beta\}$. Meanwhile, the observation density is $p(y|\lambda) = \lambda^{y}e^{-\lambda}/y!$, so the score of the data is $\partial l(y|\lambda)/\partial\lambda = \{[y/\lambda]-1\}$. Whence, the posterior score is $\partial l(\lambda|y)/\partial\lambda = \{[(\alpha_1-1)/\lambda]-\beta_1\}$, where $\alpha_1 = \alpha + y$ and $\beta_1 = \beta + 1$. Since this has the same form as the prior score, we can exponentiate and integrate it to prove that the posterior is also gamma. Thus, Bayes' theorem tells us to keep score in the gamma-Poisson model by adding y to α and 1 to β.

Suppose now that we have a series of Poisson observations y_t with prior distribution for λ_t of $\Gamma(\alpha_{t|t-1}, \beta_{t|t-1})$. Then keeping score with Bayes' theorem tells us that the posterior is $\Gamma(\alpha_{t|t}, \beta_{t|t})$ with $\alpha_{t|t} = \alpha_{t|t-1}+y_t$ and $\beta_{t|t} = \beta_{t|t-1}+1$. Let's model a possible migration over time in $\lambda = \lambda_t$ with a discount factor $\theta(0 < \theta < 1)$, as $\alpha_{t+1|t} = \theta\alpha_{t|t}$ and $\beta_{t+1|t} = \theta\beta_{t|t}$. Thus, $\alpha_{t+1|t} = \theta(\alpha_{t|t-1} + y_t) = \theta y_t + \theta^2 y_{t-1} + \cdots$, and $\beta_{t+1|t} = \theta(\beta_{t|t-1} + 1) = \theta + \theta^2 + \cdots \cong \theta/(1-\theta)$, if $t = 0$ is sufficiently far in the past to be irrelevant. In that case, $\beta_{t+1|t}$ is constant, and $\alpha_{t+1|t} = \theta\tilde{y}_t/(1-\theta)$, where $\tilde{y}_t = \theta\tilde{y}_{t-1} + (1-\theta)\, y_t$ = an exponentially weighted moving average (EWMA) of the observations y_t. In essence, Bayes' theorem tells us to track the gamma scale parameter α by keeping score with an EWMA. For an EWMA application with a somewhat different gamma-Poisson model, see [24].

Example 2: *Beta-Binomial.*
Consider the beta-binomial pair with observation $y \sim \text{bin}(\pi, m)$ and prior $\pi \sim \text{beta}\,(\alpha, \beta)$. The same logic as for gamma-Poisson tells us that keeping score with Bayes' theorem produces a posterior that is beta (α_1, β_1) with $\alpha_1 = \alpha + y$ and $\beta_1 = \beta + m - y$. With a sequence $y_t \sim \text{bin}(\pi_t, m_t)$, and prior $\pi_t \sim \text{beta}(\alpha_{t|t-1}, \beta_{t|t-1})$, we keep score with $\alpha_{t+1|t} = \theta(\alpha_{t|t-1} + y_t)$ and $\beta_{t+1|t} = \theta[\beta_{t|t-1} + (m_t - y_t)]$. If $m_t = m$ is constant and $t = 0$ is sufficiently far in the past to be negligible, then $\alpha_{t+1|t} = \theta\tilde{y}_t/(1-\theta)$, where $\tilde{y}_t$ is the EWMA of the observations as before, and $\beta_{t+1|t} = \theta\,(m - y_t) + \theta\beta_{t|t-1} = \theta(m - \tilde{y}_t)/(1-\theta)$. Yousry et al. [44] discuss the use of this kind of EWMA in manufacturing.

Example 3: *Conjugate Updating an Exponential Dispersion Model.*
Examples 1 and 2 can be generalized to an arbitrary exponential family or exponential dispersion model [19], with

$$p(y|\boldsymbol{\mu}, \phi) = \exp\{\phi[\boldsymbol{y}'\boldsymbol{\mu} - b(\boldsymbol{\mu})] - c(\boldsymbol{y}, \phi)\}, \tag{7.3}$$

for some $\phi > 0$. The multinomial distribution with $(k+1)$ categories can be written in this form, with the k-vector $\boldsymbol{\mu}$ being the logistic transformation of the probabilities with $\pi_0 = 1/\{1 + \sum \exp(\mu_i)\}$ and $\pi_i = \pi_0 \exp(\mu_i)$, $i = 1, \ldots, k$, and with $\phi \boldsymbol{y}$ being nonnegative integers whose sum never exceeds another integer N [so $(N - \phi y)$ is the number in category 0]; with $k = 1$, this is the binomial distribution.

For a distribution of the form (7.3), consider a conjugate prior, CP$(\boldsymbol{\alpha}, s)$, on the natural parameter $\boldsymbol{\mu}$ with density

$$p(\boldsymbol{\mu}) = \exp\{s[\boldsymbol{\alpha}'\boldsymbol{\mu} - b(\boldsymbol{\mu})] - g(\boldsymbol{\alpha}, s)\}, \tag{7.4}$$

where $b(\boldsymbol{\mu})$ is the same as in (7.3), and $s > 0$ and $\boldsymbol{\alpha}$ are known.

The gamma-Poisson model of Example 1 can be written in the form (7.3)-(7.4) with $\mu = \log(\lambda)$. The beta-binomial of Example 2 can also be expressed in this form with $\mu = \log[\pi/(1-\pi)]$; this is called the "logit," "log(odds)" or "logistic" transformation. If the two possible outcomes of the beta-binomial are further subdivided binomially into $(k+1) > 2$ possible outcomes, we get a Dirichlet-multinomial model.

The "scores" required for (7.2) are simple:

$$dl(\boldsymbol{y}|\boldsymbol{\mu})/d\boldsymbol{\mu} = \phi[\boldsymbol{y} - db(\boldsymbol{\mu})/d\boldsymbol{\mu}],$$

and

$$dl(\boldsymbol{\mu})/d\boldsymbol{\mu} = s[\boldsymbol{\alpha} - db(\boldsymbol{\mu})/d\boldsymbol{\mu}] \tag{7.5}$$

The sum of these two gives us the posterior score, which we write as follows:

$$dl(\boldsymbol{\mu})/d\boldsymbol{\mu} = s_1[\boldsymbol{\alpha}_1 - db(\boldsymbol{\mu})/d\boldsymbol{\mu}],$$

where

$$s_1 = s + \phi, \tag{7.6}$$

and

$$\boldsymbol{\alpha}_1 = \boldsymbol{\alpha} + \kappa(\boldsymbol{y} - \boldsymbol{\alpha}) \quad \text{with} \quad \kappa = \phi/s_1.$$

The integral of the posterior score has the same form as the logarithm of the prior (7.4); moreover, the constant of integration must match $g(\boldsymbol{\alpha}_1, s_1)$ because (7.4) must integrate to 1.[1]

For an exponential family with a conjugate prior that can be written in the form (7.3)-(7.4), these results can be obtained from standard exponential family properties without "keeping score" in this way. Specifically, the product of (7.3) and (7.4) gives us the joint distribution, also in exponential

[1] This establishes that the posterior is CP$(\boldsymbol{\alpha}_1, s_1)$, conjugate to the prior.

family form:

$$p(y|\mu)p(\mu) = \exp\{(s\alpha + \phi y)'\mu - (s + \phi)b(\mu) - c(y, \phi) - g(\alpha, s)\}. \quad (7.7)$$

Since the prior density (7.4) must integrate to 1 for any $s > 0$ and any α, it must also integrate to 1 for α_1 and s_1 per (6). This property allows us to easily integrate out μ to get the predictive distribution:

$$p(y) = \exp\{g(\alpha_1, s_1) - g(\alpha, s) - c(y, \phi)\}. \quad (7.8)$$

This predictive distribution can be used to evaluate the consistency of each new observation with this model. New observations that seem implausible relative to this predictive distribution (7.8) should trigger further study to determine if these observations (*a*) might suggest improvements to the model or to the data collection methodology or (*b*) are honest rare events that deserve to be incorporated into the posterior with other observations or (*c*) are outliers that should not be incorporated into the posterior.

The standard application of Bayes' theorem in this context proceeds by dividing the joint density (7.7) by this predictive density $p(y)$ to get a posterior of the form (7.4) with parameters (7.6). However, if we use anything other than a conjugate prior like (7.4), the posterior might not be obtained so easily. It is precisely for such situations that more general tools like keeping score using (7.2) are most useful; see also [20].

Before leaving this example, suppose we have a series of observations y_t with density (7.3) and prior CP($\alpha_{t|t-1}$, $s_{t|t-1}$). Then the posterior is CP($\alpha_{t|t}$, $s_{t|t}$), where

$$\alpha_{t|t} = \alpha_{t|t-1} + \kappa_t(y_t - \alpha_{t|t-1}),$$

with $s_{t|t} = s_{t|t-1} + \phi$ (with ϕ constant) and $\kappa_t = \phi/s_{t|t}$. Similar to examples 1 and 2, we model a possible change in μ_t between the current and the next observations with a discount factor θ on s:

$$s_{t+1|t} = \theta s_{t|t} = \theta(s_{t|t-1} + \phi) = \theta[\phi + \theta(s_{t-1|t-2} + \phi)]. \quad (7.9)$$

If $t = 0$ is sufficiently remote to be negligible, we substitute this expression into itself repeatedly to get $s_{t+1|t} \cong \phi\theta/(1-\theta)$, which makes it essentially constant over time. If we further assume that $\alpha_{t+1|t} = \alpha_{t|t}$, we get the following:

$$\alpha_{t+1|t} = \alpha_{t|t-1} + \kappa(y_t - \alpha_{t|t-1}),$$

where

$$\kappa = 1 - \theta. \quad (7.10)$$

In sum, a standard EWMA of the random variable y_t of an exponential family (7.3) estimates the prior location parameter $\alpha_{t|t-1}$ of a standard conjugate prior (7.4) of the location μ_t of $\mathbf{y}_t$ as μ_t evolves over time as modeled by the discount factor θ on the prior information parameter $s_{t|t-1}$ per (7.9) and (7.10).

The gamma-Poison and beta-binomial models of Examples 1 and 2 are special cases of this.

This exponential family EWMA has been discussed, applied, and generalized by West and Harrison [43, sec. 14.2], Grigg and Spiegelhalter [16], Klein [20] and others. We will interpret κ in (10) using "Bayes' rule of Information" in the next section. Before that, however, we note that this exponential family EWMA can be applied in a quasi-likelihood context [25], not assuming a complete distribution for the observations but only that (a) the logarithm of (7.3) times some constant provides an adequate approximation to the log(density) of the observations and (b) prior and posterior are adequately approximated by (7.4) and (7.6). We could check the adequacy of these assumptions using Markov Chain Monte Carlo (MCMC) (see Chapter 2) with a sample of such data. This could be quite valuable in engineering applications where MCMC might be used during engineering design to evaluate whether a much cheaper EWMA would be adequate for routine use where MCMC would not be feasible.

7.3 Bayes' Rule of Information

We return now to (7.2) and take another derivative to get the following:

$$\frac{\partial^2 l(\boldsymbol{\mu}|\boldsymbol{y})}{\partial\boldsymbol{\mu}\partial\boldsymbol{\mu}'} = \frac{\partial^2 l(\boldsymbol{y}|\boldsymbol{\mu})}{\partial\boldsymbol{\mu}\partial\boldsymbol{\mu}'} + \frac{\partial^2 l(\boldsymbol{\mu})}{\partial\boldsymbol{\mu}\,\partial\boldsymbol{\mu}'}. \tag{7.11}$$

In this chapter, we let $\mathbf{J}(.)$ denote the *observed information*, which we define here as the negative of the matrices of second partials in (7.11). Then (7.11) becomes

$$\begin{array}{ccccc} \mathbf{J}(\boldsymbol{\mu}|\boldsymbol{y}) & = & \mathbf{J}(\boldsymbol{y}|\boldsymbol{\mu}) & + & \mathbf{J}(\boldsymbol{\mu}) \\ \begin{pmatrix}\text{posterior}\\ \text{information}\end{pmatrix} & = & \begin{pmatrix}\text{information from}\\ \text{observation(s)}\end{pmatrix} & + & \begin{pmatrix}\text{prior}\\ \text{information}\end{pmatrix}. \end{array} \tag{7.12}$$

We call this "Bayes' Rule of Information", as it quantifies in many applications the accumulation of information via Bayes' theorem. If $\boldsymbol{y} \sim N_k(\boldsymbol{\mu}, \mathbf{V})$, we get $\mathbf{J}(\boldsymbol{y}|\boldsymbol{\mu}) = \mathbf{V}^{-1}$. Since $\mathbf{J}(\boldsymbol{y}|\boldsymbol{\mu})$ is constant independent of $\boldsymbol{\mu}$ in this case, it is also the Fisher (expected) information, though that is not true in other applications. Similarly, with a prior $\boldsymbol{\mu} \sim N_k(\boldsymbol{\theta}_0, \boldsymbol{\Sigma}_0)$, we have $\mathbf{J}(\boldsymbol{\mu}) = \boldsymbol{\Sigma}_0^{-1}$. Then (7.12) tells us that $\mathbf{J}(\boldsymbol{\mu}|\boldsymbol{y}) = \mathbf{V}^{-1} + \boldsymbol{\Sigma}_0^{-1}$. Since we know from other arguments [e.g., exponential family conjugacy (7.3)-(7.6)] that the posterior is also normal, this gives us the posterior variance in the form of its inverse, the "information".

In the normal case, the information terms in (7.12) are also called precision parameters [4], being the inverse of variances (or covariance matrices); this case is considered further in Section 7.5. In Section 7.4, we assume that the prior is normal and the observed information can be adequately approximated by a constant over probable variations in μ (though it may depend on the observation y). This will support using a normal approximation for the posterior, which is discussed further in Section 7.6.

An apparent exception that often supports the rule involves multimodal distributions. In such cases, the observed information from the observation(s) $[-\partial^2 l(y|\mu)/\partial\mu\partial\mu']$ can even have negative eigenvalues in a certain region between modes. Fortunately, many such examples are still sufficiently regular that standard results can be used to show that observations with indefinite or even negative definite information are so rare that their impact on the posterior vanishes almost surely as more data are collected. If this is not adequate, we could handle mixtures by computing the posterior as a mixture, deleting components with negligible posterior mixing probabilities as suggested by West and Harrison [43, ch. 12]. (For more on finite mixtures, see [42] and [28].)

Example 3 (**cont**.): EWMA for Exponential Dispersion Data.
What does "Bayes' Rule of Information" tell us about processing data from a (possibly overdispersed) generalized linear model (7.3) with a conjugate prior (7.4)? To find out, we differentiate (7.5):

$$\mathbf{J}(y|\mu) = \phi\left[\frac{d^2 b(\mu)}{d\mu\, d\mu'}\right], \quad \text{and} \quad \mathbf{J}(\mu) = s\left[\frac{d^2 b(\mu)}{d\mu\, d\mu'}\right] \tag{7.13}$$

To help build our intuition about this, we use dimensional analysis assuming y has "y units", and μ has "μ units". Then $b(\mu)$ has $(y\mu)$ units. If the exponent in (7.3) is dimensionless, ϕ must have $(y\mu)^{-1}$ units. For a normal distribution, "μ units" are "y units", so ϕ has y^{-2} units. For a Poisson distribution, y is counts of events, and μ is in log(counts). Then ϕ has $\left(\text{count} \times \log(\text{count})\right)^{-1}$ units. A similar analysis applies to binomial or multinomial observations, where μ is in logits and y is either counts or proportions; in the latter case, ϕ is in $(\text{count} \times \text{logit})^{-1}$. Thus, $d\,b(\mu)/d\mu$ has "y units", which it must since a standard exponential family property makes $Ey = db\,(\mu)/d\mu$. Similarly, $d^2b(\mu)/(d\mu\, d\mu')$ has $(y\mu^{-1})$ units. Then by (7.13), $\mathbf{J}(y|\mu)$ has μ^{-2} units, which it must have, because the inverses of observed and Fisher information approximate the variance of μ.

Another standard exponential family property has

$$\text{var}(y|\mu) = \phi^{-1}\left[\frac{d^2 b(\mu)}{d\mu d\mu'}\right].$$

This is the same as $\mathbf{J}(y|\mu)$ except that the scale factor ϕ is inverted, which change the units from μ^{-2} to y^2, as required for var $(y|\mu)$.

In Section 7.4, we will assume that the posterior information is always positive (or nonnegative definite) and can be adequately approximated by a constant in a region of sufficiently high probability near the posterior mode. In this case, with a normal prior, a normal posterior also becomes a reasonable approximation. Before turning to that common case, we first illustrate pathologies possible with irregular likelihood when the range of support depends on a parameter of interest.

Example 4: *Exponential - Uniform.*
Pathologies with likelihood often arise with applications where the range of support of a distribution involves parameter(s) of interest. For example, consider $y \sim \text{Uniform}(0, e^{\mu})$. We take as a prior for μ a 2-parameter exponential with mean ν^{-1} and support on (μ_0, ∞). We denote this by $\text{Exp}(\nu^{-1}, \mu_0)$; its density is as follows:

$$f(\mu) = \nu \exp[-\nu(\mu - \mu_0)]I(\mu > \mu_0),$$

where $I(A)$ is the indicator function of the event A. Then the log(density) is as follows:

$$l(\mu) = \ln(\nu) - \nu(\mu - \mu_0), \quad \text{for } (\mu > \mu_0). \tag{7.14}$$

Also, the density for y is as follows:

$$f(y|\mu) = e^{-\mu}I(0 < y < e^{\mu}),$$

so

$$l(y|\mu) = (-\mu), \quad \text{for } (0 < y < e^{\mu}). \tag{7.15}$$

Therefore, the support for the joint distribution has $\mu > \max\{\mu_0, \ln(y)\}$. To keep score with Bayes' theorem, we need the prior score and the score from the data. We get the prior score by differentiating (14):

$$\frac{\partial l(\mu)}{\partial \mu} = (-\nu), \quad \text{for } (\mu > \mu_0). \tag{7.16}$$

For the data, by differentiating (7.15) we see that the score function is a constant (-1):

$$\frac{\partial l(y|\mu)}{\partial \mu} = (-1), \quad \text{for } (0 < y < e^{\mu}), \quad \text{i.e., } \{\ln(y) < \mu\}. \tag{7.17}$$

We add this to (7.16) to get the posterior score:

$$\frac{\partial l(\mu|y)}{\partial \mu} = (-1 - \nu) = (-\nu_1), \quad \text{for } (\mu > \mu_1 = \max\{\mu_0, \log(y)\}),$$

where $\nu_1 = \nu + 1$. By integrating the posterior score over $(\mu > \mu_1)$, the range of support for μ, we find that the posterior is $\text{Exp}(\nu_1^{-1}, \mu_1)$. Thus, the 2-parameter exponential is a conjugate prior for the uniform distribution considered here. With repeated data collection, ν_1 increases by 1 with each observation pulling $E(\mu) = \mu_1 + \nu_1^{-1}$ ever closer to the lower limit μ_1.

To get the observed information, we differentiate (7.16) and (7.17) a second time to get

$$J(\mu) = J(\mu|y) = J(y|\mu) = 0.$$

Thus, in this example, the observed information from prior, data, and posterior are all 0. Clearly, the posterior gets sharper with additional data collection. Each observation reflects an accumulation of knowledge, and the score equation (7.2) helps us quantity that, even though in this example, there is no "observed information" in anything!

The problems in this case arise because the range of support for the distribution of y depends on the parameter, which means that many of the standard properties of "regular likelihood" do not hold. However, the score and information equations (7.2) and (7.12) are valid wherever the densities are defined. If we change the parameterization, we get different pathologies. For example consider $y \sim U(0, \beta)$ with β following a Pareto distribution [36]. Then the score from the data is $(-1/\beta)$ if $0 < y < \beta$, so the Fisher information defined as the variance of the score is 0. The observed information, however, is not zero; it's negative $= (-\beta^{-2})$! The usual equality between the Fisher information and the expected observed information assumes that the order of differentiation and expectation can be interchanged, which does not hold in this case. Fisher information may not be useful in such irregular situations, but we can still keep score and accumulate observed information using (7.2) and (7.12). If instead of exponential or Pareto priors and posteriors, we used a truncated normal, we would reach similar conclusions; the analysis would still include pathologies, though different from what we've just discussed.

A primary area for application of Bayes' Rule of Information (7.12) and the companion scoring rule (7.2) is for Kalman filtering, especially nonlinear extended Kalman filtering and for more general Bayesian sequential updating ([43]; [31]; [15]). Such cases involve repeated applications of Bayes' theorem, where the information from the data arriving with each cycle accumulates in the posterior, summarizing all the relevant information in the data available at that time, which then with a possible transition step becomes the prior for the next cycle.

Another important area of application is for deriving importance weighting kernels for Monte Carlo integration with random effects and/or Bayesian mixed effect models outside of the normal linear paradigm. Beyond providing a first order approximation, which may not be adequate, they provide a tool for handling relatively easily the "curse of dimensionality," which says roughly that almost everything is sparse in high enough dimensions. For example, Evans and Schwartz [8] note that the volume of a k-dimensional unit sphere as a proportion of the circumscribing unit cube, $[-1, 1]^k$, goes to zero as k increases without bounds. Thus, if we try to estimate the volume of this sphere via Monte Carlo sampling from a uniform distribution on $[-1, 1]^k$, we would need ever larger Monte Carlo samples as k increases just to maintain a fixed probability of getting at least one observation in this sphere!

However, if most of the mass of the distribution is close to the center of an appropriate normal approximation, most of the k-dimensional pseudo-random normal variates generated will also be relevant to the non-normal distribution of interest. This makes importance sampling a simple yet valuable tool for evaluating the adequacy of a normal approximation and for improving upon it when it is not adequate.

7.4 Normal Prior and Posterior

We assume in this and the next sections that the prior and posterior are both adequately approximated by normal distributions, $N_p(\boldsymbol{\theta}_0, \boldsymbol{\Sigma}_0)$ and $N_p(\boldsymbol{\theta}_1, \boldsymbol{\Sigma}_1)$, respectively. Then

$$l(\boldsymbol{\mu}) = c_0 - \frac{1}{2}(\boldsymbol{\mu} - \boldsymbol{\theta}_0)'\boldsymbol{\Sigma}_0^{-1}(\boldsymbol{\mu} - \boldsymbol{\theta}_0),$$

and

$$l(\boldsymbol{\mu}|\boldsymbol{y}) = c_1 - \frac{1}{2}(\boldsymbol{\mu} - \boldsymbol{\theta}_1)'\boldsymbol{\Sigma}_1^{-1}(\boldsymbol{\mu} - \boldsymbol{\theta}_1),$$

where c_0 and c_1 are appropriate constants (relative to $\boldsymbol{\mu}$). We'd like to use (7.12) to compute $\boldsymbol{\Sigma}_1$ and (7.2) to get $\boldsymbol{\theta}_1$. For this, we need following:

$$\frac{\partial l(\boldsymbol{\mu})}{\partial \boldsymbol{\mu}} = \left[-\boldsymbol{\Sigma}_0^{-1}(\boldsymbol{\mu} - \boldsymbol{\theta}_0)\right]; \quad \frac{\partial l(\boldsymbol{\mu}|\boldsymbol{y})}{\partial \boldsymbol{\mu}} = \left[-\boldsymbol{\Sigma}_1^{-1}(\boldsymbol{\mu} - \boldsymbol{\theta}_1)\right], \tag{7.18}$$

and

$$\mathbf{J}(\boldsymbol{\mu}) = \left[-\frac{\partial^2 l(\boldsymbol{\mu})}{\partial \boldsymbol{\mu} \partial \boldsymbol{\mu}'}\right] = \boldsymbol{\Sigma}_0^{-1}; \quad \mathbf{J}(\boldsymbol{\mu}|\boldsymbol{y}) = \left[-\frac{\partial^2 l(\boldsymbol{\mu}|\boldsymbol{y})}{\partial \boldsymbol{\mu} \partial \boldsymbol{\mu}'}\right] = \boldsymbol{\Sigma}_1^{-1}. \tag{7.19}$$

To keep things simple, we substitute (7.19) into (7.12) evaluating $\mathbf{J}(\boldsymbol{y}|\boldsymbol{\mu})$ at the prior mode $\boldsymbol{\mu} = \boldsymbol{\mu}_0$ to get the following (provided only that the likelihood for $\boldsymbol{y}$ is regular):

$$\boldsymbol{\Sigma}_1^{-1} = \mathbf{J}(\boldsymbol{y}|\boldsymbol{\mu} = \boldsymbol{\theta}_0) + \boldsymbol{\Sigma}_0^{-1}. \tag{7.20}$$

We assume in this section that variations in $\mathbf{J}(\boldsymbol{y}|\boldsymbol{\mu})$ are so small that a normal approximation with mean at the posterior mode $\boldsymbol{\theta}_1$ and "information" $\boldsymbol{\Sigma}_1^{-1}$ per (7.20) provides an adequate approximation to the posterior. This approximation is often quite accurate when the prior summarizes many previous observations and the information provided by $\boldsymbol{y}$ is relatively modest by comparison. With relatively noninformative priors, this may be less adequate. In some cases, replacing $\boldsymbol{\theta}_0$ by $\boldsymbol{\theta}_1$ in (7.20) would improve the approximation to the posterior; this would require an iteration as discussed in the appendix.

updating for both the normal mean (vector) and a scalar parameter factoring out of the information matrices, which are otherwise assumed known. The predictive distributions become (multivariate) t after integrating out first the normal mean and then the gamma/chi-square information parameter. Similarly, the marginal prior and posterior for the mean are also t. Graves, Bisgaard and Kulahci [14] used this to describe the obvious heteroscedasticity of measured deficits in angular acceleration, used to detect misfires in an automobile engine.

The resulting theory is relatively simple, given the inherent complexity of the applications, with substantial utility for many applications. Its major deficiency is that it does not generalize easily with multiple variance and covariance parameters. For this, we replace the gamma/chi-square distributions by normal approximations for the logarithms of variances and a tanh for various parameters relating to the correlations that are constrained to lie between -1 and $+1$, as discussed by Pinheiro and Bates [30]. The score and information equations (7.2) and (7.11) can be used to derive Bayesian updating equations, thereby supporting Kalman filters for monitoring variance components. For most purposes, there is little loss of accuracy in approximating a chi-square by a lognormal, except when the number of degrees of freedom is quite small, because most properties of the two distributions are so similar that substantial quantities of data would be required to tell them apart.

7.6 Bayes and the Central Limit Theorem

The use of normal approximations for prior and posterior as discussed in Section 7.4 relates to more general results regarding the sampling distribution of maximum likelihood estimators (MLEs). Standard proofs of asymptotic normality and results on rates of convergence to normality involve key steps that could be written recursively using (7.2) to "keep score" and (7.12) to "accumulate information" with succeeding observations. The results work, roughly speaking, because the "score", being stochastic, grows as the square root of n, the number of observations, while the information and higher cumulants grow linearly with n. To get a standard normal, the score is divided by the square root of the information, which makes the kth cumulant of this ratio roughly proportional to $n^{1-k/2}$, so the skewness declines as $n^{-1/2}$, the fourth cumulant (kurtosis) as n^{-1}, etc. With standard Edgeworth expansions, the first correction term involves the skewness, $O(n^{-1/2})$, the second involves the kurtosis and the square of the skewness, both $O(n^{-1})$, etc., as discussed, e.g., in [35], [38], [39], and [11]. The same math including a prior with a sufficiently flat dominating measure would produce a normal approximation to the posterior with information Σ_1^{-1} and mean $\boldsymbol{\theta}_1$ computed via (7.20) and (7.21). Under suitable regularity conditions, as more information is accumulated into the posterior, it becomes more nearly normal ([1]; [33]).

Central limit convergence of MLEs has been proven with otherwise adequately behaved multimodal distributions, even though the observed information $\mathbf{J}(y|\mu)$ is sometimes negative (or negative definite) for μ in certain regions. Fortunately, the regions of negative information vanish almost surely with increasing numbers of observations. If that is not adequate, finite mixtures in prior and observation distributions can often be adequately approximated by the obvious finite mixtures in the posterior, dropping all but the dominant components as described by West and Harrison [43, ch. 12].

7.7 Monitoring in Manufacturing Integrated Circuits

Modern manufacturing, especially of integrated circuits (ICs or "computer chips"), includes collecting substantial quantities of data on process parameters and product performance. IC production uses photolithography, printing images for successive layers for a computer chip on top of previous layers to produce a circuit. Many chips, from tens to thousands, are printed simultaneously on a disk called a wafer. They are tested, the good kept and the bad discarded. Data on electrical characteristics and the failure modes of malfunctioning chips and accompanying test circuits help managers and engineers manage and improve the production processes and thereby the yield (defined as the percent of chips that are good). The tools described in this chapter can help develop better ways of extracting useful information from the data collected. Examples 2 and 3 in Section 7.2 above show that a simple exponentially weighted moving average (EWMA) of yield, $\tilde{y}_t$, tracks the evolution of expected yield via the relationship of $\tilde{y}_t$ to the parameters of a beta distribution conjugate to the assumed binomial distribution of yield.

However, other priors seem to match the physics of wafer fab better than a beta distribution. In particular, the number of defects per chip is often approximately Poisson, and the yield (ignoring possible repair circuitry) is the probability of zero defects. Thus, $Ey = e^{-DA}$, where D is the "fail rate" = number of defects per unit area of the approximating Poisson distribution and A is the area of the chip (or the "critical area" of a particular layer). We call DA the "defect rate", to distinguish it from the "fail rate" D. Thus, DA has units of defects/chip. The area A is typically measured in cm^2, in which case D has units of defects/cm^2. To a first order of approximation, different products produced in the same factory (called a "Fab") often have approximately the same defects/cm^2, D; different products typically differ more in the area A of the chip than in the fail rate D.

Moreover, D exhibits both systematic and random variations across the wafer and between wafers within a lot, as well as random variations between lots. An important aid in modeling and decomposing these different sources of variability is simply to create appropriate probability plots. Figure 7.1 is one example, depicting the distribution of yield by lot for 755 lots of a product with 324 chips per wafer averaging 7864 chips per lot. This image is fairly typical of wafers k_i in a lot may vary from one lot to the next due, e.g., to breakage and other major processing problems, where

$$E(y_{ij}|\gamma_0, \gamma_i, \gamma_{ij}) = \pi_{ij} = \exp\left\{-\exp(\gamma_0 + \gamma_i + \gamma_{ij})A\right\},$$

where γ_0, γ_i, and γ_{ij} follow normal distributions with means Γ_0, 0, and 0, and with variances σ_0^2, σ_{lot}^2, and σ_{wafer}^2, respectively. We integrate out the wafer-level parameters γ_{ij} as follows:

$$f(y_{ij}|\gamma_i, \boldsymbol{\mu}) = \binom{n}{ny_{ij}} \int_{-\infty}^{\infty} \pi_{ij}^{n\,y_{ij}}(1-\pi_{ij})^{n(1-y_{ij})}\phi\left(\frac{\gamma_{ij}}{\sigma_{\text{wafer}}}\right)\frac{d\gamma_{ij}}{\sigma_{\text{wafer}}}, \tag{7.30}$$

where $\phi(\cdot)$ is the standard normal density and $\boldsymbol{\mu}$ is a vector of parameters including γ_0, σ_{wafer}^2, and other parameters. For the entire lot, we similarly compute the marginal likelihood as follows:

$$f(\boldsymbol{y}_i|\boldsymbol{\mu}) = \int_{-\infty}^{\infty} f(y_{ij}|\gamma_i)\phi\left(\frac{\gamma_i}{\sigma_{\text{lot}}}\right)\frac{d\gamma_i}{\sigma_{\text{lot}}}. \tag{7.31}$$

State of the art software for this kind of application could be produced as a relatively easy modification of the "lmer" function in the "lme4" package [2] in *R* [34] to (*a*) allow input of an appropriate prior for the parameters including γ_0, σ_{lot}^2, and σ_{wafer}^2, (*b*) compute the posterior (rather than the maximum of the likelihood) using (7.20) and (7.21) and the appendix, and (*c*) include the posterior in the returned object. Required derivatives of the log of this likelihood could be computed either by numerical differentiation or by numerical integration of the derivatives in (7.30) and (7.31).

For this kind of application, the "lmer" function supports using the Laplace method, which approximates the integrals (7.30) and (7.31) by replacing the logarithm of the $(k_i + 1)$-dimensional integrand by its second-order Taylor series expansion about the maximum of the integrand. This approximate integrand has the form of a normal density, which can then be integrated to produce an approximation to the integral. If either $\sigma_{\text{lot}}^2 = 0$ or $\sigma_{\text{wafer}}^2 = 0$, "lmer" allows the use of "AGQ" = adaptive Gauss-Hermite quadrature. One type of AGQ finds the maximum of the integrand, then uses Gauss-Hermite quadrature relative to the normal distribution defined by the second order Taylor expansion at the posterior mode or maximum of the likelihood. For quadrature of order r, this computes a weighted average of the ratio of the integrand to the approximating normal density at r points, with the quadrature points and weights chosen to make the integral exact when the integrand is a polynomial of order $(2r - 1)$ times the approximating normal density. The approximating normal distribution is found by balancing the score and computing the information essentially equivalent to (7.2) and (7.12), iteratively as described in the appendix. With $r = 1$, this AGQ is the Laplace approximation. Quadrature points and weights are easily obtained from "gauss.quad" in the "statmod" package in R [40].

Good importance sampling algorithms also use approximating normal distributions, averaging the ratio of the integrand to the approximating normal density over samples of pseudo-random normal deviates. Both methods average numbers that should be close to 1 unless the posterior is quite non-normal. Hartford [18] reported excellent results from applying this type of AGQ to maximum likelihood estimation for nonlinear mixed models and good results from using this same distribution with importance sampling.

AGQ works well for any integrand for which the derivative of order $2r$ is bounded by some number that is moderately small. It's primary deficiency for multivariate applications like this is the need for repeated evaluations of inner integrals. For "lot i" with k_i wafers, we must first evaluate the product (or the sum of the logarithms) of k_i integrals like (7.30), and the integrand of each must be evaluated at r points for a total of $k_i r$ evaluations. The outer integral (7.31) requires us to do this r times, for a total of $(k_i r)^r$ evaluations. Moreover, we must do this repeatedly to find the posterior mode (or the maximum likelihood) for, e.g., a normal approximation to the distribution of $[\gamma_0, \log(\sigma^2_{\text{lot}}), \log(\sigma^2_{\text{wafer}})]$. When the Laplace approximation is not adequate, AGQ is generally preferred for low-dimensional integrals, while importance sampling is recommended for higher dimensions [8].

Monitoring procedures keeping score (7.20) and accumulating information (7.21) with likelihoods like (7.30) and (7.31) have many applications. For example, in a Fab running multiple products, iterations like this could be used to refine an estimate of a different $\gamma_0 = \gamma_{0p}$ for each product p. These numbers could then be plotted with Student's t confidence bounds as described near the end of Section 7.5; if the confidence bounds for product p consistently fail to include the γ_0 for all products combined, it suggests that the fail rate (in defects/cm^2) for product p is different from the average across multiple products. Moreover, the difference $(\gamma_{0p} - \gamma_0)$ could be used to estimate the amount of money represented by this difference.

Similarly, we could compute a different $\gamma_0 = \gamma_{0su}$ for each different piece of equipment u used at step s in the process. A typical Fab has between 500 and 1500 steps and uses several superficially equivalent machines at many of them. It would be a waste of time, if not physically impossible, to have someone look at a different chart for each step. We need simple summaries to tell us which few charts to examine. One such summary might be the log(posterior odds) for a difference at a step s compared to the assumption of no difference. This is the log(prior odds) plus the log(likelihood ratio), where the likelihood under both H_0 and H_1 has the form (7.31), except that under H_1, a separate parameter γ_{0su} is estimated for each machine u used at step s, while under H_0, $\gamma_{0su} = \gamma_0$ for all u. We typically expect the condition of equipment to change over time. Immediately after maintenance, the condition of equipment used at step s will often be consistent with H_0. However, after an appropriate passage of time, the condition will likely change to H_1. A one-sided cumulative sum of log(likelihood ratio) using the predictive distribution provides a monitoring rule that is similar to computing the posterior log(odds) with a floor giving by the log(hazard odds); see [12]. These cusums can then be used to distinguish between random variations in the numbers and steps with honest equipment differences. This is similar to "Multiple Model Adaptive Estimation" (MMAE), used (at least in simulations) to diagnose malfunctions in the F-16 fighter aircraft (e.g., [17]). For steps with substantive differences, differences like $|\gamma_{0su} - \gamma_{0sv}|$ for machines u and v used at step s could be combined with production forecasts produce a "Forecasted Pareto" to help prioritize alternative improvement efforts in terms of the estimated monetary value of the apparent yield losses at different steps.

7.8 Alternative Definitions of Information in Statistics

Several different types of "information" have been defined and used in statistical work (see, e.g., [41]). Fisher and observed information are both used to develop approximate sampling distributions for maximum likelihood estimates. However, (7.12) is similar to but distinct from the traditional frequentist result that the Fisher information for the joint distribution of two independent random variables is the sum of the Fisher information for each marginal [32, sec. 5a.4].

Shannon [37] argued that the information contained in a "message" (observation) $\boldsymbol{y}$ is the number of "bits" required to produce the equivalent reduction in uncertainty, which is $E\{-\log_2[f(y)]\}$. For example, if $\boldsymbol{y}$ is the outcome of the toss of coin with probability of success p, then $E\{-\log_2[f(y)]\} = p[-\log_2(p)] + (1-p)[-\log_2(1-p)]$. When $p = 0.5$, this is 1, and declines monotonically to 0 as p goes to either 1 or 1 is 0. Thus, knowing the outcome of a single toss of a fair coin is 1 "bit" of information, while knowing the outcome of a toss of a biased coin is between 0 and 1. Important results in modern communication theory are based on Shannon's concept of information.

Using natural rather than base 2 logarithms, Kullback and Leibler [21] (see also [27]) quantified the mean information in an observation $\boldsymbol{y}$ for discriminating a probability density $f(y)$ from $g(y)$ as $E\{\log[f(y)/g(y)]|f\}$; they called this a measure of "distance" between f and g. Kullback and Leibler related their information distance to the Fisher (expected) information:

$$I(\boldsymbol{\mu}, \boldsymbol{\mu}+\boldsymbol{\delta}) = E\{\log[f(\boldsymbol{y}|\boldsymbol{\mu})/f(\boldsymbol{y}|\boldsymbol{\mu}+\boldsymbol{\delta})]|\boldsymbol{\mu}\} \cong 0.5\,\boldsymbol{\delta}'E[J(\boldsymbol{y}|\boldsymbol{\mu})|\boldsymbol{\mu}]\boldsymbol{\delta}.$$

To help educate our intuition about this, consider $\boldsymbol{y} \sim N(\boldsymbol{\mu}, \mathbf{V})$. Then

$$\begin{aligned} I(\boldsymbol{\mu}, \boldsymbol{\mu}+\boldsymbol{\delta}) &= 0.5\,E\{[\boldsymbol{y}-(\boldsymbol{\mu}+\boldsymbol{\delta})]'\mathbf{V}^{-1}[\boldsymbol{y}-(\boldsymbol{\mu}+\boldsymbol{\delta})] - [\boldsymbol{y}-\boldsymbol{\mu}]'\mathbf{V}^{-1}[\boldsymbol{y}-\boldsymbol{\mu}]\} \\ &= E\boldsymbol{\delta}'\mathbf{V}^{-1}[0.5\boldsymbol{\delta}+\boldsymbol{\mu}-\boldsymbol{y}] = 0.5\boldsymbol{\delta}'\mathbf{V}^{-1}\boldsymbol{\delta}. \end{aligned}$$

This matches the general result, since the Fisher information in this context is $\mathbf{V}^{-1}$.

In sum, several different concepts of "information" have been discussed in the statistics literature, with each serving different purposes. The focus of this chapter has been Fisher's efficient score and the observed information, which provide powerful tools for deriving exact and approximate posterior distributions. For a more general review of these and other types of "information" used in statistics, see [41], [10], [23], and [9].

7.9 Summary

We discussed Bayes' rule of information generally in (7.12) and in approximate and exact normal applications in (7.20), (7.24) and (7.26) with manufacturing applications in Section 7.7. We also showed how keeping score with Bayes' theorem provides easy derivations of the posterior for the gamma-Poisson, beta-binomial, and exponential-uniform conjugate pairs. These tools have long been used when prior and observations are normal (e.g., [29] and [22]), but without substantive consideration of their more general utility. Yousry et al. [44] describe the use in quality control of an EWMA for binomial data with a beta prior. Their derivation is similar to the discussion in Example 2, Section 7.2 above, but without the convenience of using the concept of Fisher's efficient score or of Bayesian sequential updating, promoted as a general foundation for monitoring [15].

In many cases, a normal distribution provides an adequate approximation to the posterior, even with nonlinear or non-normal likelihood. When it is not convenient to compute derivatives analytically, the score function and information from the data can be estimated by numerical differentiation. After the posterior mode and information $(\boldsymbol{\theta}_1, \Sigma_1^{-1})$ are found, the adequacy of the normal approximation might be checked using importance sampling, which averages the ratio of $\exp[l(\boldsymbol{\mu}|\mathbf{y})]$ to the $N(\boldsymbol{\theta}_1, \Sigma_1)$ density over a random sample from $N(\boldsymbol{\theta}_1, \Sigma_1)$. If the posterior is exactly $N(\boldsymbol{\theta}_1, \Sigma_1)$, this ratio will always be 1, so the sample standard deviation will be 0. If the posterior is close to normality, the ratio will be close to 1, and the sample standard deviation will be small ([8], [18]). Of course, we must also assure ourselves that the posterior does not have another substantive mode that might be completely missed with this importance sampling. If substantive discrepancies are found, they can be reported with profile confidence intervals [30], highlighting the discrepancies between the profile and the normal approximation. Certain likelihoods (e.g., mixtures; see [42] or [28]) are known to have potential difficulties. These cases might be identified by excessive variability in the observed information from the data. Once identified, special procedures can be developed appropriate to the situation.

Appendix: Non-Constant Observed Information

In this appendix, we develop an iteration to an approximate normal posterior $N_p(\boldsymbol{\theta}_1, \Sigma_1)$ from a normal prior $N_p(\boldsymbol{\theta}_0, \Sigma_0)$ and either non-normal data or data with normal errors nonlinearly related to parameters of interest $\boldsymbol{\theta}$. We shall not prove here anything about the convergence of our iteration; such a proof would follow the lines of comparable results on convergence to MLEs.

The iteration will ultimately require keeping score at the posterior mode $\boldsymbol{\theta} = \boldsymbol{\theta}_1$, rather than the prior mode as with (7.21), substituting (7.18) into (7.2) to obtain the following:

$$\mathbf{0} = \left[\frac{\partial l(\boldsymbol{y}|\boldsymbol{\theta} = \boldsymbol{\theta}_1)}{\partial \boldsymbol{\theta}}\right] - \boldsymbol{\Sigma}_0^{-1}(\boldsymbol{\theta}_1 - \boldsymbol{\theta}_0). \tag{7.32}$$

Since $\boldsymbol{\theta}_1$ is initially unknown, we expand the score from the data in a Taylor approximation about an arbitrary point $\boldsymbol{\theta} = \boldsymbol{\xi}_j$, beginning from $\boldsymbol{\xi}_0 = \boldsymbol{\theta}_0$, as follows:

$$\left[\frac{\partial l(\boldsymbol{y}|\boldsymbol{\theta} = \boldsymbol{\theta}_1)}{\partial \boldsymbol{\theta}}\right] \cong \left[\frac{\partial l(\boldsymbol{y}|\boldsymbol{\theta} = \boldsymbol{\xi}_j)}{\partial \boldsymbol{\theta}}\right] - \mathbf{J}(\boldsymbol{y}|\boldsymbol{\theta} = \boldsymbol{\xi}_j)(\boldsymbol{\theta}_1 - \boldsymbol{\xi}_j).$$

Substituting this into (7.32) produces the following:

$$0 \cong \left[\frac{\partial l(\boldsymbol{y}|\boldsymbol{\theta} = \boldsymbol{\xi}_j)}{\partial \boldsymbol{\theta}}\right] - \mathbf{J}(\boldsymbol{y}|\boldsymbol{\theta} = \boldsymbol{\xi}_j)(\boldsymbol{\theta}_1 - \boldsymbol{\xi}_j) - \boldsymbol{\Sigma}_0^{-1}\{(\boldsymbol{\theta}_1 - \boldsymbol{\xi}_j) + (\boldsymbol{\xi}_j - \boldsymbol{\theta}_0)\}. \tag{7.33}$$

Iteration $(j+1)$ begins by evaluating (7.12) at $\boldsymbol{\theta} = \boldsymbol{\xi}_j$ using (7.19) as follows:

$$\boldsymbol{\Sigma}_{1j}^{-1} = \mathbf{J}(\boldsymbol{y}|\boldsymbol{\theta} = \boldsymbol{\xi}_j) + \boldsymbol{\Sigma}_0^{-1}.$$

By substituting this into (7.33) while replacing the unknown $\boldsymbol{\theta}_1$ with an improved estimate $\boldsymbol{\xi}_{j+1}$, we get the following:

$$\boldsymbol{\Sigma}_{1j}^{-1}(\boldsymbol{\xi}_{j+1} - \boldsymbol{\xi}_j) = \left[\frac{\partial l(\boldsymbol{y}|\boldsymbol{\theta} = \boldsymbol{\xi}_j)}{\partial \boldsymbol{\theta}}\right] - \boldsymbol{\Sigma}_0^{-1}(\boldsymbol{\xi}_j - \boldsymbol{\theta}_0),$$

so

$$\boldsymbol{\xi}_{j+1} = \boldsymbol{\xi}_j + \boldsymbol{\Sigma}_{1j}\left\{\left[\frac{\partial l(\boldsymbol{y}|\boldsymbol{\theta} = \boldsymbol{\xi}_j)}{\partial \boldsymbol{\theta}}\right] - \boldsymbol{\Sigma}_0^{-1}(\boldsymbol{\xi}_j - \boldsymbol{\theta}_0)\right\},$$

if $\boldsymbol{\Sigma}_{1j}^{-1}$ is nonsingular. By (7.32), when $\boldsymbol{\xi}_j$ is sufficiently close to the desired posterior mode $\boldsymbol{\theta}_1$, the change between $\boldsymbol{\xi}_j$ and $\boldsymbol{\xi}_{j+1}$ will be negligible. An appropriate algorithm based on this mathematics would include a check to confirm that $\boldsymbol{\xi}_{j+1}$ produces a number in (7.32) closer to $\mathbf{0}$ than with $\boldsymbol{\xi}_j$ and would reduce the step size if necessary to achieve this.

Acknowledgments

The author wishes to express appreciation to the PDF Solutions management team for their support and especially to George Cheroff, whose assistance with library research has been quite valuable. In addition, Olivia Grigg, the editors and anonymous referees for this volume provided comments that improved the quality of the discussion.

References

1. Bernardo, J. M., and Smith, A.F. M. (2000) *Bayesian Theory* (NY: Wiley, prop. 5.14).
2. Bates, Douglas (2005) "Fitting linear mixed models in R", *R News*, 5, (1) 27–30.
3. Box, G., and Luceño, A. (1997) *Statistical Control by Monitoring and Feedback Adjustment* (NY: Wiley, ch. 10–11).
4. DeGroot, M. H. (1970) *Optimal Statistical Decisions* (NY: McGraw-Hill, p. 39).
5. Dey, D. K, Ghosh, S. K., and Mallick, B. K. (2000) *Generalized Linear Models: A Bayesian Perspective* (NY: Marcel Dekker, esp. ch. 3, p. 50, by Ibrahim and Chen).
6. Durbin, J. (2004) "Introduction to State Space Time Series Analysis", ch. 1 in A. Harvey, S. J. Koopman, and N. Shephard, *State Space and Unobserved Component Models* (Cambridge, UK: Cambridge U. Pr., pp. 3-25, esp. p. 22).
7. Durbin, J., and Koopman, S. J. (2002) *Time Series Analysis by State Space Methods*, corrected ed. (Oxford, UK: Oxford U. Pr., sec. 8.2, p. 157).
8. Evans, M., and Schwartz, T. (2000) *Approximating Integrals via Monte Carlo and Deterministic Methods* (Oxford, UK: Oxford U. Pr.).
9. Goel, P. K., and M. H. DeGroot (1979) "Comparison of Experiments and Information Measures", *Annals of Statistics*, 7: 1066–1077.
10. Good, I. J. (1960) "Weight of Evidence, Corroboration, Explanatory Power, Information and the Utility of Experiments", *Journal of the Royal Statistical Society, series B*, 22: 319–331.
11. Graves, S. B. (1983) *Edgeworth Expansions for Discrete Sums and Logistic Regression* (Ph.D. Dissertation, University of Wisconsin-Madison).
12. ——, Bisgaard, S., and Kulahci, M. (2002) "A Bayes-Adjusted Cusum" (technical report, www.prodsyse.com, 2005/11/24).
13. ——, Bisgaard, S., and Kulahci, M. (2002) "Designing Bayesian EWMA Monitors Using Gage R & R and Reliability Data" (technical report, www.prodsyse.com, 2005/11/24).
14. ——, Bisgaard, S., and Kulahci, M. (2002) "A Bayesian EWMA for Mean and Variance" (technical report, www.prodsyse.com, 2005/11/24).
15. ——, Bisgaard, S., Kulahci, M., Van Gilder, J., Ting, T., Marko, K., James, J., Zatorski, H., Wu, C. (2001) *Foundations of Monitoring Dynamic Systems* (technical report, from www.prodsyse.com, 2005/11/24).
16. Grigg, O. A., and Spiegelhalter, D. J. (2005) "A Simple Risk-Adjusted Exponentially Weighted Moving Average", MRC Biostatistics Unit: Technical report 2005/2, Medical Research Council of the Laboratory of Molecular Biology, Cambridge, UK (http://www.mrc-bsu.cam.ac.uk/BSUsite/Publications/pp+techrep.shtml, 2005/08/01).

Part III

Process Control and Time Series Analysis

8

A Bayesian Approach to Signal Analysis of Pulse Trains

Melinda Hock
Naval Research Laboratory, Washington, DC

Refik Soyer
George Washington University, Washington, DC

CONTENTS

ABSTRACT In this chapter, a Bayesian framework is presented for analysis of pulse trains that are corrupted by noise and missing pulses at unknown locations. The existence of missing pulses at unknown locations complicates the analysis and model selection process. This type of hidden "missingness" in the pulse data is different from the usual missing observations problem that arises in time-series analysis where standard methodology is available. We develop a Bayesian methodology for dealing with the hidden missingness. Bayesian analysis of pulse trains with hidden missingness presents a structure similar to the hidden Markov models considered in the literature. Our development is based on Markov Chain Monte Carlo (MCMC) methods and involves both inference and model selection. Analysis of the pulse trains also requires formal treatment of correlated noise terms. The presented framework deals with this issue via the use of MCMC methods and it allows for sequential processing of data by using a Bayesian dynamic linear model (DLM) setup.

8.1 Introduction

In electronic warfare (EW) applications, many threat radars are designed so that the time interval between pulses or the frequency of pulses varies in a cyclical manner. Classifying and predicting these variations in real-time is important so that the threat can be tracked and protective measures can be taken as necessary. Time interval between two pulses emitted by a threat radar is defined as a *pulse repetition interval* (PRI). The time-series of the PRIs is referred to as a *pulse train*. PRI tracking is an important problem in naval EW applications because knowledge of the PRI is used to defend ships against radar-guided missiles by performing deceptive jamming. The purpose of a PRI tracking algorithm is to predict the PRIs or, equivalently, the pulse time of arrivals (TOA) for various PRI types.

The pulse/signal environment will generally be very complex with many different radars (emitters) transmitting simultaneously. A different set of signal parameters will characterize each emitter in the environment. The electronic support (ES) receiver has the capability of separating each signal from all the others in the environment by sorting pulses with similar parameters. The ES receiver provides data to the PRI tracker from a single designated emitter. In many applications due to the complexity of the environment, sensitivity of the receiver might be reduced during jamming, and this results in missing pulses in the observed data. For tracking to be performed identifying the PRI type of the pulse train in the presence of missing pulses is essential. The pulse train available to the tracker possibly has up to 15% missing pulses. The presence of missing pulses complicates the identification of the PRI modulation type because never known is whether a pulse has actually been missed by the ES receiver.

The presence of missing pulses complicates the analysis and identification of the PRI modulation type because the location of missing pulses is unknown. As noted by Hock and Soyer [14], this type of hidden "missingness" in the pulse data is different from the usual missing observations problem that arises in time-series analysis where standard methodology is available [16,19].

Another source of difficulty in identifying the location of missing pulses is the presence of noise in the pulse train. When the ES receiver records the TOAs of pulses, these readings are subjected to noise. The major type of noise that contaminates the TOAs will be referred to as *jitter noise*. Jitter noise is added to the data at the missile radar transmitter, and it can be either unintentional or intentional. Unintentional jitter is caused by imperfections in the transmitting equipment that result in oscillator instability. Intentional jitter is a deliberate distortion of the pulse timing for the purpose of making the pulse train more difficult to track. The TOAs are also affected by measurement noise. The ES receiver adds measurement noise to the data. The magnitude of the measurement noise is determined by the resolution of the receiver. It is a quantity that can be measured and is typically small, but its presence introduces correlated noise components for the PRIs and affects the analysis.

In this Chapter, we develop a formal Bayesian approach to describe hidden missingness for the analysis of pulse trains and for identification of the type of PRI modulation present on an isolated pulse train. Several possible types of PRI modulations exist. We consider staggered PRI modulation (several different PRIs in a repeating pattern) and jittered PRI modulation (random variation in PRI about a mean value) and develop a Bayesian analysis of these with hidden missingness using Markov Chain Monte Carlo (MCMC) methods. We also present a Bayesian model selection approach for identifying PRI modulation types and develop the methodologies for analysis of pulse trains with correlated noise.

A synopsis of this chapter is as follows: Section 8.2 begins with a definition of the jitter and stagger modulation types and a discussion of hidden missingness in pulse trains. Notation and preliminaries are introduced and a discrete time Markov chain model is presented to describe the missingness structure and motivation for considering correlated noise (error) is discussed. In Section 8.3, Bayesian inference for the model using the Gibbs sampler is developed. An alternate Bayesian approach is introduced in Section 8.4 using a dynamic linear model (or a Kalman filter) setup. This setup allows for sequential processing of data and avoids matrix inversions during the implementation of the Gibbs sampler. A model comparison approach, based on marginal likelihood, is introduced in Section 8.5. The developed methodology is illustrated using simulated data in Section 8.6 and concluding remarks are given in Section 8.7.

8.2 A Hidden Markov Model for PRIs

8.2.1 Notation and Preliminaries

The data available to the PRI tracker consists of a batch of observed PRI values. This pulse train of n observed PRI are obtained by taking the difference between consecutive measured TOAs. Let $\{\tau_0, \tau_1, \ldots, \tau_n\}$ denote the observed TOA sequence and $\{z_1, \ldots, z_n\}$ denote the corresponding observed PRI sequence such that $z_i = \tau_i - \tau_{i-1}$ for $i = 1, \ldots, n$. Both the TOA values and the observed PRI values are affected by missing pulses.

To distinguish between the transmitted PRI values by the threat radar and the observed values by the ES receiver due to missingness, we denote the transmitted PRI values by the sequence $\{y_{k_i}\}$, $i = 1, \ldots, n$. In $\{y_{k_i}\}$, k_i denotes the transmitted pulse index, whereas i denotes the observed pulse index such that $k_i \geq i$ with equality holding only for the case of no missing observations in the pulse train. Similarly, the TOA sequence associated with the transmitted pulses will be denoted by $\{t_{k_i}\}$, $i = 1, \ldots, n$, where $y_{k_i} = t_{k_i} - t_{k_i - 1}$. If no missing pulses are in a given interval, the observed PRI value of the interval will be equivalent to the corresponding PRI value from the transmitted pulse train, that is, $z_i = y_{k_i}$.

However, if a pulse is missing in an interval from the incoming data stream, the next observed PRI value represents the sum of two consecutive PRI values in the transmitted sequence rather than a single PRI value. Similarly, if two consecutive pulses are missing from the incoming data stream, the following observed PRI value represents the sum of three consecutive PRI values. In general, when missing pulses corrupt the data set, the observed PRIs are expressed as

$$z_i = \sum_{j=0}^{m_i} y_{k_i - j} \tag{8.1}$$

where m_i is the number of missing pulses between the $(i-1)$st and ith observed pulses. A sum of two or more transmitted PRI values will be referred to as aggregate data. For example, in Figure 8.1, the transmitted sequence consists of $\{y_1, y_2, y_3, y_4, y_5\}$. The observed sequence has a missing pulse in the third interval and as a result, the observed sequence is obtained as $z_1 = y_1$, $z_2 = y_2$, $z_3 = y_3 + y_4$, and $z_4 = y_5$.

In the observed data $\{z_1, \ldots, z_n\}$, each data point can represent either a single PRI value or an aggregate PRI value. The number of PRI values included in a data point will be referred to as its state. The states are unobservable latent variables associated with each data point. Following Hock and Soyer [14], we

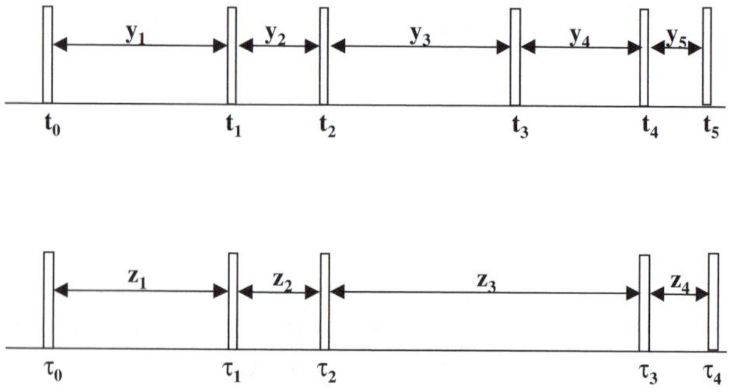

FIGURE 8.1
Effect of a single missing pulse on the observed data.

denote the states by $\{s_1, \ldots, s_n\}$. For the data point i, the state $s_i = 1$ if the ith data point represents a single PRI value, that is, no missing pulse is in the ith observed pulse interval. The state $s_i = 2$ if the ith data point represents the sum of two consecutive PRI values from the transmitted pulse sequence, that is, if a missing pulse is in the ith observed pulse interval. Thus, we can write the observed pulse index as $k_i = \sum_{j=1}^{i} s_j$. For example, in Figure 8.1, $s_1 = s_2 = s_4 = 1$ and $s_3 = 2$, implying that $k_3 = 4$.

In general, if missing pulses are m_i in the ith observed pulse interval, then $s_i = m_i + 1$. As noted by [14], in many applications the value of m is small, typically 1 or 2. We note that when a sequence of n PRIs are observed by the ES receiver and if the data is corrupted by missing pulses, the actual transmitted sequence will consist of $\sum_{i=1}^{n} s_i$ PRI values, where $n \leq \sum_{i=1}^{n} s_i$.

8.2.2 Noisy PRI Modulations and Hidden Missingness

As pointed out in Section 8.1, the two most commonly observed types of PRI modulation transmitted by the threat radar are the staggered and jittered PRI modulation.

Under the staggered PRI modulation it is assumed that the transmitted PRIs form a discrete time-series generated by a unknown periodic function. The Fourier representation theorem states that any periodic function can be expressed as a linear combination of sine and cosine terms plus a constant. The frequencies of the sine and cosine terms correspond to the different harmonics present in the function. A discrete time series can be represented by a finite number of harmonics. Specifically, if a discrete time series has period p, it can be described as

$$a_0 + \sum_{i=1}^{q} a_i cos\left(2\pi \frac{i}{p} k\right) + b_i sin\left(2\pi \frac{i}{p} k\right), \tag{8.2}$$

where $k = 1, 2, \ldots$ is the time index, a_is and b_is are the unknown Fourier coefficients, and

$$q = \begin{cases} p/2 & \text{for } p \text{ even} \\ (p-1)/2 & \text{for } p \text{ odd.} \end{cases}$$

In Equation (8.2), the first harmonic has frequency $1/p$ and completes its cycle in p time periods. The second harmonic has frequency $2/p$ and completes its cycle in $p/2$ time periods. If p is even, at most $p/2$ harmonics are required to represent the time series because the period corresponding to the $(p/2)$th harmonic is 2, which is the shortest possible cycle length. If p is odd, at most $(p-1)/2$ harmonics are needed.

As discussed in Section 8.1, two different types of noise can contaminate the PRIs: jitter noise and measurement noise. Jitter noise is added to the data at the missile radar transmitter. It is a deliberate distortion of the pulse timing for the purpose of making the pulse train more difficult to track, and its magnitude is unknown. The ES receiver on the victim ship introduces the measurement noise. The magnitude of the measurement noise is determined by the resolution of the receiver.

A staggered PRI sequence that is subjected only to jitter noise can be represented as follows:

$$y_k = a_0 + \sum_{i=1}^{q} a_i cos\left(2\pi \frac{i}{p} k\right) + b_i sin\left(2\pi \frac{i}{p} k\right) + w_k, \tag{8.3}$$

where w_k is the jitter noise term that is normally distributed with mean 0 and unknown variance σ_w^2, and the last sine term reduces to zero when p is even. In what follows, we will refer to Equation (8.3) as the stagger model. The jittered PRI modulation is a special case of Equation (8.3) where the transmitted PRIs randomly vary around a constant mean, that is,

$$y_k = a_0 + w_k. \tag{8.4}$$

We will refer to Equation (8.4) as the jitter model. Because the jitter model is a special case of the stagger model, our development will focus on the latter.

For an observed PRI sequence $\{z_1, \ldots, z_n\}$, corrupted with missing pulses, if the sequence of states $\{s_1, \ldots, s_n\}$ are known for all $i = 1, \ldots, n$, then using Equation (8.1) the time-series of the observed PRIs can be aggregated as

$$z_i = \sum_{j=0}^{m_i} y_{k_i - j} = \sum_{j=0}^{s_i - 1} y_{k_i - j}. \tag{8.5}$$

For example, in the jitter model, Equation (8.5) reduces to

$$z_i = \sum_{j=0}^{s_i - 1} (a_0 + w_{k_i - j}) = s_i a_0 + u_i, \tag{8.6}$$

where $u_i = \sum_{j=0}^{s_i-1} w_{k_i-j}$. Note that the u_i's do not share any common jitter noise terms and thus they are independent Gaussian terms with mean 0 and variance $s_i \sigma_w^2$. Similarly, for the stagger model with period p, we can write

$$z_i = \sum_{j=0}^{s_i - 1} \left[a_0 + \sum_{l=1}^{q} a_l cos\left(2\pi \frac{l}{p}(k_i - j)\right) + b_l sin\left(2\pi \frac{l}{p}(k_i - j)\right)\right] + u_i, \tag{8.7}$$

where the jitter model is obtained as the special case with $q = 0$.

To consider the presence of measurement noise at the ES receiver, we define τ_i^m as the ith observed TOA that is corrupted by the measurement noise v_i and write

$$\tau_i^m = \tau_i + v_i, \tag{8.8}$$

where τ_i denotes the measurement noise free TOA. The measurement noise sequence $\{v_i\}$ is a zero-mean independent normal sequence whose variance σ_v^2

will be known when the resolution of the receiver is known. Furthermore, the measurement noise sequence $\{v_i\}$ is independent of the jitter noise sequence $\{w_{k_i-j}\}$.

It follows from the above that the ith PRI is given by

$$z_i^m = \tau_i^m - \tau_{i-1}^m = z_i + v_i - v_{i-1}, \tag{8.9}$$

where $z_i = \tau_i - \tau_{i-1}$ and

$$z_i^m = \sum_{j=0}^{s_i-1}\left[a_0 + \sum_{l=1}^{q} a_l cos\left(2\pi\frac{l}{p}(k_i - j)\right) + b_l sin\left(2\pi\frac{l}{p}(k_i - j)\right)\right] + u_i^c, \tag{8.10}$$

with $u_i^c = u_i + v_i - v_{i-1}$. We can note from Equation (8.9) that the z_i^m terms that are one period apart will be correlated, that is,

$$Cov(z_i^m, z_{i-h}^m) = \begin{cases} -\sigma_v^2 & \text{if } h = 1 \\ 0 & \text{if } h > 1 \end{cases} \tag{8.11}$$

and the variance of z_i^m is given by

$$Var(z_i^m) = s_i\sigma_w^2 + 2\sigma_v^2. \tag{8.12}$$

It can be easily shown that $-0.5 < Corr(z_i^m, z_{i-1}^m) < 0$, and the correlation approaches to -0.5 as the ratio of the variances σ_w^2/σ_v^2 gets smaller.

In what follows, we will supress the dependence on the superscript m and write the correlated noise model for PRIs as

$$z_i = \sum_{j=0}^{s_i-1}\left[a_0 + \sum_{l=1}^{q} a_l cos\left(2\pi\frac{l}{p}(k_i - j)\right) + b_l sin\left(2\pi\frac{l}{p}(k_i - j)\right)\right] + u_i^c. \tag{8.13}$$

8.2.3 Modeling Hidden Missingness

Given the latent sequence $\{s_1, \ldots, s_n\}$ under the stagger model with period p, we can write the observed PRIs as

$$z_i = \sum_{j=0}^{s_i-1} \mathbf{X}_{k_i-j}\boldsymbol{\theta} + u_i^c, \tag{8.14}$$

where

$$\mathbf{X}_{k_i-j} = \left(1 \;\; \cos\left(2\pi\frac{1}{p}(k_i - j)\right) \;\; \sin\left(2\pi\frac{1}{p}(k_i - j)\right) \cdots \sin\left(2\pi\frac{q}{p}(k_i - j)\right)\right)$$

and $\boldsymbol{\theta}' = (a_0\, a_1\, b_1 \ldots a_q\, b_q)$. Thus, given the latent sequence $\{s_1, \ldots, s_n\}$, it is possible to write the observed PRIs as a linear model, that is,

$$\boldsymbol{z} = \boldsymbol{X\theta} + \boldsymbol{u}^c, \tag{8.15}$$

8.3 Bayesian Analysis of the HMM

For the stagger model with period p, the observed PRIs are given by Equation (8.14). The Bayesian approach requires that we specify our uncertainty about the unknown coefficient vector $\boldsymbol{\theta}$ and the unknown variance σ_w^2 in the linear model setup of Equation (8.15) by a prior probability distribution. Under the normality of u_i^c's, a conjugate prior for this Equation (8.15) is given by the normal-gamma distribution

$$\boldsymbol{\theta}|\phi \sim N(\boldsymbol{m}, \boldsymbol{V}/\phi), \tag{8.22}$$

$$\phi \sim \text{Gamma}\ (d/2, c/2). \tag{8.23}$$

In the above, $\phi = 1/\sigma_w^2$ is the unknown precision, $\boldsymbol{m}$ is a $p \times 1$ specified mean vector, $\boldsymbol{V}$ is a specified variance-covariance matrix of $\boldsymbol{\theta}$, and d and c are specified prior parameters for the distribution of ϕ. The joint distribution of $\boldsymbol{\theta}$ and ϕ that is obtained as the product of the Equation (8.22) and Equation (8.23) is known as the Normal-Gamma distribution, which implies that the marginal prior of $\boldsymbol{\theta}$ is a multivariate Student t density with degrees of freedom $(d + p)$, mean vector $\boldsymbol{m}$, and scale matrix $\frac{c}{d}\boldsymbol{V}$. This density will be denoted as

$$\boldsymbol{\theta} \sim T(d + p, \boldsymbol{m}, \boldsymbol{V}c/d). \tag{8.24}$$

It is assumed that a priori, $(\boldsymbol{\theta}, \phi)$ are independent of the latent variables s_i's.

For the observed PRI sequence $\boldsymbol{z}^T = (z_1 \dots z_n)$, given $(\boldsymbol{\theta}, \phi)$ and the latent variables $\boldsymbol{S}_n = (s_1, \dots, s_n)$, the distribution of $\boldsymbol{z}$ will be a multivariate normal density given by $(\boldsymbol{z}|\boldsymbol{\theta}, \phi, \boldsymbol{S}_n) \sim N(\boldsymbol{X\theta}, \boldsymbol{U}^c)$ as in Equation (8.18), where the measurement noise variance σ_v^2 is known. Given the latent sequence $\boldsymbol{S}_n$, the Bayesian analysis of the model can be made using conjugate analysis to obtain the posterior distribution $p(\boldsymbol{\theta}, \phi|\boldsymbol{S}_n, \boldsymbol{z})$. But as $\boldsymbol{S}_n$ is not observed, the Bayesian analysis requires the joint posterior distribution $p(\boldsymbol{\theta}, \phi, \boldsymbol{S}_n|\boldsymbol{z})$ to be given by

$$p(\boldsymbol{\theta}, \phi, \boldsymbol{S}_n|\boldsymbol{z}) \propto p(\boldsymbol{z}|\boldsymbol{\theta}, \phi, \boldsymbol{S}_n)\ p(\boldsymbol{\theta}, \phi)\ p(\boldsymbol{S}_n) \tag{8.25}$$

due to the prior independence of $(\boldsymbol{\theta}, \phi)$ and $\boldsymbol{S}_n$. Because $\boldsymbol{S}_n$ follows a Markov chain with unknown transition matrix $\boldsymbol{P}$, the joint posterior distribution required for the Bayesian analysis is

$$p(\boldsymbol{\theta}, \phi, \boldsymbol{S}_n, \boldsymbol{P}|\boldsymbol{z}) \propto p(\boldsymbol{z}|\boldsymbol{\theta}, \phi, \boldsymbol{S}_n)\, p(\boldsymbol{\theta}, \phi)\, p(\boldsymbol{S}_n|\boldsymbol{P})\, p(\boldsymbol{P}), \tag{8.26}$$

where $p(\boldsymbol{P})$ is the prior distribution for the transition matrix $\boldsymbol{P}$ of the Markov chain, which is assumed to be independent of $(\boldsymbol{\theta}, \phi)$. We assume that $\boldsymbol{P}_i$, the ith row of $\boldsymbol{P}$, follows a Dirichlet distribution

$$p(\boldsymbol{P}_i) \propto \prod_{j=1}^{M+1} p_{ij}^{\alpha_{ij}-1}, \quad i = 1, \dots, M+1 \tag{8.27}$$

with parameters α_{ij}, denoted as $\boldsymbol{P}_i \sim Dirichlet(\alpha_{ij};\ j = 1, \ldots, M+1)$. Also, $\boldsymbol{P}_i$'s are assumed to be independent for all i.

Under any choice of prior distributions, the joint posterior distribution $p(\boldsymbol{\theta}, \phi, \boldsymbol{S}_n, \boldsymbol{P}|z)$ in Equation (8.26) cannot be evaluated analytically. Such evaluation is also computationally infeasible due to the presence of latent sequence $\boldsymbol{S}_n$. Each latent variable s_i can belong to one of the $M+1$ categories, implying that $(M+1)^n$ possible latent sequences exist for n observed pulses. Thus in what follows, a Markov Chain Monte Carlo approach, or more specifically a Gibbs sampler, will be presented to generate samples from the joint posterior distribution Equation (8.26). The attractive feature of the Gibbs sampler is that it enables the generation of samples from the posterior distributions without having to obtain the exact distributional forms. This is achieved by successive drawings from the full conditional distributions of $\boldsymbol{\theta}$, ϕ, $\boldsymbol{S}_n$, and $\boldsymbol{P}$ given z. For example, the full conditional distribution of $\boldsymbol{\theta}$ is $p(\boldsymbol{\theta}|\phi, \boldsymbol{S}_n, \boldsymbol{P}, z)$. For a more detailed discussion of the Gibbs sampler and other related Monte Carlo methods, see Gelfand and Smith [9], Casella and George [1], and Chapter 2 in this volume.

8.3.1 Implementation of the Gibbs Sampler

In the stagger model, the implementation of the Gibbs sampler requires the full conditional distributions $p(\boldsymbol{\theta}|\phi, \boldsymbol{S}_n, \boldsymbol{P}, z)$, $p(\phi|\boldsymbol{\theta}, \boldsymbol{S}_n, \boldsymbol{P}, z)$, $p(\boldsymbol{P}|\boldsymbol{\theta}, \phi, \boldsymbol{S}_n, z)$, and $p(s_i|\boldsymbol{\theta}, \phi, \boldsymbol{S}^{-i}, \boldsymbol{P}, z)$ for $i = 1, \ldots, n$, where $\boldsymbol{S}^{-i} = \{s_j|\ j \neq i,\ j = 1, 2, \ldots, n\}$. Once the full conditional distributions are obtained the Gibbs sampler can be implemented by succesively generating from these distributions to obtain a sample from the joint posterior distribution.

From Equation (8.26), we can see that given $\boldsymbol{S}_n$, z does not depend on the transition matrix $\boldsymbol{P}$. Thus, learning about $\boldsymbol{P}$ is through the updating of $\boldsymbol{S}_n$ based on observed data z. This implies that the full conditional distribution of $\boldsymbol{P}$ is $p(\boldsymbol{P}|\boldsymbol{\theta}, \phi, \boldsymbol{S}_n, \mathbf{z}) = p(\boldsymbol{P}|\ \boldsymbol{S}_n)$. Furthermore, given $\boldsymbol{S}_n$ the full conditional distributions of $\boldsymbol{\theta}$ and ϕ are not dependent on $\boldsymbol{P}$, that is, $p(\boldsymbol{\theta}|\phi, \boldsymbol{S}_n, \boldsymbol{P}, z) = p(\boldsymbol{\theta}|\phi, \boldsymbol{S}_n, z)$ and $p(\phi|\ \boldsymbol{\theta}, \boldsymbol{S}_n, \boldsymbol{P}, z) = p(\phi|\ \boldsymbol{\theta}, \boldsymbol{S}_n, z)$.

Note that the latent sequence $\boldsymbol{S}_n$ follows a Markov chain with a state space of dimension $M+1$, where M is the maximum possible number of missing pulses in an interval. We can write down the full conditional distribution of s_i as

$$p(s_i|\boldsymbol{\theta}, \phi, \boldsymbol{S}^{-i}, \boldsymbol{P}, z) \propto p(z|s_i, \boldsymbol{\theta}, \phi, \boldsymbol{S}^{-i}, \boldsymbol{P})\, p(s_i|\boldsymbol{\theta}, \phi, \boldsymbol{S}^{-i}, \boldsymbol{P}). \tag{8.28}$$

As pointed out in the above, the first term on the right-hand side of Equation (8.28) does not depend on the transition matrix $\boldsymbol{P}$, that is, $p(z|s_i, \boldsymbol{\theta}, \phi, \boldsymbol{S}^{-i}, \boldsymbol{P}) = p(z|s_i, \boldsymbol{\theta}, \phi, \boldsymbol{S}^{-i})$. It follows from Equation (8.14) that only the $z_i, z_{i+1}, \ldots, z_n$ will depend on s_i, and thus

$$p(s_i|\boldsymbol{\theta}, \phi, \boldsymbol{S}^{-i}, \boldsymbol{P}, z) \propto p(z_i, \ldots, z_n|\boldsymbol{\theta}, \phi, \boldsymbol{S}_n)\, p(s_i|\ \boldsymbol{S}^{-i}, \boldsymbol{P}), \tag{8.29}$$

where $p(z_i, \ldots, z_n|\boldsymbol{\theta}, \phi, \boldsymbol{S}_n)$ is a multivariate normal density as implied by Equation (8.18). In other words, $(z_i, \ldots, z_n)$ will have a mean vector that is the product of a submatrix consisting of rows $i, i+1, \ldots, n$ of $\boldsymbol{X}$, and $\boldsymbol{\theta}$ and will have a variance-covariance matrix consisting of rows and columns $i, i+1, \ldots, n$ of $\boldsymbol{U}^c$. Using the Markov property, for $2 \leq i \leq n-1$

$$p(s_i \mid \boldsymbol{S}^{-i}, \boldsymbol{P}) \propto p(s_i | s_{i-1}, \boldsymbol{P})\, p(s_{i+1} | s_i, \boldsymbol{P}) \tag{8.30}$$

and Equation (8.29) reduces to

$$p(s_i | \boldsymbol{\theta}, \phi, \boldsymbol{S}^{-i}, \boldsymbol{P}, \boldsymbol{z}) \propto p(s_i | s_{i-1}, \boldsymbol{P})\, p(s_{i+1} | s_i, \boldsymbol{P})\, p(z_i, \ldots, z_n | \boldsymbol{\theta}, \phi, \boldsymbol{S}_n). \tag{8.31}$$

Since the s_i's are discrete random variables, Equation (8.31) can be easily normalized by summing over all possible values of s_i. Also, for the special cases of $i = 1$ and $i = n$,

$$p(s_i | \boldsymbol{\theta}, \phi, \boldsymbol{S}^{-i}, \boldsymbol{P}, \boldsymbol{z}) \propto \begin{cases} p(s_1 | \boldsymbol{P})\, p(s_2 | s_1, \boldsymbol{P})^n\, p(z_1, \ldots, z_n | \boldsymbol{\theta}, \phi, \boldsymbol{S}_j), \; i = 1 \\ p(s_n | s_{n-1}, \boldsymbol{P})\, p(z_n | \boldsymbol{\theta}, \phi, \boldsymbol{S}_n), \; i = n, \end{cases}$$

where $p(s_1 | \boldsymbol{P})$ can be chosen as the stationary distribution of the Markov chain with transition matrix $\boldsymbol{P}$. For the case of the jitter model where $\boldsymbol{\theta} = a_0$ is a scalar, the full conditional of s_i reduces to

$$p(s_i | a_0, \phi, \boldsymbol{S}^{-i}, \boldsymbol{P}, \boldsymbol{z}) \propto p(s_i | s_{i-1}, \boldsymbol{P})\, p(s_{i+1} | s_i, \boldsymbol{P})\, p(z_i | a_0, \phi, s_i), \tag{8.32}$$

where $p(z_i | a_0, \phi, s_i)$ is a normal density. Unlike the stagger model, in this case full conditional of s_i depends only on the neighboring latent variables s_{i-1} and s_{i+1}.

As previously pointed out, the full conditional distribution of the transition matrix $\boldsymbol{P}$ is dependent only on $\boldsymbol{S}_n$, that is, $p(\boldsymbol{P} | \boldsymbol{\theta}, \phi, \boldsymbol{S}_n, z) = p(\boldsymbol{P} | \boldsymbol{S}_n)$. Given the independent Dirichlet priors on each row $\boldsymbol{P}_i \sim Dirichlet(\alpha_{ij};\, j = 1, \ldots, M+1)$, the full conditional distribution of each row is given by

$$p(\boldsymbol{P}_i | \boldsymbol{S}_n) \propto \prod_{k=1}^{n} p(s_{k+1} | s_k = i)\, p(\boldsymbol{P}_i),$$

where $p(\boldsymbol{P}_i)$ is the Dirichlet prior given in Equation (8.27). Thus, the above can be written as

$$p(\boldsymbol{P}_i | \boldsymbol{S}_n) \propto \prod_{j=1}^{M+1} p_{ij}^{\alpha_{ij}-1}\, p_{ij}^{\sum_{k=1}^{n} 1(s_{k+1}=j, s_k=i)}, \tag{8.33}$$

where $1(\cdot)$ is an indicator function. From the above, the full conditional distribution of $\boldsymbol{P}_i$ is Dirichlet of the form

$$(\boldsymbol{P}_i | \boldsymbol{S}_n) \sim Dirichlet\left(\alpha_{ij} + \sum_{k=1}^{n} 1(s_{k+1} = j, s_k = i)\,;\; j = 1, \ldots, M+1 \right). \tag{8.34}$$

Given $\boldsymbol{S}_n$, the $\boldsymbol{P}_i$'s, $i = 1, \ldots M+1$, are independent.

The full conditional distributions of $\boldsymbol{\theta}$ and ϕ do not depend on $\boldsymbol{P}$ given $\boldsymbol{S}_n$, that is, $p(\boldsymbol{\theta}|\phi, \boldsymbol{S}_n, \boldsymbol{P}, z) = p(\boldsymbol{\theta}|\phi, \boldsymbol{S}_n, z)$ and $p(\phi|\,\boldsymbol{\theta}, \boldsymbol{S}_n, \boldsymbol{P}, z) = p(\phi|\,\boldsymbol{\theta}, \boldsymbol{S}_n, z)$. In obtaining $p(\boldsymbol{\theta}|\phi, \boldsymbol{S}_n, z)$, we can write

$$p(\boldsymbol{\theta}|\phi, \boldsymbol{S}_n, z) \propto p(z|\boldsymbol{\theta}, \phi, \boldsymbol{S}_n)\; p(\boldsymbol{\theta}|\phi), \tag{8.35}$$

where $p(\boldsymbol{\theta}|\phi)$ is given by Equation (8.22) and it does not depend on the latent vector $\boldsymbol{S}_n$ due to the prior independence assumptions. The first term on the right-hand side of Equation (8.35) is the conditional likelihood function of $\boldsymbol{\theta}$ given $(\phi, \boldsymbol{S}_n, z)$, which is given by the multivariate normal form in Equation (8.18). Thus, Equation (8.35) can be written as

$$p(\boldsymbol{\theta}|\phi, \boldsymbol{S}_n, z) \propto exp\left(-\frac{1}{2}[(z - \boldsymbol{X\theta})'(\boldsymbol{U}^c)^{-1}(z - \boldsymbol{X\theta}) + (\boldsymbol{\theta} - \boldsymbol{m})'\phi \boldsymbol{V}^{-1}(\boldsymbol{\theta} - \boldsymbol{m})]\right).$$

Using standard Bayesian conjugate analysis [7, p. 251], From the above that $(\boldsymbol{\theta}|\phi, \boldsymbol{S}_n, z) \sim N(\boldsymbol{m}^*, \boldsymbol{V}^*)$ where the posterior mean and variance are given by

$$\boldsymbol{m}^* = (\phi \boldsymbol{V}^{-1} + \boldsymbol{X}'(\boldsymbol{U}^c)^{-1}\boldsymbol{X})^{-1}(\phi \boldsymbol{V}^{-1}\boldsymbol{m} + \boldsymbol{X}'(\boldsymbol{U}^c)^{-1}z) \tag{8.36}$$

and

$$\boldsymbol{V}^* = (\phi \boldsymbol{V}^{-1} + \boldsymbol{X}'(\boldsymbol{U}^c)^{-1}\boldsymbol{X})^{-1} \tag{8.37}$$

Note that the above results are similar to Bayesian analysis presented by Gelman et al. [11, p. 255] for regression models with unequal variances or with heteroskedasticity.

The full conditional distribution of ϕ, $p(\phi|\,\boldsymbol{\theta}, \boldsymbol{S}_n, z)$ is given by

$$p(\phi|\,\boldsymbol{\theta}, \boldsymbol{S}_n, z) \propto p(z|\boldsymbol{\theta}, \phi, \boldsymbol{S}_n)\; p(\phi|\boldsymbol{\theta}) \propto p(z|\boldsymbol{\theta}, \phi, \boldsymbol{S}_n)\; p(\boldsymbol{\theta}|\phi) p(\phi). \tag{8.38}$$

Given $\boldsymbol{\theta}$ and $\boldsymbol{S}_n$, the first term on the right-hand side of Equation (8.38), $p(z|\boldsymbol{\theta}, \phi, \boldsymbol{S}_n)$, is the conditional likelihood of ϕ, which is a multivariate normal form, and the remaining terms are the components of the Normal-Gamma prior given by Equation (8.22) and Equation (8.23). Thus, Equation (8.38) can be written as

$$p(\phi|\,\boldsymbol{\theta}, \boldsymbol{S}_n, z) \propto \left[|\,\boldsymbol{U}^c\,|^{-1/2} exp\left(-\frac{1}{2}[(z - \boldsymbol{X\theta})'(\boldsymbol{U}^c)^{-1}(z - \boldsymbol{X\theta})]\right)\right]$$
$$\times \left[\phi^{p/2} exp\left(-\frac{\phi}{2}[(\boldsymbol{\theta} - \boldsymbol{m})'\boldsymbol{V}^{-1}(\boldsymbol{\theta} - \boldsymbol{m})]\right)\right] \times \phi^{(d/2)-1}\, exp\left(-\phi\frac{c}{2}\right), \tag{8.39}$$

where $\boldsymbol{U}^c = \boldsymbol{J} + \boldsymbol{U}/\phi$ is a function of ϕ. In this case, the full conditional distribution of ϕ does not have a known form. However, in drawing from the full conditional of Equation (8.39), a Metropolis (see for example [2]) step can be used at each iteration of the Gibbs sampler. In so doing, a Gamma distribution can be used as the proposal density for $p(\phi|\,\boldsymbol{\theta}, \boldsymbol{S}_n, z)$. This is a very reasonable choice because $\boldsymbol{U}^c$ approaches $\boldsymbol{U}$ for small values of the measurement noise variance σ_v^2 and in this case Equation (8.39) will approach to a gamma density with parameters $d^*/2$ and $c^*/2$, that is, $(\phi|\,\boldsymbol{\theta}, \boldsymbol{S}_n, z) \sim$ Gamma $(d^*/2, c^*/2)$, where $d^* = n + p + d$ and

$$c^* = [(z - \boldsymbol{X\theta})'\boldsymbol{U}^{-1}(z - \boldsymbol{X\theta}) + (\boldsymbol{\theta} - \boldsymbol{m})'\boldsymbol{V}^{-1}(\boldsymbol{\theta} - \boldsymbol{m}) + c]. \tag{8.40}$$

In this case, the proposal density is not symmetric and an independence chain — where the proposed transition is independent of previous position — can be used.

Alternatively, a reparameterization of the model can be used and the full conditional distribution of ϕ can be obtained as a gamma density. It is not unreasonable to assume that the measurement noise variance is proportional to the jitter noise variance, that is, we can assume that the measurement noise variance is given by $\sigma_v^2 \sigma_w^2 = \sigma_v^2/\phi$ in Equation (8.17). In this case, we can define (8.19) as $\boldsymbol{U}^c = (\boldsymbol{J} + \boldsymbol{U})/\phi$ and show that $(\phi|\,\boldsymbol{\theta}, \boldsymbol{S}_n, z) \sim$ Gamma $(d^*/2, c^*/2)$, where $d^* = n + p + d$ and c^* is given by Equation (8.40) by replacing $\boldsymbol{U}$ with $(\boldsymbol{J} + \boldsymbol{U})$.

We note that the implementation of the Gibbs sampler requires inversion of the variance covariance matrix at each iteration. This is not an issue in the special case of no measurement noise, where $\boldsymbol{U}^c$ approaches to the diagonal matrix $\boldsymbol{U}$ in Equation (8.19). However, in the general case with measurement noise, as n gets large the matrix inversions can become computationally infeasible. An alternate inference strategy for the general model can be developed by using a Kalman filter-type setup for the correlated noise problem. Implementation of the Gibbs sampler in this setup requires sequential processing of the data and this procedure has certain advantages in developing inference because it avoids matrix inversions of $\boldsymbol{U}^c$. The state-space framework that uses the Kalman filtering to avoid such matrix inversions is not uncommon in signal processing literature; for example, see Doucet and Duvaut [6] for a different application. Furthermore, for the PRI tracker to have the ability to process data in a sequential manner is desirable and therefore the Kalman filter setup is attractive for real time implementation of the methodology. The Kalman filter setup will be presented next.

8.4 Kalman Filter Setup for the Model

The Kalman filter setup is also referred to as the *dynamic linear model* (DLM) setup in Bayesian time-series literature, where many well-known models can be represented as special cases; for example, see Harrison and Stevens [13] and West and Harrison [19]. In our case, the main reason for using the DLM setup is to generate from the full posterior conditional distributions via sequential processing of data. As pointed out in the previous section, the DLM setup enables us to avoid matrix inversions in the implementation of the Gibbs sampler.

Given the latent sequence S_n, we consider the stagger model given by Equation (8.14) and rewrite it as

$$z_i = \sum_{j=0}^{s_i-1}(\mathbf{X}_{k_i-j}\boldsymbol{\theta} + w_{k_i-j}) + v_i - v_{i-1} \tag{8.41}$$

by substituting $u_i^c = u_i + v_i - v_{i-1} = \sum_{j=0}^{s_i-1} w_{k_i-j} + v_i - v_{i-1}$. We can represent the above as

$$z_i = [\boldsymbol{F}_i' \;\; 1 \;\; -1]\boldsymbol{\Theta}_i + u_i, \tag{8.42}$$

where $\boldsymbol{F}_i' = \sum_{j=0}^{s_i-1}\mathbf{X}_{k_i-j}$, $u_i = \sum_{j=0}^{s_i-1} w_{k_i-j}$, and $\boldsymbol{\Theta}_i$ is a state vector that includes $\boldsymbol{\theta}$ as well as the noise terms v_i and v_{i-1}, that is,

$$\boldsymbol{\Theta}_i = \begin{bmatrix} \boldsymbol{\theta}_i \\ v_i \\ v_{i-1} \end{bmatrix} \tag{8.43}$$

with $\boldsymbol{\theta}_i = \boldsymbol{\theta}$ for all i. In the Kalman filter model setup, Equation (8.42) is referred to as the observation equation (see Chapter 1). The system, or state, equation of the Kalman filter setup for the correlated noise model is given by

$$\boldsymbol{\Theta}_i = \mathbf{G}\boldsymbol{\Theta}_{i-1} + \mathbf{v}_i \tag{8.44}$$

where

$$\mathbf{G} = \begin{bmatrix} \boldsymbol{I}_{p\times p} & \mathbf{0}_{p\times 1} & \mathbf{0}_{p\times 1} \\ \mathbf{0}_{1\times p} & 0 & 0 \\ \mathbf{0}_{1\times p} & 1 & 0 \end{bmatrix}, \quad \mathbf{v}_i = \begin{bmatrix} \mathbf{0}_{p\times 1} \\ v_i \\ 0 \end{bmatrix}, \tag{8.45}$$

$\boldsymbol{I}_{p\times p}$ is the p−dimensional identity matrix and $\mathbf{0}_{p\times 1}$ is $p \times 1$ vector of zeros. The Kalman filter setup of the above equations is similar to the dynamic linear model setup for the moving average process with known coefficients given in [13].

We note that given the latent sequence $\mathbf{S}_n$, the above setup satisfies all the assumptions of the normal dynamic linear models of West and Harrison [19]. More specifically, in the observation Equation (8.42), u_i represents the observation noise of the model and u_i's are independent normally distributed random variables with mean 0 and variance $s_i\sigma_w^2$, where σ_w^2 is the unknown jitter noise variance as before. In the system Equation (8.44), v_i represents the system noise vector of the DLM that is normally distributed with zero mean vector and the diagonal variance-covariance matrix W_i, where all the diagonal elements are zero except the $(p+1, p+1)$ element, which is the

of Equation (8.56). Generation from the second distribution, that is, from $p(\boldsymbol{\Theta}_1, \ldots \boldsymbol{\Theta}_n|\phi, \mathbf{S}_n, D_n)$ can be achieved by iteratively generating from the full conditional distributions $p(\boldsymbol{\Theta}_i|\phi, \boldsymbol{\Theta}^{-i}, \mathbf{S}_n, D_n)$ for $i = 1, \ldots, n$, where $\boldsymbol{\Theta}^{-i} = \{\boldsymbol{\Theta}_j | \, j \neq i, \, j = 1, 2, \ldots, n\}$. However, this is not very efficient and it is more desirable to generate directly from $p(\boldsymbol{\Theta}_1, \ldots \boldsymbol{\Theta}_n|\phi, \mathbf{S}_n, D_n)$ whenever possible. In what follows, we will present a scheme to directly generate from the joint posterior $p(\boldsymbol{\Theta}_1, \ldots \boldsymbol{\Theta}_n|\phi, \mathbf{S}_n, D_n)$. This is based on the forward filtering backward sampling algorithm of Fruhwirth-Schnatter [8] that is given in West and Harrison [19]. After we present the general algorithm, we will illustrate the adoption of this algorithm to the correlated noise model and discuss the implementation details. In our presentation, for notational convenience we will suppress dependence on ϕ and $\mathbf{S}_n$.

As pointed out in [19], using the Markov structure of the DLM setup, the joint distribution $p(\boldsymbol{\Theta}_1, \ldots \boldsymbol{\Theta}_n| \, D_n)$ can be written as

$$p(\boldsymbol{\Theta}_1, \ldots, \boldsymbol{\Theta}_n|D_n) = p(\boldsymbol{\Theta}_n|D_n) \, p(\boldsymbol{\Theta}_{n-1}|\boldsymbol{\Theta}_n, D_{n-1}) \cdots p(\boldsymbol{\Theta}_1|\boldsymbol{\Theta}_2, D_1), \quad (8.57)$$

where the first term $p(\boldsymbol{\Theta}_n|D_n)$ is available from the DLM updating Equation (8.53) for $i = n$. Thus, we can start the sampling from $\boldsymbol{\Theta}_n$ and then sequentially sample $\boldsymbol{\Theta}_{n-1}, \ldots, \boldsymbol{\Theta}_1$ using densities $p(\boldsymbol{\Theta}_{i-1}|\boldsymbol{\Theta}_i, D_{i-1})$ for $i = n-1, \ldots, 2$. The required distributions can be obtained using the DLM setup. Using Bayes rule, we can write

$$p(\boldsymbol{\Theta}_{i-1}|\boldsymbol{\Theta}_i, D_{i-1}) \propto p(\boldsymbol{\Theta}_i|\boldsymbol{\Theta}_{i-1}, D_{i-1}) \, p(\boldsymbol{\Theta}_{i-1}|D_{i-1}), \quad (8.58)$$

where

$$(\boldsymbol{\Theta}_i|\boldsymbol{\Theta}_{i-1}, D_{i-1}) \sim N(\mathbf{G}\boldsymbol{\Theta}_{i-1}, \mathbf{W}_i/\phi)$$

and

$$(\boldsymbol{\Theta}_{i-1}|D_{i-1}) \sim N(\boldsymbol{m}_{i-1}, \boldsymbol{\Sigma}_{i-1}/\phi).$$

It follows from the above that,

$$(\boldsymbol{\Theta}_{i-1}|\boldsymbol{\Theta}_i, D_{i-1}) \sim N(\boldsymbol{h}_{i-1}, \boldsymbol{H}_{i-1}/\phi), \quad (8.59)$$

where

$$\boldsymbol{h}_{i-1} = \boldsymbol{m}_{i-1} + \boldsymbol{\Sigma}_{i-1}\mathbf{G}'\mathbf{R}_i^{-1}(\boldsymbol{\Theta}_i - \mathbf{G}\boldsymbol{m}_{i-1}) \quad (8.60)$$

$$\boldsymbol{H}_{i-1} = \boldsymbol{\Sigma}_{i-1} - \boldsymbol{\Sigma}_{i-1}\mathbf{G}'\mathbf{R}_i^{-1}\mathbf{G}\boldsymbol{\Sigma}_{i-1}. \quad (8.61)$$

Note that because all the distributions are of normal form, generating from the joint posterior distribution Equation (8.57) is straightforward in a general DLM. However, implementation of the algorithm in special cases like we have requires consideration of certain technical issues. In our case, the state vectors $\boldsymbol{\Theta}_i$ and $\boldsymbol{\Theta}_{i-1}$ share common components as $\boldsymbol{\Theta}_i' = (\boldsymbol{\theta} \; v_i \; v_{i-1})$ and $\boldsymbol{\Theta}_{i-1}' = (\boldsymbol{\theta} \; v_{i-1} \; v_{i-2})$. As a result, the variance-covariance matrix $\boldsymbol{H}_{i-1}$ in

Equation (8.61) is not positive definite and the distribution $p(\Theta_{i-1}|\Theta_i, D_{i-1})$ cannot be sampled. As pointed out in West and Harrison [19], this issue arises in many problems where consecutive state vectors share common components and the algorithm must be modified to alleviate such difficulties.

In our case, going from Θ_i to Θ_{i-1} in $p(\Theta_{i-1}|\Theta_i, D_{i-1})$, all but one of the components of the state vector Θ_{i-1} will be known, and therefore we only need to draw from a univariate distribution to update Θ_{i-1}. Thus, we can generate from Equation (8.57) using the following algorithm:

(i) Sample Θ_n from $p(\Theta_n|D_n)$ using, Equation (8.53) with $i = n$,

(ii) for $i = n-1, n-2, \ldots, 2$, sample the value of v_{i-1} from $p(v_{i-1}|\boldsymbol{\theta}, v_{i+1}, v_i, D_i)$ and update the remaining elements of Θ_i using the common components from Θ_{i+1}.

In implementing the algorithm, the only distribution we need then is the univariate density $p(v_{i-1}|\boldsymbol{\theta}, v_{i+1}, v_i, D_i)$. Using Bayes rule and the fact that $D_i = (z_i, D_{i-1})$, we can write

$$p(v_{i-1}|\boldsymbol{\theta}, v_{i+1}, v_i, D_i) \propto p(z_i|\boldsymbol{\theta}, v_{i+1}, v_i, v_{i-1}, D_{i-1})\, p(v_{i-1}|\boldsymbol{\theta}, v_{i+1}, v_i, D_{i-1}).$$

In the above, we note that using the observation equation the first term reduces to

$$p(z_i|\boldsymbol{\theta}, v_{i+1}, v_i, v_{i-1}, D_{i-1}) = p(z_i|\Theta_i, v_{i+1}, D_{i-1}) = p(z_i|\Theta_i)$$

where $(z_i|\Theta_i) \sim N(\mathbb{F}_i\Theta_i, s_i\phi)$ and $\mathbb{F}_i\Theta_i = \boldsymbol{F}_i'\boldsymbol{\theta} + v_i - v_{i-1}$. To obtain the second term, we note that given D_{i-1}, v_{i-1} is independent of $v_i, v_{i+1}, \ldots, v_n$. Thus,

$$p(v_{i-1}|\boldsymbol{\theta}, v_{i+1}, v_i, D_{i-1}) = p(v_{i-1}|\boldsymbol{\theta}, D_{i-1})$$

and $p(v_{i-1}|\boldsymbol{\theta}, v_{i+1}, v_i, D_i)$ reduces to

$$p(v_{i-1}|\boldsymbol{\theta}, v_i, D_i) \propto p(z_i|\Theta_i)\, p(v_{i-1}|\boldsymbol{\theta}, D_{i-1}) \tag{8.62}$$

where the density $p(v_{i-1}|\boldsymbol{\theta}, D_{i-1})$ can be obtained as a normal density from the posterior distribution of $(\Theta_{i-1}|D_{i-1})$ using Normal distribution theory. More specifically, we have

$$(\Theta_{i-1}|D_{i-1}) \sim N(\boldsymbol{m}_{i-1}, \Sigma_{i-1}/\phi)$$

where $\Theta_{i-1}' = (\boldsymbol{\theta}\ v_{i-1}\ v_{i-2})$, implying that

$$(v_{i-1}, \boldsymbol{\theta}|D_{i-1}) \sim N\left[\begin{pmatrix} m_{v_{i-1}} \\ \boldsymbol{m_\theta} \end{pmatrix}, \begin{pmatrix} \Sigma_{v_{i-1}} & \Sigma_{\theta v_{i-1}} \\ \Sigma_{\theta v_{i-1}} & \Sigma_\theta \end{pmatrix}\right]$$

with dependence on ϕ suppressed.

We note that in general (see Chapter 1 and its appendix), if we have

$$\begin{pmatrix} \boldsymbol{X}_1 \\ \boldsymbol{X}_2 \end{pmatrix} \sim N\left[\begin{pmatrix} \boldsymbol{\mu}_1 \\ \boldsymbol{\mu}_2 \end{pmatrix}, \begin{pmatrix} \boldsymbol{V}_{11}\ \boldsymbol{V}_{12} \\ \boldsymbol{V}_{21}\ \boldsymbol{V}_{22} \end{pmatrix}\right]$$

then

$$(\mathbf{X}_1|\mathbf{X}_2) \sim N(\boldsymbol{\mu}_1(\mathbf{X}_2), \mathbf{V}_1(\mathbf{X}_2)),$$

where

$$\boldsymbol{\mu}_1(\mathbf{X}_2) = \boldsymbol{\mu}_1 + \mathbf{V}_{12}\mathbf{V}_{22}^{-1}(\mathbf{X}_2 - \boldsymbol{\mu}_2) \tag{8.63}$$

and

$$\mathbf{V}_1(\mathbf{X}_2) = \mathbf{V}_{11} - \mathbf{V}_{12}\mathbf{V}_{22}^{-1}\mathbf{V}_{21}. \tag{8.64}$$

Now letting $\mathbf{X}_1 = v_{i-1}$ and $\mathbf{X}_2 = \boldsymbol{\theta}$, from the above,

$$(v_{i-1}|\boldsymbol{\theta}, D_{i-1}) \sim N(\mu_{v_{i-1|\theta}}, V_{v_{i-1|\theta}}) \tag{8.65}$$

where $\mu_{v_{i-1|\theta}}$ and $V_{v_{i-1|\theta}}$ are obtained from Equation (8.63) and Equation (8.64) accordingly. Given Equation (8.65) and the fact that

$$(z_i|\boldsymbol{\Theta}_i) \sim N(\mathbb{F}_i\boldsymbol{\Theta}_i, s_i/\phi),$$

using Equation (8.62), we can show that

$$(v_{i-1}|\boldsymbol{\theta}, v_i, D_i) \sim N(\mu_{v_{i-1}}, V_{v_{i-1}}), \tag{8.66}$$

where

$$\mu_{v_{i-1}} = \left(\frac{s_i/\phi}{s_i/\phi + V_{v_{i-1|\theta}}}\right)\mu_{v_{i-1|\theta}} + \left(\frac{V_{v_{i-1|\theta}}}{s_i/\phi + V_{v_{i-1|\theta}}}\right)(\mathbf{F}_i'\boldsymbol{\theta} + v_i - z_i) \tag{8.67}$$

and

$$V_{v_{i-1}} = \frac{(s_i/\phi)\, V_{v_{i-1|\theta}}}{s_i/\phi + V_{v_{i-1|\theta}}}. \tag{8.68}$$

We note that in writing Equation (8.66), dependence on the latent variable s_i and the precision ϕ has been suppressed, that is, Equation (8.66) gives us $p(v_{i-1}|\phi, \mathbf{S}_n, \boldsymbol{\theta}, v_i, D_i)$, and given $\boldsymbol{\theta}$, we can write $p(v_{i-1}|\phi, \mathbf{S}_n, \boldsymbol{\theta}, v_i, D_i) = p(v_{i-1}|\phi, s_i, \boldsymbol{\theta}, v_i, D_i)$. To summarize, using the DLM setup in generating the full conditional distribution $p(\boldsymbol{\Theta}_1, \ldots, \boldsymbol{\Theta}_n|\phi, \mathbf{S}_n, D_n)$, we start with sampling $\boldsymbol{\Theta}_n$ from $p(\boldsymbol{\Theta}_n|\phi, \mathbf{S}_n, D_n)$, and then for $i = n-1, n-2, \ldots, 1$, we sample the value of v_{i-1} from Equation (8.66) and update the remaining elements of $\boldsymbol{\Theta}_i$ using the common components from $\boldsymbol{\Theta}_{i+1}$. At a given iteration of the Gibbs sampler, this gives us a sample for $\boldsymbol{\theta}, v_n, \ldots, v_0$ and ϕ is sampled directly from $p(\phi|\mathbf{S}_n, D_n)$ given by Equation (8.56).

As in Section 8.3.1, the full conditional distribution of $\boldsymbol{P}$ is given by the Dirichlet form of Equation (8.34). In obtaining the full conditional distribution of latent variables $\mathbf{S}_n$, we now consider $p(s_i|\boldsymbol{\theta}, \phi, \mathbf{S}^{-i}, \boldsymbol{P}, z, v)$ where $v = (v_0, v_1, \ldots, v_n)$ and $z = D_n$. Thus in Equation (8.28), we need to obtain

$p(z|s_i, \boldsymbol{\theta}, \phi, \mathbf{S}^{-i}, v)$, which can be written as

$$p(z|s_i, \boldsymbol{\theta}, \phi, \mathbf{S}^{-i}, v) = \prod_{j=1}^{n} p(z_j|s_i, \boldsymbol{\theta}, \phi, \mathbf{S}^{-i}, v_j, v_{j-1}) \tag{8.69}$$

due to the independence of z_i's given $\boldsymbol{\Theta}_i$ as implied by the observation Equation (8.46). Using a similar development to that in Section 8.3.1, we can show that the full conditional distribution of s_i is obtained as

$$p(s_i|\boldsymbol{\theta}, \phi, \mathbf{S}^{-i}, \mathbf{P}, z, v) \propto p(s_i|s_{i-1}, \mathbf{P})\, p(s_{i+1}|s_i, \mathbf{P}) \prod_{j=i}^{n} p(z_j|\boldsymbol{\Theta}_j, \phi, \mathbf{S}_j) \tag{8.70}$$

where $\boldsymbol{\Theta}'_j = (\boldsymbol{\theta}\ v_j\ v_{j-1})$.

8.5 Model Comparison

In Bayesian paradigm, a formal model comparison can be made by also describing uncertainty about models probabilistically, that is, specify prior probabilities describing our uncertainty about the candidate models and update these to posterior model probabilities after data is observed. For example, assume that we are considering two alternative models, say 1 and 2. Prior to observing the data, we can describe our uncertainty about these models via probabilities p_1 and p_2, then after observing data z, the posterior model probabilities are given by

$$p(i|z) \propto p(z|i)\, p_i$$

for $i = 1, 2$, where $p(z|i)$ denotes the marginal likelihood under model i. Note that if we take the ratio of the posterior model probabilities, we obtain the posterior odds in favor of model 1 written as

$$\frac{p(1|z)}{p(2|z)} = \frac{p(z|1)}{p(z|2)} \times \frac{p_1}{p_2},$$

where the first ratio, that is,

$$\frac{p(z|1)}{p(z|2)} \tag{8.71}$$

is known as the Bayes factor. In Bayesian paradigm, model comparison is typically based on Bayes factors that are obtained as the ratio of marginal likelihoods under two competing models; see Kass and Raftery [15] for a comprehensive review.

In many problems, $p(z|i)$ is not available in an analytical form and its evaluation using posterior Monte Carlo samples is not a trivial task. Thus, various alternatives to marginal likelihoods have been suggested in the literature for model selection using Monte Carlo samples; see Gelfand [10] for a recent review.

However, in certain problems where a Gibbs sampler is used and all the full conditional distributions are known, it is possible to approximate the marginal likelihoods from the posterior samples using a method introduced by Chib [3]. As discussed in Section 8.3.1 and Section 8.4 in Bayesian analyses of the stagger model with hidden missingness, all the full conditionals are known. Thus, in what follows, the approach proposed by Chib will be adopted to compare models with different periods.

8.5.1 Marginal Likelihood Computation for the Model

The approach presented by Chib [3] is based on two ideas. First, the marginal likelihood for a particular model is expressed as

$$p(z) = \frac{p(z|\Theta)\,p(\Theta)}{p(\Theta|z)}, \tag{8.72}$$

where Θ is a vector of parameters. Secondly, as pointed out by Chib, the above holds for any value of Θ, say Θ^*, and the value of posterior density $p(\Theta^*|z)$ can be estimated by $\hat{p}(\Theta^*|z)$ using Monte Carlo samples. Because $p(z|\Theta^*)$ and $p(\Theta^*)$ can be evaluated at Θ^*, the log marginal likelihood can be estimated as

$$ln\ \hat{p}(z) = ln\ p(z|\Theta^*)\ + ln\ p(\Theta^*) - \hat{p}(\Theta^*|z). \tag{8.73}$$

In Equation (8.73), the only term that is not readily available is $\hat{p}(\Theta^*|z)$, but as shown in [3], this can be obtained using the outputs from the Gibbs sampler. In our case, Θ consists of $(\boldsymbol{\theta}, \phi, S_n, \boldsymbol{P})$ and using independence assumptions, we can write the marginal likelihood as

$$p(z) = \frac{p(z|\boldsymbol{\theta}, \phi, S_n)\ p(\boldsymbol{\theta}|\phi)\,p(\phi)\,p(S_n|\boldsymbol{P})\ p(\boldsymbol{P})}{p(\boldsymbol{\theta}, \phi, S_n, \boldsymbol{P}|z)}. \tag{8.74}$$

All the terms in the numerator of Equation (8.74) can be evaluated at $(\boldsymbol{\theta}, \phi, S_n, \boldsymbol{P}) = (\boldsymbol{\theta}^*, \phi^*, S_n^*, \boldsymbol{P}^*)$. Thus to approximate $p(z)$, we need to obtain $p(\boldsymbol{\theta}^*, \phi^*, S_n^*, \boldsymbol{P}^*|z)$. Using the multiplication rule and the conditional independence of $(\boldsymbol{\theta}, \phi)$ with transition matrix $\boldsymbol{P}$ given S_n, $p(\boldsymbol{\theta}^*, \phi^*, S_n^*, \boldsymbol{P}^*|z)$ is given by

$$p(\boldsymbol{\theta}^*, \phi^*, S_n^*, \boldsymbol{P}^*|z) = p(S_n^*|z)\ p(\boldsymbol{P}^*|S_n^*)\,p(\phi^*|\,S_n^*, z)\ p(\boldsymbol{\theta}^*|\phi^*, S_n^*, z), \tag{8.75}$$

where the term $p(\boldsymbol{P}^*|S_n^*)$ is the product of independent Dirichlet densities of Equation (8.34) and $p(\boldsymbol{\theta}^*|\phi^*, S_n^*, z)$ is the full conditional of $\boldsymbol{\theta}$, which is normal with mean and variance given by Equation (8.36) and Equation (8.37). The third term on the right-hand side of Equation (8.75), $p(\phi^*|\,S_n^*, z)$, can be obtained as

$$p(\phi^*|\, S_n^*, z) = \int p(\phi^*|\, \boldsymbol{\theta}, S_n^*, z)\; p(\boldsymbol{\theta}|S_n^*, z)\, d\boldsymbol{\theta}. \tag{8.76}$$

Note that the posterior samples obtained using the Gibbs sampler are from the posterior density $p(\boldsymbol{\theta}|z)$ and not from $p(\boldsymbol{\theta}|S_n^*, z)$. However, as suggested in [3], if we continue to sample for additional G' iterations using conditional densities

$$p(\phi|\boldsymbol{\theta}, S_n^*, z) \quad \text{and} \quad p(\boldsymbol{\theta}|\phi, S_n^*, z),$$

then we can obtain a Monte Carlo estimate as

$$p(\phi^*|S_n^*, z) \approx \frac{1}{G'} \sum_{g=1}^{G'} p(\phi^*|\boldsymbol{\theta}^{(g)}, S_n^*, z) \tag{8.77}$$

where $\boldsymbol{\theta}^{(g)}$ are samples from $p(\boldsymbol{\theta}|\phi, S_n^*, z)$. In our particular case, this step can be avoided by updating $\boldsymbol{\theta}$ and ϕ as a block to obtain $p(\boldsymbol{\theta}, \phi|S_n, z)$, since $(\boldsymbol{\theta}, \phi|S_n, z)$ follows a normal-gamma density. We note that this is applicable by using the reparameterization of the measurement error variance as discussed in Section 8.3.1 and Section 8.4. Otherwise, if a Metropolis step is needed within the Gibbs, then the method of Chib and Jeliazkov [5] can be used. In our development, we assume that the reparameterization is used and the full conditionals are available. Thus, the only term we need to evaluate is $p(S_n^*|z)$. Using the multiplication rule, we can write

$$p(S_n^*|z) = p(s_1^*|z)\, p(s_2^*|s_1^*, z) \cdots p(s_i^*|S_{i-1}^*, z) \cdots p(s_n^*|S_{n-1}^*, z), \tag{8.78}$$

where the first term $p(s_1^*|z)$ can be estimated from the draws available from the Gibbs sampler as

$$p(s_1^*|z) \approx \frac{1}{G} \sum_{g=1}^{G} p(s_1^*|\boldsymbol{\theta}^{(g)}, \phi^{(g)}, (S^{-i})^{(g)}, \mathbf{P}^{(g)}, z). \tag{8.79}$$

Evaluation of the remaining densities requires additional sampling. For a general term $p(s_i^*|S_{i-1}^*, z)$ that is given by

$$p(s_i^*|S_{i-1}^*, z) = \int p(s_i^*|\boldsymbol{\theta}, \phi, S_{l>i}, \mathbf{P}, S_{i-1}^*, z) dp(\boldsymbol{\theta}, \phi, S_{l>i}, \mathbf{P}|S_{i-1}^*, z), \tag{8.80}$$

where $S_{l>i} = \{s_l\,;\, l > i\}$, we need to continue sampling from full conditionals of $(\boldsymbol{\theta}, \phi, s_i, S_{l>i}, \mathbf{P})$ given (S_{i-1}^*, z). In other words, additional sampling will use the full conditional distributions

$$p(\boldsymbol{\theta},\, \phi|s_i, S_{l>i}, S_{i-1}^*, z),\; p(\mathbf{P}|s_i, S_{l>i}, S_{i-1}^*),\; p(s_i|\boldsymbol{\theta}, \phi, S_{l>i}, \mathbf{P}, S_{i-1}^*, z),$$

and

$$p(s_j|\boldsymbol{\theta}, \phi, s_i, S_{l>i}^{-j}, \mathbf{P}, S_{i-1}^*, z), \quad j = i+1, \ldots, n.$$

TABLE 8.1

Posterior Distributions of Selected s_i's Under Stagger Model with Period 3

i	Actual s_i	$p(s_i = 1 \mid z)$	$p(s_i = 2 \mid z)$	$p(s_i = 3 \mid z)$
4	1	0.9980	0.0020	0.0000
12	2	0.0322	0.9637	0.0041
18	2	0.1221	0.8778	0.0001
45	3	0.0002	0.0937	0.9061
54	2	0.1168	0.8828	0.0004
65	2	0.1281	0.8718	0.0001
71	2	0.0063	0.9806	0.0131
80	2	0.0391	0.9574	0.0035

We note that the performance of the approach is dependent on its ability to infer the unknown locations of the missing pulses. In Table 8.1, for the stagger model with period 3 and $\sigma_w = 25$, we illustrate the posterior distributions of the latent variables s_i's associated with missing pulse locations $i = 12, 18, 45, 54, 65, 71,$ and 80. The posterior distribution of s_4 associated with a no missing pulse location is also shown for comparison purposes. Each row in Table 8.1 represents the posterior distribution associated with the particular pulse location. We note that for s_4, which is the state variable for location 4 with no missing pulse, the posterior probability of no missing pulse ($s_4 = 1$) is 0.998. For locations $i = 12$, 71, and 80, where there is one missing pulse in each case, the corresponding posterior probabilities $p(s_i = 2 \mid z)$ are all higher than 0.95. Note that these are the PRI values that are large in magnitude as can be seen from Figure 8.1. For locations $i = 18, 54,$ and 65, with one missing pulse the probabilities $p(s_i = 2 \mid z)$ are not as large as in the previous group but they are still larger than 0.87. Note that these are the locations with PRI values that are not large in magnitude but the procedure is still able to infer the missing pulse with high probability. Finally, location 45 has two missing pulses and the corresponding posterior probability $p(s_{45} = 3 \mid z) = 0.9061$. In summary, the inference procedure seems to be identifying the location of missing pulses.

Using the above priors, the Gibbs sampler was run for each candidate model following our development in Section 8.3. From the Gibbs sampler output, for each candidate model the marginal likelihoods were computed using the procedure presented in Section 8.5. The analyses were repeated for each model using simulated data with $\sigma_w = 1, 5, 10, 15, 20, 25,$ and 30. In each simulation, the data were generated from the model with period 3 and there were eight missing pulses. In Figure 8.3, we present the log marginal likelihoods of the four models for the different values of the σ_w (as shown on the x-axis). The figure illustrates that our approach identifies the stagger model with period 3 as the correct model in all cases except the last one, where $\sigma_w = 30$. As expected, performance deteriorates as σ_w gets large as shown by the decreasing value of the log marginal likelihood for stagger model with

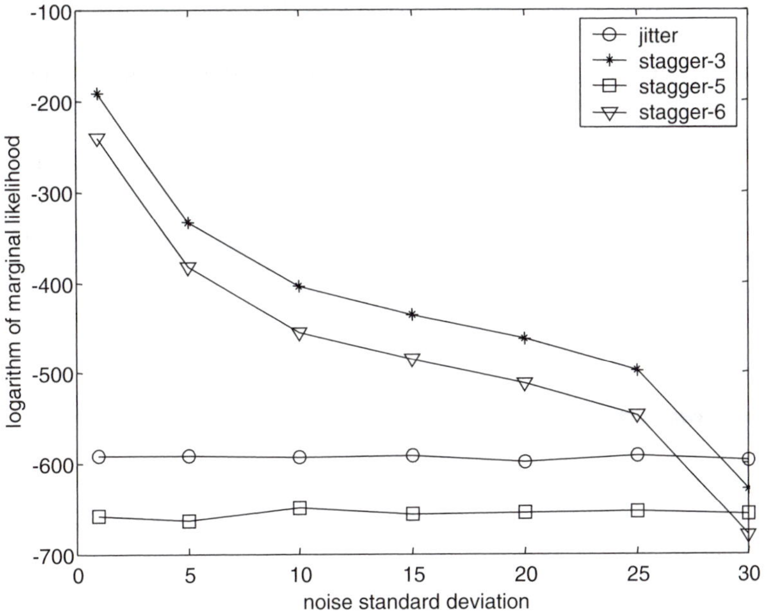

FIGURE 8.3
Plots of the log marginal likelihoods versus σ_w.

period 3. Similarly, the posterior distribution of σ_w under the correct model was more concentrated around the actual values of σ_w compared to the other models.

8.7 Conclusion

In this chapter, a Bayesian approach was developed for analysis of pulse trains corrupted by missing pulses at unknown locations. The development was motivated by electronic warfare applications where it is of interest to infer the locations of missing pulses and to identify the PRI modulation type.

The presented Bayesian approach is based on Markov Chain Monte Carlo methods and enables us to make posterior and predictive inferences about model parameters, location of missing pulses, and future PRIs. It also allows us to evaluate the marginal likelihoods of models with different periods and thus provides us with a Bayesian model comparison tool. An alternative setup of the model using the Bayesian DLM framework allows for sequential processing of data and avoids matrix inversions during the implementation of the Gibbs sampler.

9

Bayesian Approaches to Process Monitoring and Process Adjustment

Rong Pan
Department of Industrial Engineering, Arizona State University

CONTENTS

ABSTRACT In this chapter, some Bayesian algorithms for detecting a persistent process mean shift and for adjusting the process back to target are presented. We discuss the connection between the Bayesian algorithm and the cumulative sum (CUSUM) algorithm, a popular tool for detecting small process mean shifts. The process adjustment method is based on a Kalman filter technique, which provides a sequential adjustment strategy along with process measurements. The integration of this sequential adjustment method with different control charts is evaluated through simulations.

9.1 Introduction

Quality control of a manufacturing process consists of two distinct functions: detecting any abnormal process change and adjusting process after the change has been identified. Using control charts to monitor manufacturing process outputs is considered to be the specific function of statistical process control (SPC). In SPC, the root cause of process change should be isolated and removed after the change is detected. This typically requires production stoppage and thorough analysis and diagnosis of the manufacturing process. If some controllable process variables exist and the influence of these controllable variables on process outputs has been quantified, we can adjust the process by varying these controllable variables, which is known as process adjustment and is particularly desirable on automated processes.

However, SPC techniques do not provide an explicit process adjustment method. Process adjustment is usually regarded as function pertaining to engineering process control (EPC), an area that has traditionally belonged to process engineers rather than to quality engineers. Integrating EPC and SPC techniques for process quality control has been discussed by several authors in the recent literature [4, 24, 26, 29, 40, 41]. This chapter will derive both process monitoring and process adjustment methods from Bayes theory and compare them with conventional methods.

We assume a univariate process that consists of a measurable quality characteristic y and a single controllable factor x. The process mean is defined as the expectation of y. The process is initially in a stationary and uncorrelated state, but a random disturbance can shift the process mean off-target. A control chart is in use to monitor this process and the chart is applied to individual samples because the production lot size assumed is small. Whenever the control chart signals an "out-of-control" alarm, we suspect that a mean shift has occurred and proceed with adjusting the process. This chapter focuses on studying methods for sequential process adjustment based on the Kalman filter (KF) technique. We define a sequential adjustment strategy as a finite number of adjustments implemented after a shift disturbance on a process is detected.

In the following sections, some commonly used control charts are reviewed and a Bayesian algorithm for detecting process change is derived. In particular, we discuss the connection between the Bayesian algorithm and the cumulative sum (CUSUM) algorithm. A sequential process adjustment method based on the KF technique is then proposed. Finally, the performance of various combinations of this adjustment method and different control charts is evaluated for a mean shift type of process change as the one represented in Figure 9.1.

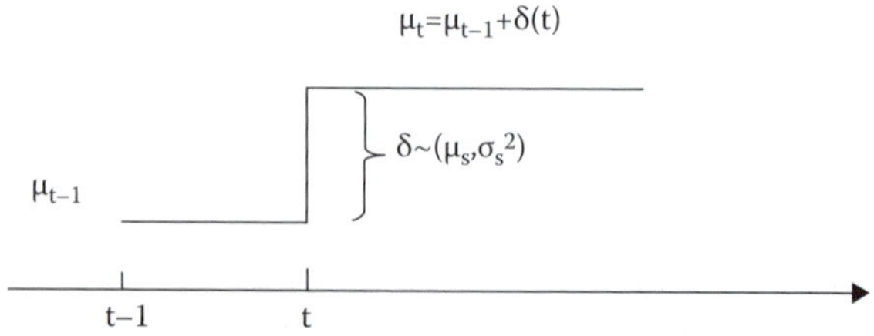

FIGURE 9.1
Step-type disturbance on the process mean.

9.2 Bayesian Algorithm for Detecting Process Change

Assume that a process quality characteristic y is under surveillance of a process monitoring scheme. Process measurements, $y_1, y_2, \ldots,$ are originally generated from a Normal distribution $N(\mu_0, \sigma^2)$; however, a process disturbance will shift the process mean to μ_1 at a random time. For any process monitoring scheme, minimizing the number of false alarms that occur before the appearance of the process disorder is desirable. Alternately, we must guarantee a small delay in the time taken to detect a disorder.

Shewhart control charts with $\pm 3\sigma$ control limits are the simplest type of process monitoring scheme, but it is well known that the Shewhart chart is insensitive to small or moderate persistent mean shifts [27]. In order to detect small shifts more quickly, CUSUM and EWMA (exponentially weighted moving average) charts are usually recommended. In particular, a CUSUM chart can be shown to be the generalized sequential probability ratio test (SPRT) for the hypothesis $H_0 : \mu = \mu_0$ versus $H_1 : \mu = \mu_1$, where μ_1 is a predetermined out-of-control mean [23]. Considering the single-sided monitoring scheme for $\mu_1 > \mu_0$, the test statistics of the CUSUM algorithm are

$$c_t^+ = max\{0, y_t - k + c_{t-1}^+\}, \tag{9.1}$$

where t is considered here as discrete time and $k = (\mu_1 + \mu_0)/2$ is the middle between the original process mean value and the shifted process mean value that one wishes to detect [44]. The control limit of the CUSUM statistics is defined as $H = h\sigma$, where h is another design parameter. Whenever c^+ exceeds H, an out-of-control alarm is signaled.

EWMA charts use the EWMA smoothing method to predict the process mean. This utilizes not only the current measurement but the discounted historical data as well. The EWMA statistic is defined as

$$z_t = \lambda y_t + (1 - \lambda) z_{t-1}, \quad 0 < \lambda < 1, \tag{9.2}$$

where λ is the smoothing parameter or weight, and the starting value z_0 is the process target. The EWMA chart control limits are $\pm L\,\sigma\sqrt{\frac{\lambda}{(2-\lambda)}[1 - (1 - \lambda)^{2t}]}$.

Bayesian approach to a quality control model was first attempted in [13]. Shiryaev [37] and Roberts [34] independently proposed the same Bayesian algorithm for detecting process mean shift. This algorithm also appears in Basseville and Nikiforov [2]. Here we describe this algorithm in detail.

At a discrete time t, suppose the manufacturing process could be in one of two states, 0 or 1, where state 0 is a normal state modeled by a distribution f_0 and state 1 is the abnormal state modeled by a distribution f_1. An intrinsic transition probability of process change exists, ξ, such that the normal process can change to abnormal at the next time step with probability ξ and then stay abnormal until it is detected. In other words, the random disturbance that can change the process state from 0 to 1 strikes on the process with a rate of occurrence as ξ. Therefore, the current state of the process depends on its immediate previous state, i.e., it is a Markov chain. We must estimate the posterior probability of process in state 1, given all process observations.

Let π_t be the probability of the process in state 1 at time t; then π_0 is the prior probability of process change when no process measurement is available. By Bayes' law, the posterior probability of the process change at time 1, π_1, is updated by

$$\begin{aligned}\pi_1 &= P(S_1 = 1|\pi_0, \xi, y_1)\\ &= \frac{P(S_1 = 1, y_1|\pi_0, \xi)}{P(y_1|\pi_0, \xi)}\\ &= \frac{\pi_0 f_1(y_1) + (1-\pi_0)\xi f_1(y_1)}{\pi_0 f_1(y_1) + (1-\pi_0)\xi f_1(y_1) + (1-\pi_0)(1-\xi) f_0(y_1)}.\end{aligned} \tag{9.3}$$

Similarly, π_t is given by

$$\pi_t = \frac{\pi_{t-1} f_1(y_t) + (1-\pi_{t-1})\xi f_1(y_t)}{\pi_{t-1} f_1(y_t) + (1-\pi_{t-1})\xi f_1(y_t) + (1-\pi_{t-1})(1-\xi) f_0(y_t)}. \tag{9.4}$$

If π_t exceeds a certain threshold, then a signal of process change will be issued.

In dealing with this algorithm is convenient using a log-likelihood ratio form. Let $g_t = ln\frac{\pi_t}{1-\pi_t}$, then

$$g_t = ln(\xi + e^{g_{t-1}}) - ln(1-\xi) + ln\frac{f_1(y_t)}{f_0(y_t)}. \tag{9.5}$$

A signal will be issued at time t when $g_t \geq s$, where s is the threshold value.

To calculate g_t, the state transition probability ξ, as well as f_1, the probability density function of the abnormal process, must be known. These assumptions are not realistic for most manufacturing processes. This seems to be the reason that the Bayesian algorithm has not been widely applied in manufacturing quality control practice.

A popular algorithm used in quality control is CUSUM algorithm proposed by Page [28]. Equation (9.1) is equivalent to a sequential probability ratio test, which is

$$g_t = max\left\{g_{t-1} + ln\frac{f_1(y_t)}{f_0(y_t)}, 0\right\} \tag{9.6}$$

and $g_0 = 0$. The alarm signal is issued when $g_t \geq s$.

Notice that in the CUSUM algorithm, the decision function g_t can be reset to 0 and the measurement data used in detecting process change are within a moving time window. In other words, the action involved in the CUSUM algorithm can be one of the following three decisions:

$g_t < 0$: The process is deemed to be normal and the monitoring scheme is reinitiated by setting $g_t = 0$.

$0 \leq g_t < s$: The process cannot be judged and the current monitoring scheme is continued.

$g_t \geq s$: The process is deemed to be abnormal and signal an out of control alarm.

This is not the feature of the Bayesian algorithm. However, when $\lambda = 0$, the Bayesian decision function will degenerate to a similar form as the CUSUM algorithm except with multiple decisions. In the next section, the connection between the CUSUM algorithm and Bayesian decision theory will be discussed.

Suppose a normal (in-control) process can be modeled by a normal distribution $N(\mu_0, \sigma^2)$ and σ^2 is known. A random disturbance will shift the process mean from μ_0 to μ_1. Then we can easily derive the probability ratio

$$\frac{f_1(y_t)}{f_0(y_t)} = e^{-\frac{(y_t-\mu_1)^2-(y_t-\mu_0)^2}{2\sigma^2}} \tag{9.7}$$

and the log ratio

$$ln\frac{f_1(y_t)}{f_0(y_t)} = \frac{b}{\sigma}(y_t - k), \tag{9.8}$$

where $b = (\mu_1 - \mu_0)/\sigma$ and $k = (\mu_1 + \mu_0)/2$.

Assume at the beginning of process monitoring, no prior information of the state of the process is known, i.e., $\pi_0 = 0.5$; then $g_0 = 0$. If the chance of process disturbance is so small that $\xi \approx 0$, then

$$g_t = g_{t-1} + \frac{b}{\sigma}(y_t - k), \tag{9.9}$$

and clearly

$$g_t = \frac{b}{\sigma}\sum_{i=1}^{t}(y_i - k). \tag{9.10}$$

Given one decision threshold s, an alarm of process mean change will be signaled when $g_t \geq s$, or

$$\sum_{i=1}^{t}(y_i - k) > H, \tag{9.11}$$

where $H = h\sigma$ and $h = s/b$.

By the CUSUM algorithm, two decision thresholds exist — 0 and s. When $g_t < 0$, the current process is deemed to be in-control and the monitoring

decreases. It is reasonable to assume that the distance between a_t and s_t will decrease in proportion to the inverse of the precision of posterior estimation. A convenient choice is to let $s_t = k$, $a_0 = m_0$, and $(k - a_t) = \frac{n_0}{n_t}(k - a_0)$; then by combining Equation (9.13) and Equation (9.14), we have

$$
\begin{aligned}
\sum_{i=1}^{t}(y_i - k) &< 0, && \textit{reinitiate;} \\
0 \le \sum_{i=1}^{t}(y_i - k) &< n_0(k - m_0), && \textit{continue;} \\
\sum_{i=1}^{t}(y_i - k) &\ge n_0(k - m_0), && \textit{signal.}
\end{aligned} \tag{9.15}
$$

This is the same as the CUSUM algorithm, where the threshold for the out-of-control alarm is set at $n_0(k - m_0)$. Let $m_0 = \mu_0$ and comparing Equation (9.15) with Equation (9.12), it is easy to find that s in Equation (9.12) is equivalent to $(\mu_1 - \mu_0)^2/(2\tau^2)$, i.e., half of the ratio of the square of process mean shift size to the variance in prior distribution. Therefore, the value of the signal threshold in the CUSUM algorithm reflects one's prior belief of the in-control process distribution model.

The CUSUM algorithm can be formulated as a sequential probability ratio test (SPRT) for the exponential distribution family. However, a major weakness of SPRT is its failure to signal outliers and huge jumps in quality for nonexponential family models such as the t distribution and distributions with inverse polynomial tails. Harrison and Lai [17] discussed this problem in detail. They called Equation (9.15) the "popular decision scheme" (PDS) and demonstrated that it can be applied on more complex process models. Recent development of this scheme for detecting other types of process change can be found in [12, 36].

9.4 Kalman Filter Approach to Process Mean Adjustment

After detecting a shift in process mean, adjusting the process to bring back the mean level to an acceptable value is necessary. This normally requires a root cause diagnosis for the process change and the estimation of a current process mean. Many change point identification and estimation algorithms can be employed for this purpose in the retrospective study of SPC. Chen and Elsayed [6] applied a Bayesian method on identification and estimation of process mean change, and Kelton et al. [18] use the aggregated mean for estimation. However, they are one-time adjustment strategies, that is, no adjustment will be made until the current process mean value can be precisely estimated. Another adjustment strategy is on line adaptive adjustment, which is based on the feedback principle that has become an important resource in the toolkit used by quality engineers [3, 5, 35]. This section presents a formulation that unifies several well-known process adjustment schemes and shows how several extensions can be obtained using standard methods from control engineering, initially discussed in del Castillo et al. [9, 10].

To motivate the type of adjustment problems considered in this chapter and to introduce some necessary notation, consider the setup adjustment problem first studied by Grubbs [14]. Suppose, without loss of generality, that process measurements y_t correspond to the deviations from target of some quality characteristic of the items as they are produced at discrete points in time $t = 1, 2, \ldots$, i.e., $y_t \sim N(0, \sigma^2)$ if the process was properly set up. Here σ^2 is the variance of random process error, or measurement error. In some manufacturing processes such as machining, an incorrect setup operation can result in drastic consequences in the quality of the parts produced thereafter. Grubbs proposed a method for the adjustment of the machine to bring the process back to target if at start-up it was off-target by d units.

This is analogous to Deming's "funnel experiment" when the funnel is initially off-target. If the adjustment cost is negligible, a significant cost is associated with running the process off-target. That the process should be adjusted back to the target is evident. Figure 9.2 illustrates this experiment at process on-target, off-target, and adjusted scenarios.

Consider a manufacturing process has a setup error d, which is a random variable with known mean $\bar{d}$ and known variance P_0. Assume a controllable process variable x will directly impact on process measurement y. So the full process model can be formulated as

$$y_t = d + x_{t-1} + \varepsilon_t, \tag{9.16}$$

where $d \sim D(\bar{d}, P_0)$, D is any distribution with first and second moments finite, and $\varepsilon_t \sim N(0, \sigma^2)$.

Let us first assume d is a unknown constant, then for the first manufactured part, the expected quality characteristics is

$$\mu_1 = d + x_0, \tag{9.17}$$

where x_0 is the initial setting of the controllable variable. After the first process measurement y_1, an adjustment $\nabla x_1 = x_1 - x_0$ is made on the process, which results in a new process mean of

$$\mu_2 = \mu_1 + \nabla x_1 = d + x_1. \tag{9.18}$$

Because in fact d is a random variable, the objective of process adjustment is to find the process adjustments $\nabla x_1, \nabla x_2, \ldots$ that minimize

$$E\left[\sum_{t=1}^{n} \mu_t^2\right]. \tag{9.19}$$

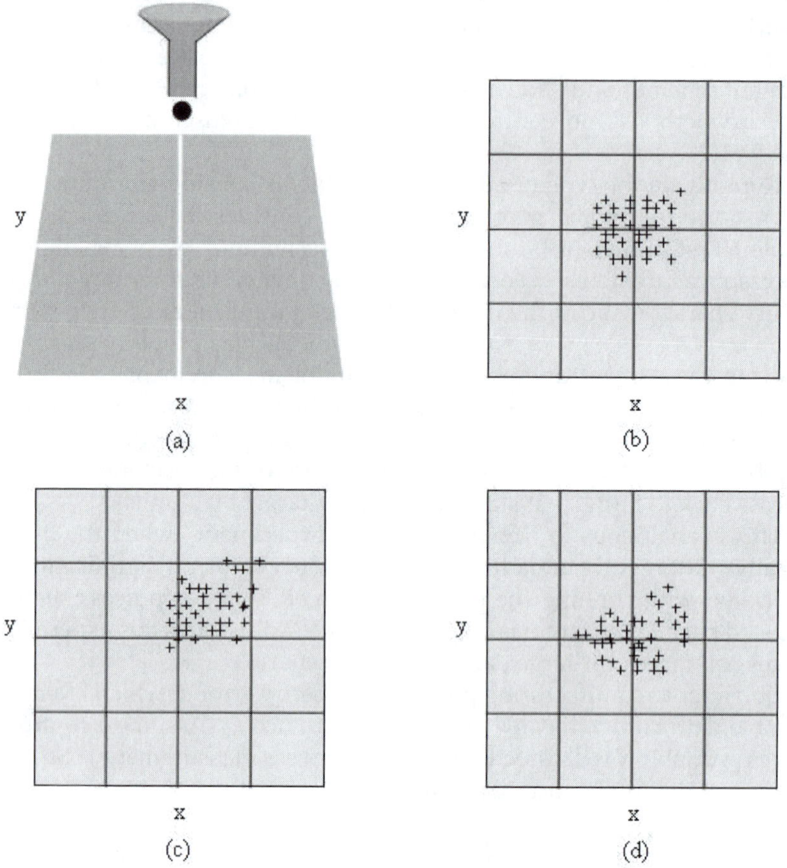

FIGURE 9.2
Deming's funnel experiments with a process setup error and process adjustments. (a) Funnel apparatus; (b) Simulated locations of dropping balls when no setup error exists; (c) Simulated locations of dropping balls when a one-unit setup error on both x- and y-axes exists; (d) Simulated locations of dropping balls when five sequential adjustments based on Grubbs' harmonic rule are taken on the off-target process.

In other words, quadratic off-target costs are assumed but no cost is incurred when performing the adjustments.

Optimization of this type of criterion is based on the *separation principle* [1]. For the setup adjustment problem under consideration, this principle indicates that the optimal solution can be found by solving separately the problem of estimating the μ_t's (process means) from the problem of finding the best adjustments $\{\nabla x_t\}$. If the optimal adjustment equation obtained through this separation is identical to what would have been obtained if the process were deterministic, the controller is said to be a certainty equivalence controller. This essentially means that the parameter estimates are used in the control equation as if they were the true parameters. In our case, if the means μ_t were known, evidently the best adjustment would be simply to set $\nabla x_t = -\mu_t$, which yields a minimum variance process. In what follows, we will derive the posterior distribution of μ_t and then the process adjustment formula by using the Kalman filter technique.

The estimation problem of μ_t is solved in a Bayesian framework using a simple Kalman filter [25]. Given the model of Equation (9.16), define

$$y^t|\mu_t \sim N(\mu_t, \sigma^2).$$

Let m_{t-1} and P_{t-1} be the posterior mean and variance of the parameter μ_{t-1}, respectively. Then after the adjustment ∇x_{t-1}, the prior distribution of μ_t becomes

$$\mu_t|y^{t-1} \sim N(m_{t-1} + \nabla x_{t-1}, P_{t-1}),$$

where $y^t = \{y_1, y_2, \ldots, y_t\}$ are all available data at time t. The first mean has a prior distribution $\mu_1 \sim N(\overline{d} + x_0, P_0)$ where $\overline{d} = E[d]$, $P_0 = Var(d)$, and x_0 is the initial setpoint of the machine. Note that if $\overline{d}$ is known, then we should set $x_0 = -\overline{d}$ and get $\mu_1 \sim N(0, P_0)$.

Given this setup, we have that the posterior distribution of μ_t given the observations is

$$\mu_t|y^t \sim N(m_t, P_t)$$

with

$$m_t = m_{t-1} + \nabla x_{t-1} + \frac{P_{t-1}}{P_{t-1} + \sigma^2}[y_t - (m_{t-1} + \nabla x_{t-1})] \tag{9.20}$$

and

$$P_t = \frac{P_{t-1}\sigma^2}{\sigma^2 + P_{t-1}}, \tag{9.21}$$

which is a recurrence equation easily solved by iteration, yielding

$$P_t = \frac{\sigma^2 P_0}{\sigma^2 + t P_0}, \tag{9.22}$$

where P_0 is known. The Kalman filter estimate of the process mean given the data is

$$m_t = E[\mu_t|y^t] = m_{t-1} + \nabla x_{t-1} + K_t(y_t - (m_{t-1} + \nabla x_{t-1})), \tag{9.23}$$

where the quantities

$$K_t = \frac{P_{t-1}}{P_{t-1} + \sigma^2} = \frac{1}{t + \frac{\sigma^2}{P_0}} \tag{9.24}$$

are the "Kalman weights." Under the stated assumptions of normality, m_t is the minimum mean square error (MMSE) estimator of μ_t. As shown by

Duncan and Horn [11], if the normality assumptions are relaxed, m_t is the MMSE *linear* estimator, i.e., among all estimators that are linear combinations of the observations, it has smallest MSE, but better nonlinear estimators might be found.

To minimize $E[\sum_{t=1}^{n} \mu_t^2]$, one can argue as follows. Conditioning on all available data at time $t-1$, we have that

$$E\left[\mu_t^2|y^{t-1}\right] = \mathrm{Var}(\mu_t|y^{t-1}) + [E(\mu_t|y^{t-1})]^2$$
$$= \frac{\sigma^2 P_0}{\sigma^2 + (t-1)P_0} + (m_{t-1} + \nabla x_{t-1})^2, \tag{9.25}$$

which is minimized by taking $\nabla x_{t-1} = -m_{t-1}$. From our earlier discussion, this is a certainty equivalence controller. Applying this adjustment rule at every point in time also minimizes the sum of the squared mean reported in Equation (9.19). The minimum of the expected sum of the squared mean is

$$E\left[\sum_{t=1}^{n} \mu_t^2\right] = \sum_{t=1}^{n} \frac{\sigma^2 P_0}{\sigma^2 + P_0(t-1)}. \tag{9.26}$$

Substituting the control rule into the process mean estimate, we get

$$m_t = \frac{1}{t + \frac{\sigma^2}{P_0}} y_t = K_t y_t, \tag{9.27}$$

and the adjustment rule is

$$\nabla x_t = \frac{-y_t}{t + \frac{\sigma^2}{P_0}}, \tag{9.28}$$

an expression identical to Grubbs' extended rule [14].

We conclude that Grubbs' extended rule minimizes the expected sum of squared deviations provided the setup error mean and variance are known. If the errors are all normally distributed, Grubbs' extended rule is the optimal solution for the criterion of Equation (9.19). If the errors are not normal, Grubbs' extended rule is the best linear control law that minimizes Equation (9.19) [1, 19]. These additional facts can also be proved using LQ or LQG (linear control gaussian) theory.

Besides the aforementioned equivalence with Grubbs' extended rule, the Kalman filter of Equation (9.27) together with Equation (9.28) provide three other particular cases of interest:

Grubbs harmonic rule and Robbins and Monro stochastic approximation. If $P_0 \to \infty$, which implies lack of any a priori information on the offset d, Grubbs' "harmonic rule" is obtained because, under these conditions, the Kalman weights are

$$K_t \to \frac{1}{t}.$$

The mean estimates become

$$m_t = \frac{1}{t} y_t$$

and

$$x_t = x_{t-1} - \frac{1}{t} y_t, \tag{9.29}$$

which is exactly Grubbs' harmonic rule. Grubbs obtained Equation (9.29) by solving a constrained optimization problem — i.e., min $Var(Y_{t+1})$, subject to $E[Y_{t+1}] = 0$ — under the assumption the setup error d was an unknown constant (a machine offset). Del Castillo et al. [9] also shows that Grubbs' harmonic rule is a special case of Robbins and Monro's stochastic approximation algorithm for the sequential estimation of the offset d [33].

Recursive least squares. If σ^2 is set equal to one, the recursive least squares estimate of d is obtained [22, 45]:

$$\widehat{d}_t = \widehat{d}_{t-1} + \frac{1}{\frac{1}{P_0} + t} y_t.$$

In this case, we are again assuming the first case studied by Grubbs, that of an unknown constant setup error d.

Unreliable measurements. If $\sigma^2 \to \infty$, i.e., if the measurements are completely unreliable, this implies that $K_t \to 0$ and $\nabla x_t = 0$ for $t = 2, 3, \ldots$. In such case, it is optimal to let $x = -\overline{d}$. If σ^2 is large but finite, the KF adjustment method will eventually bring the process to target but the convergence will be very slow.

EWMA controller. Although not a particular case of the KF formulation, also of interest is to consider the case when $K_t = \lambda$, in which case a discrete integral controller [5], also called an EWMA controller [8, 35], is obtained:

$$\nabla x_t = -\lambda y_t.$$

This controller has the main advantage of compensating against sudden shifts that can occur at any point in time besides the initial offset d. This means that the controller remains "alert" to compensate for shifts or other disturbances. A disadvantage is that it is not clear what value of λ to use. Because of this, some attempts have been made at developing adaptive techniques that modify λ as the control session evolves [31]. In particular, Guo et al. [15] proposed to apply a "time varying" EWMA controller such that it minimizes the mean square deviation of the quality characteristic after a sudden shift occurs. Not surprisingly, the optimal weights once again obey Grubbs' harmonic rule:

$$\lambda_t^* = \frac{1}{t - \tau + 1},$$

where τ is the point in time when a shift in the mean occurs. Hence, if $\tau = 1$ we have the initial setup error case, and the time varying EWMA controller exactly equals Grubbs' harmonic rule.

9.5 Extensions to More Complex Setup Adjustment Problems

It was mentioned in the previous section that the Kalman filter model that yields Grubbs' extended rule is optimal for the criterion $E[\sum_{t=1}^{n} \mu_t^2]$ and this constitutes a simple instance of a linear quadratic control problem. Recent literature has made use of the more general LQ formulation to derive optimal adjustment rules for more complex setup adjustment problems. Del Castillo et al. [9] provided KF solutions for three specific problems — errors in adjustment, quadratic adjustment costs, and multiple-input-multiple-output — that might be of interest to quality engineers. Problems such as errors in adjustments and cost of adjustments were also considered by Trietsch [39], Pan and del Castillo [30], and Lian and del Castillo [20].

The previous section considered a single manufacturing process with an unknown process setup error. Colosimo et al. [7] proposed a different approach for process adjustments over a set of batches when no previous knowledge on the parameters of the distribution of the setup offsets and on the process variability is available, i.e., when μ, P_0, and σ^2 are unknown. Their approach is based on hierarchical Bayesian models and uses Markov Chain Monte Carlo (MCMC) to derive estimates used to compute the adjustments at each observation. Further development of the MCMC approach to process adjustments can be found in Lian et al. [21]

9.6 Integration of Process Monitoring and Process Adjustment

The KF solution to optimal process adjustment cannot only be applied on correcting process setup errors but also on compensating process mean shifts that could be caused by random disturbance. Parameter d in the model of Equation (9.16) can be viewed as the size of mean shift. When an on line process monitoring scheme is employed, this shift will be detected and estimated by the control chart algorithm. Then using this information as the prior belief of a mean shift and start a sequence of adjustments is natural.

Pan and del Castillo [29] proposed an integrated process monitoring and adjustment scheme that consists of three steps: monitor the process using a control chart, estimate the shift size when a shift in the process mean is

TABLE 9.1

Six Methods of Integrating Control Charts and Sequential Adjustments

Method	Shift detection	Shift size estimation	Adjustment
1	Shewhart chart for individuals (3σ limits)	Last observation (Taguchi's method)	One adjustment after an out-of-control alarm
2	Shewhart chart for individuals (3σ limits)	Maximum likelihood estimate (Wiklund's method)	One adjustment according to the MLE value
3	CUSUM chart for individuals (k=0.5 h=5)	CUSUM estimate	One adjustment according to the CUSUM estimate
4	Shewhart chart for individuals (3σ)	last observation (Taguchi's method)	Five sequential adjustments
5	Shewhart chart for individuals (3σ)	MLE (Wiklund's method)	Five sequential adjustments
6	CUSUM chart for individuals (k=0.5 h=5)	CUSUM estimate	Five sequential adjustments

detected, and finally apply an adjustment procedure to bring the process mean back to target. They compared the performance of six combinations of control charts and adjustment methods, which are listed in Table 9.1.

On the shift size estimation, the last observation method, as suggested by Taguchi [38], utilizes the last value of y_t that exceeds the control limit of a Shewhart chart as the current estimation of process mean. This estimation always gives a large shift size, and thus is significantly biased when the actual shift size is small. The Wiklund method [42, 43] is a maximum likelihood estimation (MLE) of the process mean based on a truncated normal probability density function. The argument relies on the fact that the estimation of the process mean is made on the condition that one finds a point exceeding the control limit of the Shewhart chart. The CUSUM estimate is taken from Montgomery [27]. For a positive mean shift, the CUSUM estimate of the mean after the CUSUM chapter issues a out-of-control signal is

$$\hat{\mu} = k + \frac{c_t^+}{N^+}, \tag{9.30}$$

where N^+ is the number of periods in which a run of nonzero values of c^+ were observed.

After the shift is detected, the process is adjusted either by one single adjustment based on the shift size estimation or by sequential adjustments following the Grubbs' harmonic rule of Equation (9.29). Assume that a shift occurs at time t_0, i.e., $\mu_t = \delta$ for $t \geq t_0$; then $K_t = \frac{1}{t-t_0}$ and the sequential adjustment scheme is of the form

$$x_t = x_{t-1} - K_t y_t, \quad \text{for } t > t_0. \tag{9.31}$$

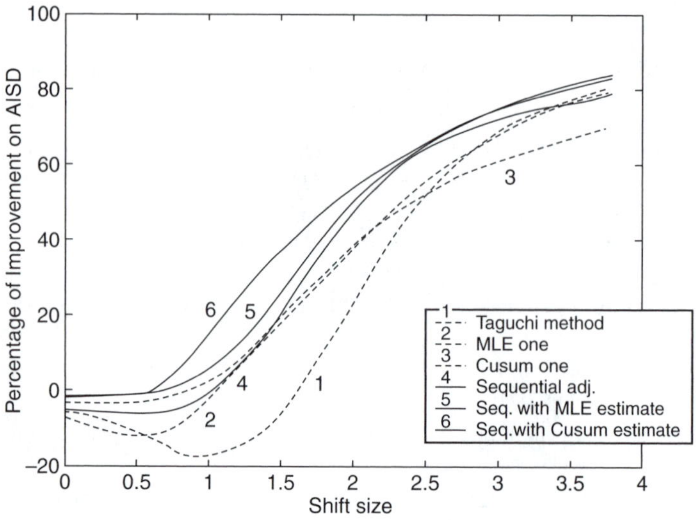

FIGURE 9.3
Performance of six integrated methods of control charts and adjustments (the process mean was shifted after the fifth observation).

The performance of an adjustment scheme is evaluated by the scaled *Average integrated squared deviation* (AISD) of the process output, which is defined as

$$\frac{\text{AISD}(n)}{\sigma^2} = \frac{1}{n\sigma^2} \sum_{t=1}^{n} y_t^2. \tag{9.32}$$

We first simulate a manufacturing process $y_t \sim N(0, \sigma^2)$ for a total of 50 observations. A process mean shift with a size varying from 0 to 4σ occurs after the fifth run. Process adjustments are conducted immediately after the shift is detected. The mean value of 10,000 simulation results are illustrated in Figure 9.3. The y-axis in the figure represents the percentage improvement in the AISD of using some adjustment method compared to the AISD without adjustment, i.e.,

$$\frac{\text{AISD}_{no\ adjust} - \text{AISD}_{method\ i}}{\text{AISD}_{no\ adjust}} \times 100,$$

so this is a "the larger the better" value. On the figure, the average percentage improvement in AISD is plotted with respect to the actual shift size, which was varied from 0 to 4σ. One can see that the sequential adjustment methods (Methods 4 to 6) are superior to the one-step adjustment methods (Methods 1 to 3) for almost all shift sizes. More specifically, using a CUSUM chart and sequential adjustments (method 6) has a significant advantage over other methods when the shift size is small or moderate, and using a Shewhart chart and sequential adjustments (method 4) is better for large shifts.

To study a general shifting process, the mean shift in another set of simulations is modeled by a stochastic process in which shifts occur randomly in time according to a geometric distribution. Specifically, the occurrence of a shift at each run is a Bernoulli trial with probability $p = 0.05$, and the shift size is normally distributed as $N(\mu_s, 1)$ and μ_s varies from 0 to 4σ. Besides the previous six methods, an EWMA controller was studied for comparison purposes. The EWMA control scheme takes the same form as Equation (9.31) except that K_t is a constant λ; here we set this control parameter at 0.2. No process monitoring is needed for the integral control scheme because the controller is always in action. The simulations were repeated 10,000 times.

Figure 9.4 shows that sequential adjustment methods still outperform any one-step adjustment method. Evidently, the EWMA controller performs better than any other sequential method when the shift size mean is small, which explains the popularity of EWMA controllers. However, one main advantage of the proposed SPC/EPC integrated methods is that they detect process changes using common SPC charts, whereas the EWMA controller alone does not have this SPC function — in other words, there is no possibility for process improvement through correction of assignable causes if only an EWMA controller is utilized. Process improvement through human intervention is facilitated by having a monitoring (SPC) mechanism that triggers the adjustment procedure and keeps a time-based record of alarms useful for process diagnostics.

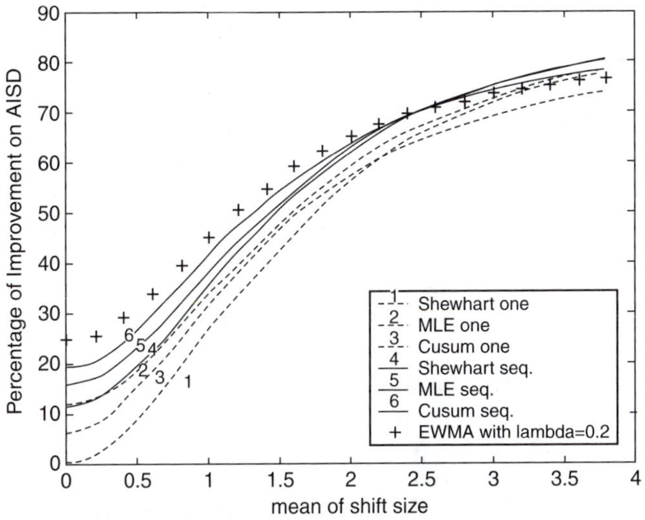

FIGURE 9.4
Performance of six integrated methods and EWMA controller for a general shift model (the shift occurs with probability $p = 0.05$ at each observation).

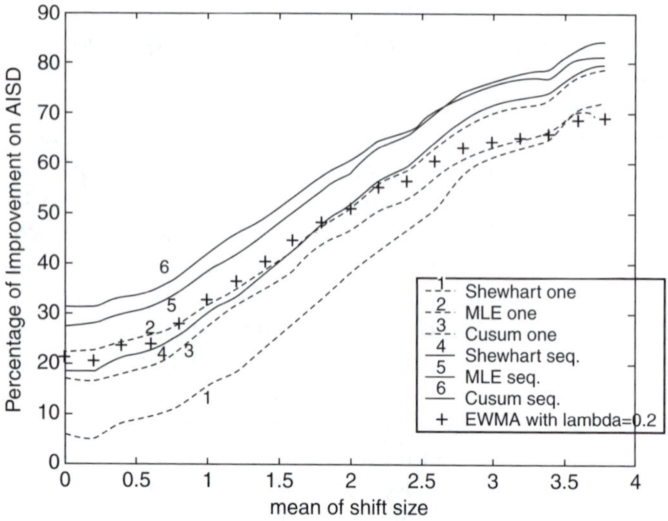

FIGURE 9.5
Performance of EPC and SPC integration for the general shift model, less frequent shifts (p=0.01).

Another drawback of the EWMA controller is that one must decide what value of the control parameter λ to use. This parameter should be small to maintain the stability of the process [35], but small parameter values might not be optimal from an AISD point of view, especially when the mean shift size is large.

Moreover, the high performance of the EWMA scheme comes from the frequent random shifts modeled in the previous simulation study (an average of 2.5 shifts per 50 runs). If the chance of shifts decreases, the inflation of variance that is caused by adjusting an on-target process will deteriorate the effectiveness of this scheme. In Figure 9.5, the simulation was conducted with the probability of random shifts p being decreased to 0.01. In this case, one can see that the EWMA method cannot compete well with the sequential adjustment methods combined with CUSUM or Shewhart chart monitoring.

More simulation results for different probabilities of shifts p are listed in Table 9.2. The EWMA adjustment method is found to be better for small shifts and Method 4 is better for large shifts when p is large; as p gets smaller ($p < 0.02$), i.e., as the process is subject to more infrequent random shocks, Method 6 gets harder to beat. Therefore, the proposed control chart-sequential adjustment integrated methods work better when p is small. Pan and del Castillo [29] studied some other process monitoring-adjustment combinations to further improve this integrated quality control scheme. Interested readers may look up their paper.

TABLE 9.2

Performance of SPC/EPC Integrated Adjustment Schemes and EWMA Scheme When Varying the Probability of a Shift

		Mean of shift size				
% improvement on AISD		**0**	**1σ**	**2σ**	**3σ**	**4σ**
p = 0.05	Method 4	11.20	36.05	62.27	**74.93**	**81.04**
		(0.30)	(0.38)	(0.36)	(0.31)	(0.29)
	Method 6	18.89	41.50	64.07	73.90	78.73
		(0.28)	(0.35)	(0.33)	(0.30)	(0.29)
	EWMA controller	**24.91**	43.11	60.47	67.76	70.71
	($\lambda = 0.1$)	(0.27)	(0.32)	(0.30)	(0.28)	(0.28)
	EWMA controller	24.51	**45.32**	65.26	73.31	76.68
	($\lambda = 0.2$)	(0.30)	(0.36)	(0.33)	(0.31)	(0.30)
	EWMA controller	21.16	44.02	**65.59**	74.21	78.38
	($\lambda = 0.3$)	(0.33)	(0.39)	(0.36)	(0.33)	(0.32)
p = 0.035	Method 4	6.65	24.31	47.76	**62.41**	**68.85**
		(0.26)	(0.37)	(0.40)	(0.39)	(0.38)
	Method 6	13.80	30.35	50.56	61.91	66.68
		(0.25)	(0.33)	(0.36)	(0.36)	(0.36)
	EWMA controller	**18.31**	32.18	48.68	56.01	59.58
	($\lambda = 0.1$)	(0.25)	(0.32)	(0.34)	(0.34)	(0.34)
	EWMA controller	16.82	**32.81**	51.21	61.09	64.35
	($\lambda = 0.2$)	(0.29)	(0.36)	(0.39)	(0.38)	(0.39)
	EWMA controller	13.13	30.33	**51.76**	60.82	65.48
	($\lambda = 0.3$)	(0.32)	(0.40)	(0.41)	(0.41)	(0.42)
p = 0.02	Method 4	1.48	11.85	28.86	41.60	**48.34**
		(0.24)	(0.32)	(0.39)	(0.43)	(0.45)
	Method 6	8.07	17.52	**32.53**	**41.94**	47.20
		(0.21)	(0.29)	(0.36)	(0.39)	(0.41)
	EWMA controller	**10.37**	**18.86**	30.68	38.05	41.06
	($\lambda = 0.1$)	(0.22)	(0.29)	(0.35)	(0.38)	(0.39)
	EWMA controller	7.35	17.09	31.40	39.49	43.57
	($\lambda = 0.2$)	(0.26)	(0.33)	(0.40)	(0.43)	(0.45)
	EWMA controller	2.16	13.03	28.90	38.19	42.28
	($\lambda = 0.3$)	(0.28)	(0.37)	(0.44)	(0.47)	(0.49)
p = 0.005	Method 4	−3.36	−1.02	3.64	9.02	12.57
		(0.18)	(0.21)	(0.28)	(0.34)	(0.37)
	Method 6	**1.32**	**3.60**	**7.88**	**11.77**	**14.37**
		(0.12)	(0.16)	(0.23)	(0.28)	(0.32)
	EWMA controller	−0.36	1.55	5.53	7.72	9.95
	($\lambda = 0.1$)	(0.13)	(0.17)	(0.24)	(0.27)	(0.30)
	EWMA controller	−5.55	−2.91	1.42	4.89	7.19
	($\lambda = 0.2$)	(0.16)	(0.21)	(0.27)	(0.32)	(0.35)
	EWMA controller	−11.25	−8.47	−2.94	0.42	3.15
	($\lambda = 0.3$)	(0.18)	(0.23)	(0.31)	(0.36)	(0.39)

Note: The numbers are the mean values and standard errors (in parenthesis) of the percentage improvement on AISD (compared to the process without adjustment) computed from 10,000 simulations. Bold numbers are the largest improvement for each p and mean shift size combination.

9.7 Summary

In this chapter, Bayesian algorithms for detecting process change and for adjusting processes were developed. We discussed the connection between the Bayesian process monitoring scheme and the CUSUM algorithm. A sequential process adjustment method was derived from the Kalman filter technique. Through simulation studies, this sequential adjustment method combined with various control charts was found effective in quality control of a manufacturing process that could be deteriorated by shifts in the process mean.

References

1. Åström, K.J., *Introduction to Stochastic Control Theory*. San Diego, CA: Academic Press, 1970.
2. Basseville, M. and Nikiforov, I.V., *Detection of Abrupt Changes: Theory and Application*, Englewood Cliffs, NJ: Prentice-Hall, 1993.
3. Box, G.E.P, Jenkins, G.M., and Reinsel, G., *Time Series Analysis, Forecasting, and Control*, 3rd ed., Englewood Cliffs, NJ: Prentice Hall, 1994.
4. Box, G.E.P. and Kramer, T., "Statistical process monitoring and feedback adjustment — A discussion," *Technometrics*, 34, 251–267, 1992.
5. Box, G.E.P. and Luceño, A., *Statistical Control by Monitoring and Feedback Adjustment*, New York: John Wiley & Sons, 1997.
6. Chen, A. and Elsayed, E.A., "An alternative mean estimator for processes monitored by SPC charts," *International Journal of Production Research*, 38, 13, 3093–3109, 2000.
7. Colosimo, B.M., Pan, R., and del Castillo, E., "A sequential Markov Chain Monte Carlo approach to setup adjustment of a process over a set of lots," *Journal of Applied Statistics*, 31, 5, 499–520, 2004.
8. Del Castillo, E., "Some properties of EWMA feedback quality adjustment schemes for drifting disturbances," *Journal of Quality Technology*, 33, 2, 153–166, 2001.
9. Del Castillo E., Pan, R., and Colosimo, B.M., "An unifying view of some process adjustment methods," *Journal of Quality Technology*, 35, 3, 286–293, 2003.
10. Del Castillo, E., Pan, R., and Colosimo, B.M., "Small sample performance of some statistical process setup adjustment methods," *Communications in Statistics—Simulation and Computation*, 32, 3, 923–941, 2003.
11. Duncan, D.B. and Horn, S.D., "Linear dynamic recursive estimation from the viewpoint of regression analysis," *Journal of the American Statistical Association*, 67, 340, 815–821, 1972.
12. Gargallo, P. and Salvador, M., "Monitoring residual autocorrelations in dynamic linear models," *Communications in Statistics — Simulation and Computation*, 32, 4, 1079–1104, 2003.

13. Girshick, M.A. and Rubin, H., "A Bayes approach to a quality control model," *The Annals of Mathematical Statistics*, 23, 1, 114–125, 1952.
14. Grubbs, F.E., "An optimum procedure for setting machines or adjusting processes," *Industrial Quality Control*, July 1954, reprinted in *Journal of Quality Technology*, 15, 4, 186–189, 1983.
15. Guo, R.S., Chen, A., and Chen, J.J., "An enhanced EWMA controller for processes subject to random disturbances," in *Run to Run Control in Semiconductor Manufacturing*, Moyne, J., Del Castillo, E., and Hurwitz, A. (Eds.), Boca Raton FL: CRC Press, 2000.
16. Harrison, P.J. and Veerapen, P.P., "A Bayesian decision approach to model monitoring and CUSUMs," *Journal of Forecasting*, 13, 29–36, 1994.
17. Harrison, P.J. and Lai, I.C.H., "Statistical process control and model monitoring," *Journal of Applied Statistics*, 26, 2, 273–292, 1999.
18. Kelton, W.D., Hancock, W.M., and Bischak, D.P., "Adjustment rules based on quality control charts," *International Journal of Production Research*, 28, 2, 365–400, 1990.
19. Lewis, F.L., *Optimal Estimation, with an Introduction to Stochastic Control Theory*, New York, NY: John Wiley & Sons, 1986.
20. Lian, Z. and del Castillo, E., "Setup adjustment under unknown process parameters and fixed adjustment cost," *Journal of Statistical Planning and Inference*, 136, 3, 1039–1060, 2006.
21. Lian, Z., Colosimo, B.M., and del Castillo, E., "Setup error adjustment: Sensitivity analysis and a new MCMC control rule," Quality & Reliability Engineering International, 22, 403–418, 2006.
22. Ljung, L. and Söderström, T., *Theory and Practice of Recursive Identification*, Cambridge, MA: The MIT Press, 1987.
23. Lorden, G., "Procedures for reacting to a change in distribution," *Annals of Mathematical Statistics*, 41, 2, 520–527, 1971.
24. MacGregor, J.F., "A different view of the funnel experiment," *Journal of Quality Technology*, 22, 255–259, 1990.
25. Meinhold, R.J. and Singpurwalla, N.D., "Understanding the Kalman Filter," *The American Statistician*, 37, 2, 123–127, 1983.
26. Montgomery, D.C., Keats, J.B., Runger, G.C., and Messina, W.S., "Integrating statistical process control and engineering process control," *Journal of Quality Technology*, 26, 79–87, 1994.
27. Montgomery, D.C., *Introduction to Statistical Quality Control*, 3rd ed., New York: John Wiley & Sons, 1996.
28. Page, E.S., "Continuous inspection schemes," *Biometrika*, 41, 100–115, 1954.
29. Pan, R. and del Castillo, E., "Integration of sequential process adjustment and process monitoring techniques," *Quality and Reliability Engineering International*, 19, 1–16, 2003.
30. Pan, R. and del Castillo, E., "Scheduling methods for the statistical setup adjustment problem," *International Journal of Production Research*, 41, 7, 1467–1481, 2003.
31. Patel, N.S. and Jenkins, S.T., "Adaptive optimization of run-to-run controllers: The EWMA example," *IEEE Transactions on Semiconductor Manufacturing*, 13, 1, 97–107, 2000.
32. Pollak, M. and Siegmund, D., "A diffusion process and its applications to detecting a change in the drift of Brownian motion," *Biometrika*, 72, 2, 267–280, 1985.

33. Robbins, H. and Monro, S., "A stochastic approximation method," *Annals of Mathematical Sciences*, 22, 400–407, 1951.
34. Roberts, S.W., "A comparison of some control chart procedures," *Technometrics*, 8, 411–430, 1966.
35. Sachs, E., Hu, A., and Ingolfsson, A., "Run by run process control: Combining SPC and feedback control," *IEEE Transactions on Semiconductor Manufacturing*, 8, 1, 26–43, 1995.
36. Salvador, M. and Gargallo, P., "Automatic selective intervention in dynamic linear models," *Journal of Applied Statistics*, 30, 10, 1161–1184, 2003.
37. Shiryaev, A.N., "On optimum method in quickest detection problems," *Theory of Probability and its Applications*, III, 1, 22–45, 1963.
38. Taguchi, G., "Quality engineering in Japan," *Commu. Statist. — Theory Meth.*, 14, 2785–2801, 1985.
39. Trietsch, D., "The harmonic rule for process setup adjustment with quadratic loss," *Journal of Quality Technology*, 30, 1, 75–84, 1998.
40. Tsung, F., Wu, H., and Nair, V., "On the efficiency and robustness of discrete proportional-integral control schemes," *Technometrics*, 40, 30, 214–222, 1998.
41. Tucker, W.T., Faltin, F.W., and Vander Weil, S.A., "Algorithmic statistical process control: An elaboration," *Technometrics*, 35, 4, 363–375, 1993.
42. Wiklund, S.J., "Estimating the process mean when using control charts," *Economic Quality Control*, 7, 105–120, 1992.
43. Wiklund, S.J., "Adjustment strategies when using shewhart charts," *Economic Quality Control*, 8, 3–21, 1993.
44. Woodall, W.H. and Adams, B.M., "The statistical design of CUSUM charts," *Quality Engineering*, 5, 4, 559–570, 1993.
45. Young, P., *Recursive Estimation and Time-Series Analysis*, New York: Springer-Verlag, 1984.

Part IV

Process Optimization and Designed Experiments

10

A Review of Bayesian Reliability Approaches to Multiple Response Surface Optimization

John J. Peterson
Statistical Sciences Department, GlaxoSmithKline Pharmaceuticals, R&D

CONTENTS

ABSTRACT This chapter is a review of recently proposed Bayesian predictive approaches to response surface optimization, in particular multiple response surface optimization. The posterior predictive distribution of a regression model is used to compute the posterior probability that a vector of response variables, $\boldsymbol{Y}$, is contained in a specified region, A, conditional on a vector of predictive factors, $\boldsymbol{x}$. Response surface optimization can then be achieved by maximizing this posterior probability with respect to $\boldsymbol{x}$. For a chosen regression model, all of the uncertainty, as well as the correlation among the response types, is accounted for through the posterior predictive distribution. Some previously published frequentist approaches had not accounted for the correlation among the response types and many have ignored some aspects of model parameter uncertainty.

Applications of this Bayesian approach to response surface optimization include the standard multivariate regression model, the seemingly unrelated regressions model, the incorporation of noise variables, the "dual response" model (for mean and standard deviation), and Bayesian model averaging. Related frequentist approaches are also reviewed. In addition, the notion of preposterior calculations are discussed as they relate to sample size augmentation that may be needed to refine a posterior predictive probability measure. Some future research possibilities are also outlined.

10.1 Introduction

In multiple response surface optimization it is important for investigators to have access to an optimization procedure that is easy to interpret and takes into account the uncertainty of all of the unknown model parameters and future (multivariate) responses. The Bayesian paradigm offers such an approach through the posterior predictive distribution of the responses. This posterior predictive distribution allows the investigator to compute a reliability measure (e.g. conformance specification probability) that can be used for process optimization. The Bayesian approach is also nice in that a preposterior analysis can be done to estimate the effect of acquiring more data to reduce model parameter uncertainty. In addition to inferences about single and multiple response types, this approach allows one to compute a Bayesian credible region of factor-level points that possess an associated reliability measure that is deemed satisfactory by the investigator.

In this section, I review two traditional approaches to multiple response surface optimization and some additional approaches found in the response surface literature. In subsequent sections of this chapter, I review recent results based upon the posterior predictive approach and briefly discuss some future modeling extensions in this area.

Two traditional approaches to multiple response surface optimization have employed "overlapping response surfaces" and desirability functions. Both of these approaches have serious problems with regard to ignoring model parameter uncertainty. Fortunately, the posterior predictive distribution approach discussed in this chapter solves these problems and can even be adapted to the above mentioned two traditional approaches in a natural way.

The earliest and simplest approach to multiple response surface optimization is the "overlapping mean response surface" method. Apparently, this dates back to Lind et al. [25]. This method involves simply looking at overlapping response surfaces (e.g. by way of contour plots) to ascertain what has been called a "sweet spot" (Anderson and Whitcomb [2]) where two or more mean response surfaces possess a region of overlap with a desirable multiple-response configuration. Further discussion of this approach can be found in Montgomery and Bettencourt [29].

Below is an example of an optimization of a HPLC assay where the "overlapping mean response surface" method was used. Here, four responses were

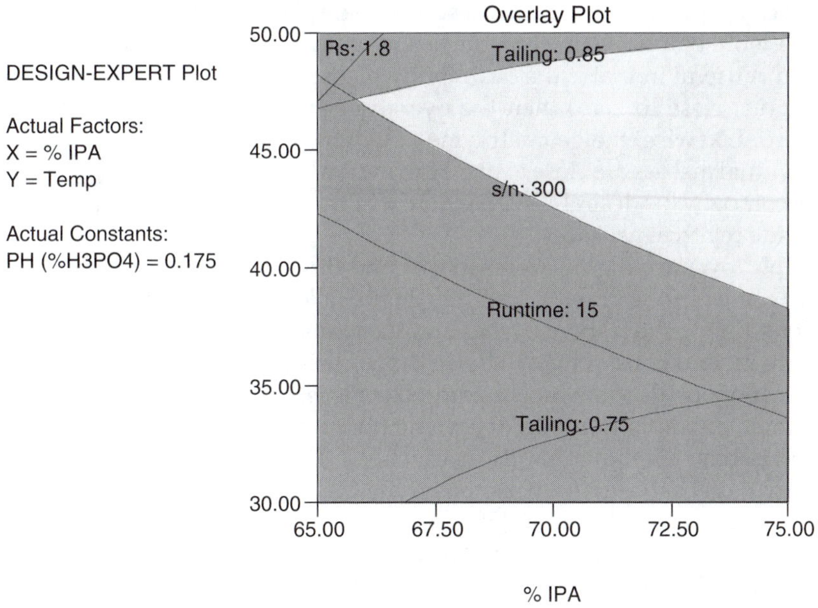

FIGURE 10.1
An overlay plot of the four mean response surfaces for the HPLC assay experiment.

considered for this optimization problem, "tailing", "run time", "Rs (critical resolution)", and "S/N (signal to noise ratio)". It was desired to configure the assay such that: Rs > 1.8, run time < 15, S/N > 300, and tailing is between 0.75-0.85. Three factors were involved in the optimization: "percent isopropyl alcohol (%IPA)", "column temperature (Temp)", and pH. With the pH set at 0.175 (halfway between the experimental upper and lower pH limits), an overlapping mean response surface plot (from the Design Expert statistical package) is given in Figure 10.1. The "sweet spot" is represented by the lighter triangular area in Figure 10.1. However, it is not clear how reliable this HPLC assay will be for factor levels within the "sweet spot" region. In other words, it is not clear what is the value of the posterior probability,

$$p(x) = \Pr(\text{Rs} > 1.8,\ \text{Run time} < 15,\ \text{S/N} > 300,\\ 0.75 < \text{tailing} < 0.85 \mid \boldsymbol{x},\ \text{data})$$

for each factor level configuration $\boldsymbol{x} = (x_1, x_2, x_3)'$, where $x_1 = \%\text{IPA}$, $x_2 = \text{Temp}$, and $x_3 = \text{pH}$. In general, this approach may be awkward to graph if there are more than two or three experimental factors.

Another traditional approach to multiple response surface optimization involves the use of what are called desirability functions. This approach was put forth by Harrington [18], and later refined by Derringer and Suich [15], and del Castillo, Montgomery, and McCarville [14]. The desirability function takes on values in [0, 1] and is the (weighted) geometric mean of individual metrics on [0, 1], one for each response type. The weighting allows one to place more emphasis on some response types versus others. The geometric mean is useful in that if the desirability metric for one response type is very poor (i.e. close to zero) then the overall desirability function will be close to zero. Likewise, the geometric mean will be close to one, only if all of the individual metrics are close to one. Harrington [18] proposed an absolute scale to accompany his desirability function, while other desirability functions only provide relative measures.

The "overlapping mean response" and desirability function approaches have some serious drawbacks. They do not take into consideration the correlations among the response types and the variability of the predictions. They only take into consideration the values of the mean response surfaces; the uncertainty of all of the model parameters is ignored.

Some attempts to model the prediction properties of multiple response surfaces have used quadratic loss functions. These approaches require some type of target value for each response type. Khuri and Conlon [22], Pignatiello [38], Ames et al. [1], Vining [50], and Ko et al. [24] have proposed various types of quadratic loss function approaches to multiple response surface optimization. Except for Ames et al. [1], these methods take into account the correlation structure of the responses types. However, they do not take into account the uncertainty of the variance-covariance matrix of the regression error vectors. In addition, the properties of a loss function based upon a multivariate quadratic form may be difficult for some experimenters to understand.

The most recent result by Ko et al. [24] provides a multivariate quadratic loss function of the form

$$(E[\hat{\boldsymbol{Y}}(\boldsymbol{x})] - \boldsymbol{T})'\boldsymbol{C}(E[\hat{\boldsymbol{Y}}(\boldsymbol{x})] - \boldsymbol{T}) + trace[\boldsymbol{C}\boldsymbol{\Sigma}_{\hat{\boldsymbol{Y}}(\boldsymbol{x})}] + trace[\boldsymbol{C}\boldsymbol{\Sigma}_{\boldsymbol{Y}(\boldsymbol{x})}], \tag{10.1}$$

where $\boldsymbol{C}$ is a "cost" matrix, $\hat{\boldsymbol{Y}}(\boldsymbol{x})$ is the predicted value of $\boldsymbol{y}$ at $\boldsymbol{x}$, $\boldsymbol{T}$ is a user-specified vector of target values, $\boldsymbol{\Sigma}_{\hat{\boldsymbol{Y}}(\boldsymbol{x})}$ is the variance-covariance matrix of the predicted response, $\hat{\boldsymbol{Y}}(\boldsymbol{x})$, at $\boldsymbol{x}$, and $\boldsymbol{\Sigma}_{\boldsymbol{Y}(\boldsymbol{x})}$ is the variance-covariance matrix of the true response, $\boldsymbol{Y}$, at $\boldsymbol{x}$. For the standard multivariate regression model, $\boldsymbol{\Sigma}_{\boldsymbol{Y}(\boldsymbol{x})} = \boldsymbol{\Sigma}$ (a constant matrix) and $\boldsymbol{\Sigma}_{\hat{\boldsymbol{Y}}(\boldsymbol{x})} = \boldsymbol{x}'(\boldsymbol{X}'\boldsymbol{X})^{-1}\boldsymbol{x}\boldsymbol{\Sigma}$. While the second term in (10.1) takes into account the uncertainty of the predicted response, it does not take into account the uncertainty of not knowing $\boldsymbol{\Sigma}$. Likewise, the third term in (10.1) also does not account for the uncertainty of not knowing $\boldsymbol{\Sigma}$. Ko et al. [24] and earlier related papers simply estimate $\boldsymbol{\Sigma}$ and treat it as fixed. As we shall see below, the Bayesian approach takes into account the uncertainty of all of the model parameters, including $\boldsymbol{\Sigma}$. In fact, as discussed below, the Bayesian approach can even average over different possible model forms to account for model form uncertainty as well.

The overlapping mean response, desirability function, and quadratic loss function approaches have the drawback that they do not completely characterize the uncertainty associated with future multivariate responses and their associated optimization measures. The danger of this is that an experimenter may use one of these methods to get an optimal factor configuration, validate it with two or three successful runs, and then begin production. For example, suppose that the probability that a future multivariate response is satisfactory is only 0.7. Even so, the chance of getting three successful, independent validation runs is 0.343, which can easily happen. Hunter [21] states that the variance of univariate response indices for multiresponse optimization "can be disturbing" and further study is needed to assess the influence of parameter uncertainty.

Del Castillo [13] proposed a multiple response surface optimization method that uses confidence regions for the optimum to address regression model parameter uncertainty. While this approach is easier to understand than a multivariate quadratic form, it requires the type of problem that can be formulated as having a primary response variable (and various secondary response variables). Furthermore, the correlation structure of the responses types are not taken into consideration.

Chiao and Hamada [12] took an important step in multiple response surface optimization. They have proposed an approach which provides a method to estimate the probability that a multivariate normal response will satisfy desired conformance conditions. Their approach is nice in that it takes into account the variance-covariance structure of the multivariate response, can accommodate heteroscedastic and noise variable regression models, and is easy for investigators to understand. However, even this approach does not take into account the uncertainty of the model parameter estimates. In some cases, this can cause the probability of conformance to appear larger than it should be.

I review in the next section the basic approach proposed in Peterson [36] and show some of its advantages over the two traditional approaches mentioned above as well and the quadratic loss function approach. In Sections 10.3 to 10.6, I discuss various generalizations of this posterior predictive approach. In Section 10.7, a summary and some future possibilities are presented.

10.2 Illustration of the Posterior Predictive Approach using the Standard Multivariate Regression Model

Let $\boldsymbol{Y} = (Y_1, Y_2, \ldots Y_r)'$ be the multivariate $(r \times 1)$ response vector and let $\boldsymbol{x} = (x_1, x_2, \ldots, x_k)'$ be the $(k \times 1)$ vector of factor variables. The standard regression model for multiple response surface modeling is the standard multivariate multiple regression model,

$$\boldsymbol{Y} = \boldsymbol{B}\boldsymbol{z}(\boldsymbol{x}) + \boldsymbol{e}, \tag{10.2}$$

where $\boldsymbol{B}$ is an $r \times q$ matrix of regression coefficients and $\boldsymbol{z}(\boldsymbol{x})$ is a $q \times 1$ vector-valued function of $\boldsymbol{x}$. The vector $\boldsymbol{e}$ has a multivariate normal distribution with mean vector $\boldsymbol{0}$ and variance-covariance matrix, $\boldsymbol{\Sigma}$. Typically in response surface analysis, the model in (10.2) is composed of a quadratic model for each mean response, but we are allowing a more general covariate structure here. The typical multivariate regression assumption, that $z(x)$ is the same for each response type, is assumed in this section.

If the experimenter is simply interested in $\boldsymbol{Y}$ being in some desirable subset of the response space, A, then he or she should consider the posterior probability,

$$p(\boldsymbol{x}) = \Pr(\boldsymbol{Y} \in A \mid \boldsymbol{x}, \text{data}), \tag{10.3}$$

where the probability measure, $p(\boldsymbol{x})$, is based upon the posterior predictive distribution of $\boldsymbol{Y}$ given the vector of factor levels, $\boldsymbol{x}$. This will then give the experimenter a measure of the reliability of $\boldsymbol{Y}$ being in A for a given $\mathbf{x}$. This measure, of course, takes into account the variance-covariance structure of the data and the uncertainty of all of the model parameters through the posterior predictive distribution of $\boldsymbol{Y}$. A search of the $\boldsymbol{x}$-space will then provide the experimenter with information on conditions for optimizing the reliability of $\boldsymbol{Y}$ being in A.

If for some reason, an investigator desires to use a some type of desirability function, $D(\boldsymbol{y})$, this posterior predictive approach can still be applied by using the probability measure, $p(\boldsymbol{x}) = \Pr(D(\boldsymbol{Y}) > D^* \mid \boldsymbol{x}, \text{data})$ for some specified lower desirability bound D^*. (An investigator may want to use a desirability function to weight various response types differently.) The event $\{D(\boldsymbol{Y}) > D^*\}$ is of course equivalent to $\{\boldsymbol{Y} \in A\}$ for an appropriate set A. A similar statement can be made about a quadratic loss function, $Q(\boldsymbol{y})$ for some upper bound Q^*. Whatever the form of the D or Q functions, the variance-covariance structure of the responses and uncertainty of all of the model parameters are accounted for through the posterior predictive distribution.

Assuming the classical noninformative prior for the model parameters, $\boldsymbol{B}$ and $\boldsymbol{\Sigma}$, which is proportional to $|\boldsymbol{\Sigma}|^{-(r+1)/2}$, it follows that the posterior predictive distribution for $\boldsymbol{Y}$ given $\boldsymbol{x}$ is multivariate t with $\nu = n - r - q + 1$ df. (Here, of course, it is assumed that n is such that $\nu \geq 1$.) This multivariate t-distribution has a location parameter vector equal to $\hat{\boldsymbol{\mu}} = \hat{\boldsymbol{B}}\boldsymbol{z}(\boldsymbol{x})$, where $\hat{\boldsymbol{B}}$ is the least squares estimate of $\boldsymbol{B}$. Furthermore, this t-distribution has a scale parameter matrix, $\boldsymbol{S}$, equal to

$$(1 + \boldsymbol{z}(\boldsymbol{x})'\boldsymbol{D}^{-1}\boldsymbol{z}(\boldsymbol{x}))\hat{\boldsymbol{\Sigma}},$$

where $\boldsymbol{D} = \sum_{i=1}^{n} \boldsymbol{z}(\boldsymbol{x}_i)\,\boldsymbol{z}(\boldsymbol{x}_i)'$ and $\hat{\boldsymbol{\Sigma}} = \frac{1}{\nu}(\boldsymbol{y}^* - (\hat{\boldsymbol{B}}\boldsymbol{Z})')'(\boldsymbol{y}^* - (\hat{\boldsymbol{B}}\boldsymbol{Z})')$. Note, $\boldsymbol{Z}$ is the $q \times n$ matrix formed by the n $\boldsymbol{z}(\boldsymbol{x}_i)$covariate vectors and $\boldsymbol{y}^*$ is the $n \times r$ matrix formed by the $n\boldsymbol{y}_i'(1 \times r)$ vectors. For more details see Press [39] (Chapter 12) and Chapter 1 in this volume.

Probabilities, like $p(\boldsymbol{x})$ in (10.3) can be computed simply by Monte Carlo simulation from the above multivariate t-distribution. One can simulate a multivariate t random variable (r.v.), $\boldsymbol{Y}$, by simulation of a multivariate normal r.v. and an independent chi-square r.v. For this particular problem the simulation is done as follows. Let $\mathbf{W}$ be a multivariate normal r.v. with zero mean vector and variance-covariance matrix equal to $\mathbf{S}$. Let U be a chi-square r.v. with ν df. that is stochastically independent of $\mathbf{W}$. Next, define

$$Y_j = (\sqrt{\nu} W_j / \sqrt{U}) + \hat{\boldsymbol{\mu}}_j, \quad \text{for } j = 1, \ldots, r \tag{10.4}$$

where Y_j is the jth element of $\mathbf{Y}$, W_j is the jth element of $\mathbf{W}$, and $\hat{\mu}_j$ is the jth element of $\hat{\boldsymbol{\mu}} = \hat{\mathbf{B}}z(x)$. It follows then that $\mathbf{Y}$ has a multivariate t-distribution with ν df. Note that $\mathbf{W}$ can be easily simulated by generating the random variable, $\boldsymbol{\Gamma}'\boldsymbol{e}$, where $\boldsymbol{\Gamma}'\boldsymbol{\Gamma} = \mathbf{S}$ and $\boldsymbol{e}$ is an $r \times 1$ vector of independent standard normal rv's.

Using Monte Carlo simulation from (4), one can approximate the reliability $p(\boldsymbol{x})$ in (10.3) for various $\boldsymbol{x}$-values in the experimental region (using large N)

$$p(\boldsymbol{x}) \approx \frac{1}{N} \sum_{s=1}^{N} I(\mathbf{Y}^{(s)} \in A),$$

where N is the number of simulations, $I(.)$ is the 0-1 indicator function, and the $\mathbf{Y}^{(s)}$ r.v.'s are simulated conditional on $\boldsymbol{x}$ and the data. For a small number of factors it is computationally reasonable to grid over the experimental region to compute values of $p(\boldsymbol{x})$ for purposes of optimization. However, even for three or more factors, it may be preferable to have a more efficient approach to optimizing $p(\boldsymbol{x})$. One approach is to maximize $p(\boldsymbol{x})$ using general optimization methods such as those discussed in Nelder-Mead [33], Price [40], or Chatterjee, Laudato, and Lynch [9].

Another approach is to compute $p(\boldsymbol{x})$ for $\boldsymbol{x}$-points in some response surface experimental design and then fit a closed-form response surface model to obtain an approximate reliability surface, $\tilde{p}(\boldsymbol{x})$. Since values of $\tilde{p}(\boldsymbol{x})$ can be computed much more quickly than $p(\boldsymbol{x})$, approximate optimization of $p(\boldsymbol{x})$ can be done. For example, a ridge analysis could be done on $\tilde{p}(\boldsymbol{x})$ to explore in an approximate fashion how $p(\boldsymbol{x})$ changes as $\boldsymbol{x}$ moves out from the center of the experimental region in an optimal way (i.e. optimally over spheres of increasing radius centered at the center of the experimental region). For a review of ridge analysis see Hoerl [20].

To evaluate the reliability of optimal results produced by the "overlapping mean responses", desirability function, and quadratic loss function approaches, Peterson [36] computes Bayesian probabilities, $p(\boldsymbol{x})$, of the form $\Pr(\mathbf{Y} \in A \mid \boldsymbol{x}, \text{data})$, $\Pr(D(\mathbf{Y}) > D^* \mid \boldsymbol{x}, \text{data})$, and $\Pr(Q(\mathbf{Y}) < Q^* \mid \boldsymbol{x}, \text{data})$ respectively. One example involves a 3-component mixture experiment with two responses, Y_1 and Y_2, where it is assumed that one desires Y_1 to be at most 234 and Y_2 to be at most 18.5, hence for this experiment $A = \{(y_1, y_2)' : y_1 \leq 234, y_2 \leq 18.5\}$. Using the standard multivariate class of regression models, the resulting estimated models were:

$$\begin{aligned} \hat{y}_1 &= 248x_1 + 272x_2 + 533x_3 - 485x_1x_3 - 424x_2x_3 \\ \hat{y}_2 &= 18.7x_1 + 14.1x_2 + 35.4x_3 - 36.7x_1x_3 + 18.0x_2x_3\ . \end{aligned} \tag{10.5}$$

Using the overlapping mean responses approach it appears that the point $x = (0.75, 0, 0.25)'$ is a good choice. (See Figure 1 in Peterson [36].) However, using the multivariate t posterior predictive distribution described above it turns out that $\Pr(Y \in A \mid x, \text{data})$ is only 0.68 which is disappointingly small even though both corresponding means satisfy the conditions described by the set A. For this example, the desirability function of Deringer and Suich [15] is also optimized to obtain an optimum value of $D_{opt} = 0.74$ obtained at $x = (0.78, 0, 0.22)'$. However, the associated Bayesian reliability $\Pr(D(Y) > D^* = 1/2\, D_{opt} \mid x, \text{data})$ is only 0.6. In other words, the posterior probability of $D(Y)$ exceeding only *one-half* of D_{opt} is only 0.6 at $x = (0.78, 0, 0.22)'$. While the more commonly used Derringer-Suich [15] desirability function does not provide an absolute quality criterion, the Harrington [18] desirability function does. Harrington provides an absolute quality scale for his desirability function (call it D_H): D_H = 0-0.37 ("very poor" to "poor"), D_H = 0.37-0.60 ("fair"), D_H = 0.60-0.80 ("good"), and D_H = 0.80-1 ("excellent"). Here, it is interesting to note that $D_H(\hat{y})$ is maximized to 0.96 at $x = (0.76, 0, 0.24)'$. However, $p_H(\mathbf{x}) = \Pr(D_H(Y) \geq 0.60 \mid \mathbf{x}) = 0.72$. This shows that the probability of obtaining a future response that is at least borderline "good" is only 0.72 despite the fact that the optimal desirability was estimated to be 0.96, which is "excellent" on the Harrington scale!

For the quadratic loss function, $Q(y)$, developed in Vining [50], the minimized value of $Q(\hat{y})$ is $Q_{opt} = 7.79$ at $x = (0.76, 0, 0.24)'$. However $\Pr(Q(Y) < 2Q_{opt} \mid x = (0.76, 0, 0.24)')$ is only 0.38, i.e. the posterior probably of the $Q(Y)$ quadratic loss being less than *twice* Q_{opt} is only 0.38.

In the situation where the posterior probability, $p(x)$, is not large enough it is still possible to estimate how $p(x)$ would change if more data were gathered or if the process variability were reduced. If the response means are all satisfactory, i.e. if $E(Y \mid x)$ is in the interior of A, then $p(x)$ should increase if the spread of the posterior predictive distribution were reduced. The spread of the posterior predictive distribution should decrease if more data are gathered or if the process variability is reduced. One can estimate the effect of reducing the process variation by using the pseudo-data, $y_j^* = \hat{y}_j + (1 - \lambda_j)\hat{e}_j$, $j = 1, \ldots, p$, (where $100\,\lambda_j$ represents a percent reduction in residual size for the jth response type) to artificially generate new process data having the same mean values but with reduced variances. This will help with understanding how the process variation affects the reliability.

An additional approach is to modify the posterior predictive distribution in such a way that one can simulate responses from this distribution as if more data had been acquired. For the posterior sampling model in (4) this is easily done as follows. Note that the multivariate t posterior predictive density depends upon the data only through the sufficient statistics, $\hat{\Sigma}$ and $\hat{B}$, the degrees of freedom, and the design matrix Z. By increasing the rows of the design matrix (to add new data points), and augmenting the df accordingly, one can simulate new data. This corresponds to adding artificial data in such a way that the sufficient statistics, $\hat{\Sigma}$ and $\hat{B}$, remain the same. This will give the experimenter an idea of how much the reliability can be increased by reducing model uncertainty. For example, the experimenter can forecast the effects of replicating the experiment a certain number of times. This idea is similar in spirit to the notion of a "preposterior" analysis as described by Raiffa and Schlaiffer [41]. For example, Peterson [36] shows that for the mixture example discussed above by reducing the process variation by 25% we should expect an increase in $\Pr(\boldsymbol{Y} \in A \mid \boldsymbol{x}, \text{data})$ from 0.68 to 0.80 (using the same amount of data). On the other hand, without changing the underlying process variation, by replicating the experiment to double the amount of data we should expect an increase in $\Pr(\boldsymbol{Y} \in A \mid \boldsymbol{x}, \text{data})$ from 0.68 to 0.88. If reducing the process variation is difficult and 0.88 is considered an adequate reliability, then it may be preferable to simply repeat the experiment to get more data. If higher reliability is needed, then one may need to both decrease the process variation and increase the size of the experiment. For example, if the process variation were reduced by 50% and the size of the experiment were doubled then we would expect $\Pr(\boldsymbol{Y} \in A \mid \boldsymbol{x}, \text{data})$ to increase to 0.995.

With regard to the choice of λ_j values, these would be calibrated to increase $p(\boldsymbol{x}) = \Pr(\boldsymbol{Y} \in A \mid data, \boldsymbol{x})$ to the required level if the preposterior analysis indicates that a reasonable increase in sample size will not increase $p(\boldsymbol{x})$ to the desired level of reliability. (As stated previously, this of course assumes that $E(\boldsymbol{Y})$ is in the interior of A.) For example, if three replications of the data increase $p(\boldsymbol{x})$ from 0.65 to 0.75, and the minimum required value for $p(\boldsymbol{x})$ is 0.9, then one may want to consider what value of λ (or λ's) would be needed to push $p(\boldsymbol{x})$ up to 0.9. This will provide an indication of how much the process variability will need to be reduced to achieve the required reliability.

It is worth pointing out that simulating from the posterior predictive distribution to get "new" data raises a subtle but important issue for Bayesian analysis. The issue is that if $p(\boldsymbol{x}) = \Pr(\boldsymbol{Y} \in A \mid data, \boldsymbol{x})$ is your optimization criterion, then the use of additional data simulated from the posterior predictive distribution does not help. A heuristic argument for this begins with the fact that $\boldsymbol{Y}$ values simulated from the posterior predictive distribution are adjusted for the uncertainty in the regression model parameters. If we use *these* $\boldsymbol{Y}$ values as additional data, then the Bayesian process to compute $p(\boldsymbol{x})$ over adjusts for model parameter uncertainty resulting in an updated value for $p(\boldsymbol{x})$ that is the same as before. This can in fact be shown mathematically as follows. To keep the notation simple, I will consider the use of only one new $\boldsymbol{Y}$ value simulated from the posterior predictive distribution and I will suppress notation for the dependence upon $\boldsymbol{x}$.

Consider $\Pr(\boldsymbol{Y} \in A \mid \boldsymbol{y}_1, \ldots \boldsymbol{y}_n, \boldsymbol{Y}_{n+1})$, where $\boldsymbol{Y}_{n+1}$ was simulated from the posterior predictive distribution conditional on $\boldsymbol{y}_1, \ldots, \boldsymbol{y}_n$. A natural preposterior update, using $\boldsymbol{Y}_{n+1}$, might be to compute (e.g. via Monte Carlo simulation),

$$E_{\boldsymbol{Y}_{n+1}|\boldsymbol{y}_1,\ldots,\boldsymbol{y}_n}\{\Pr(\boldsymbol{Y} \in A \mid \boldsymbol{y}_1, \ldots, \boldsymbol{y}_n, \boldsymbol{Y}_{n+1})\}.$$

However, it can be shown that $E_{\boldsymbol{Y}_{n+1}|\boldsymbol{y}_1,\ldots,\boldsymbol{y}_n}\{\Pr(\boldsymbol{Y} \in A \mid \boldsymbol{y}_1, \ldots, \boldsymbol{y}_n, \boldsymbol{Y}_{n+1})\} = \Pr(\boldsymbol{Y} \in A \mid \boldsymbol{y}_1, \ldots, \boldsymbol{y}_n)$, so this type of update puts you right back where you started. To see this note that

$$
\begin{aligned}
&E_{Y_{n+1}|y_1,\ldots,y_n}\{\Pr(\boldsymbol{Y} \in A|\boldsymbol{y}_1, \ldots, \boldsymbol{y}_n, \boldsymbol{Y}_{n+1})\} \\
&= \int \int_A p(y|y_1, \ldots, y_n, y_{n+1})p(y_{n+1}|y_1, \ldots, y_n)dy\,dy_{n+1} \\
&= \int \int_A p(y, y_{n+1}|y_1, \ldots, y_n)\,dy\,dy_{n+1} \\
&= \int_A \int p(y, y_{n+1}|y_1, \ldots, y_n)dy_{n+1}dy \\
&= \Pr(\boldsymbol{Y} \in A|\boldsymbol{y}_1, \ldots, \boldsymbol{y}_n).
\end{aligned}
$$

As a byproduct of Monte Carlo computation of $\Pr(\boldsymbol{Y} \in A|\boldsymbol{x}, \text{data})$, it is easy to also compute marginal probabilities involving each of the individual responses. If we denote

$$p_i(\boldsymbol{x}) = \Pr(Y_i \in A_i|\boldsymbol{x}, \ \text{data}),$$

where A_i is an interval (possibly one or two sided), then we can assess the $p_i(\boldsymbol{x})$'s simultaneously along with $p(\boldsymbol{x})$. Modification of A and the A_i's can be used to allow an investigator to directly observe economic or performance issues through the $p(\boldsymbol{x})$ and $p_i(\boldsymbol{x})$'s respectively, which can be easily related to how often the process will be invoked. Of course, other marginal (joint) probabilities concerning the Y_i's can be computed in a similar fashion.

For purposes of assessing process ruggedness, a Bayesian credible region can also be computed. For example, this can be done by using $p(\boldsymbol{x})$ for mapping out all of the $\boldsymbol{x}$-values for which $p(\boldsymbol{x})$ is at least 0.95, say. See Peterson [36] for an example.

As one can see, the posterior predictive approach described above offers a good deal of flexibility for inference about multiple response surface experiments. The Bayesian approach provides a complete way to model all of the sources of uncertainty. Alternatively, one could compute $p(\boldsymbol{x}; \boldsymbol{\mu}, \boldsymbol{\Sigma}) = \Pr(\boldsymbol{Y} \in A|\boldsymbol{x}, \boldsymbol{\mu}, \boldsymbol{\Sigma})$ and try to apply frequentist inference theory to $p(\boldsymbol{x}; \boldsymbol{\mu}, \boldsymbol{\Sigma})$, but this would be difficult as $p(\boldsymbol{x}; \boldsymbol{\mu}, \boldsymbol{\Sigma})$ may require Monte Carlo computations just to compute it for any specific values of $\boldsymbol{\mu}$ and $\boldsymbol{\Sigma}$. Optimization over $p(\boldsymbol{x}; \boldsymbol{\mu}, \boldsymbol{\Sigma})$ would somehow have to take into account the uncertainty of $\boldsymbol{\mu}$ and $\boldsymbol{\Sigma}$.

10.3 Incorporation of Noise Variables

Noise variables are factors that affect a process but are not directly controlled in typical process runs. Such factors may be related to uncontrolled conditions such as ambient temperature, humidity, process aging, variation in raw materials, or customer usage. The strategy of configuring controllable factors to reduce the influence of the noise variables often comes under the heading of "robust parameter design". The application of response surface methodology for improving processes involving noise variables dates back to the early 1990's (Box and Jones [6], Vining and Myers [49], Myers, Khuri, and Vining [30]).

The quadratic response surface model (for one response) that is often used for incorporating noise variables with controllable factors was initially proposed by Box and Jones (1990), and appears in well-known response surface texts by Khuri and Cornell [23] (p. 423) and Myers and Montgomery [32] (p. 557). It has the form

$$Y = \beta_0 + \boldsymbol{\beta}'\boldsymbol{x}_c + \boldsymbol{x}_c'\boldsymbol{B}\boldsymbol{x}_c + \boldsymbol{\gamma}'\boldsymbol{x}_n + \boldsymbol{x}_c'\boldsymbol{\Delta}\boldsymbol{x}_n + e, \tag{10.6}$$

where $\boldsymbol{x}_c$ is a $k \times 1$ vector of control factors, $\boldsymbol{x}_n$ is an $l \times 1$ vector of noise variables, and e is a random normal error term with mean zero and variance σ^2. Here, $\boldsymbol{\beta}$ is a $k \times 1$ vector of regression coefficients corresponding to $\boldsymbol{x}_c$ and $\boldsymbol{B}$ is a $k \times k$ symmetric matrix with diagonal elements equal to $\beta_{ii} (i = 1, \ldots, k)$ and off-diagonal elements equal to $^1/_2\, \beta_{ij}\ (i < j)$. Furthermore, $\boldsymbol{\gamma} = (\gamma_1, \ldots, \gamma_l)'$ and $\boldsymbol{\Delta}$ is a $k \times l$ matrix composed of elements $\delta_{ij} (i = 1, \ldots, k;\ j = 1, \ldots l)$. In the model in (10.6), it is assumed that the noise variables have a multivariate normal distribution, and have been scaled so that they have a mean vector equal to $\mathbf{0}$ and a variance-covariance matrix equal to $\sigma^2_{x_n}\boldsymbol{\Omega}$. It is also typically assumed that $\boldsymbol{\Omega}$ is equal to the identity matrix, $\boldsymbol{I}$, so the noise variables are independent. The assumption that $\boldsymbol{\Omega} = \boldsymbol{I}$ may not always hold but is appropriate for many practical situations (Borror, Montgomery, and Myers [5]).

In the presence of random noise variables, $\boldsymbol{x}_n$, we wish to do robust optimization, that is we wish to find factor levels of $\boldsymbol{x}_c$ (the controllable factors) that minimize the influence of the noise variables. Let $\boldsymbol{x} = (\boldsymbol{x}_c', \boldsymbol{x}_n')'$. Frequentist univariate approaches have considered two basic strategies. One involves computing

$$\min_{x_c \in R} Var_{x_n,e}(Y|\boldsymbol{x}),$$

$$\text{subject to } E_{x_n,e}(Y|\boldsymbol{x}) = c$$

where the expected value and variance are computed relative to both the x_n and e r.v.'s. Here, R is some specified experimental region and c is a specified constant. This approach however has the drawback requiring the experimenter to specify a (possibly artificial) constraint value for $E_{x_n,e}(Y|\boldsymbol{x})$. See Miró-Quesada and del Castillo [28] for a recent paper on a frequentist approach to this problem.

Another strategy is to minimize $E_{x_n,e}\{(Y - T)^2|\boldsymbol{x}\}$ for some specified target value, T. For robust optimization problems that involve "larger the better" (LTB) or "smaller the better" (STB) response values, other loss functions have been proposed. For the STB case the loss function is sometimes defined as above but with $T = 0$, whereas for the LTB problems, loss function forms involving $m(\boldsymbol{x}_c, \boldsymbol{x}_n; \boldsymbol{\phi})^{-2}$ or $\exp(-m(\boldsymbol{x}_c, \boldsymbol{x}_n; \boldsymbol{\phi}))$ have been proposed, where $m(\boldsymbol{x}_c, \boldsymbol{x}_n; \boldsymbol{\phi})$ represents the first five terms in (10.6) (Steiner and Hamada [48]). Myers and Montgomery [31] (p492) have also proposed loss functions for the STB and LTB cases based upon notions of upper and lower prediction limits respectively.

On the other hand, for the STB (LTB) case Peterson and Kuhn [37] propose setting

$$T = \min_{\substack{(\max) \\ \boldsymbol{x}_c \in R}} (\hat{\beta}_0 + \boldsymbol{x}_c'\hat{\boldsymbol{\beta}} + \boldsymbol{x}_c'\hat{\boldsymbol{B}}\boldsymbol{x}_c),$$

where $\hat{\boldsymbol{\beta}}$ and $\hat{\boldsymbol{B}}$ are the least squares estimates of $\boldsymbol{\beta}$ and $\boldsymbol{B}$ respectively. Suppose we wish to minimize (maximize) the mean response. The expression for the target, T, is taken as the minimum (maximum) mean response value with respect to the controllable variables, $\boldsymbol{x}_c$. This may be a more natural and direct approach than previous target forms that have been used for "smaller the better" and "larger the better" situations.

Chiao and Hamada [12], Romano et al. [44], and Miró-Quesada and del Castillo [27] have proposed non-Bayesian solutions to the multivariate-response robust parameter design problem. For consulting purposes, the latter two optimization criteria are mathematically more difficult to intuitively understand than that of Chiao and Hamada [12] who obtains a frequentist estimate of the probability of conformance, unconditional on the noise variables. Their approach, however, requires a crossed-array replicated design because the variances at each factor level combination are computed from the replications. All of these three multivariate approaches do not take into account the uncertainty of the all of the model parameter estimates.

Miró-Quesada et al. [26] extend the approach of Peterson [36] to incorporate noise variables. They compute the posterior probability,

$$p(\boldsymbol{x}_c) = \Pr(\boldsymbol{Y} \in A|\boldsymbol{x}_c, \text{data}) = E_{\boldsymbol{x}_n}(\Pr(\boldsymbol{Y} \in A|\boldsymbol{x}, \text{data}));$$

in other words the probability of conformance unconditional on the noise variables. Robust process optimization can then be done by maximizing $p(\boldsymbol{x}_c)$. In addition, $p(\boldsymbol{x}_c)$ is easier to understand than the most of the frequentist noise variable optimization objective functions. And, of course, the Bayesian approach models the uncertainty of all of the model parameters. A Monte Carlo estimate of $p(\boldsymbol{x}_c)$ can be obtained by a slight modification of the approach in Peterson [36]. Since it is typically assumed that the distribution of the noise variables are known, we simply simulate an $\boldsymbol{x}_n$ to obtain $\boldsymbol{x} = (\boldsymbol{x}_c', \boldsymbol{x}_n')'$ and then use $\boldsymbol{x}$ to simulate $\boldsymbol{Y}$ given $\boldsymbol{x}$ and the data. This process is repeated N times (for large N) to get a Monte Carlo estimate of $p(\boldsymbol{x}_c)$.

Miró-Quesada et al. [26] use the HPLC example discussed in the Introduction section to compute

$$\begin{aligned} p(\boldsymbol{x}_c) = \Pr(&\text{Rs} > 1.8,\ \text{Run time} < 15,\ \text{S/N} > 300, \\ &0.75 < \text{tailing} < 0.85|\boldsymbol{x}_c, \text{data}), \end{aligned}$$

assuming that the (coded) "%IPA" factor was a noise factor with mean 0 and standard deviation 0.1. The maximum $p(\boldsymbol{x}_c)$ value obtained over the experimental region was 0.966 (using 5,000 simulations). On the other hand, if the "%IPA" factor was not modeled as a noise factor, but instead as a control factor, the maximum $p(\boldsymbol{x})$ value obtained over the experimental region was estimated to be 1 even for 10,000 simulations. That $p(\boldsymbol{x}_c)$ is less than $p(\boldsymbol{x})$ is to be expected, because for $p(\boldsymbol{x}_c)$, the "%IPA" factor is random in a neighborhood of zero, which is not optimal when considered as a control factor.

10.4 Bayesian Model Averaging

The posterior predictive approach to response surface optimization can be taken one step further by averaging over a competing set of models. For the univariate response case (with normally distributed regression errors), this approach has been put forth by Rajagopal and del Castillo [42]. Using previous results on Bayesian model selection they obtain an expression for the posterior probability of model i, $\Pr(M_i|\text{data})$, for each of several plausible models. It follows directly that

$$p(\boldsymbol{x}) = \sum_i \Pr(Y \in A|\boldsymbol{x}, \text{data}, M_i) \Pr(M_i|\text{data}) \tag{10.7}$$

is a probability of conformance that is a weighted average over several regression models, where the weights correspond to the posterior probabilities of the models under consideration. Rajagopal and del Castillo [42] derive a computationally efficient and accurate expression for $\Pr(Y \in A|\boldsymbol{x}, \text{data}, M_i)$ that does not require Monte Carlo simulation. The expression for $\Pr(M_i|\text{data})$ is in closed form, so $p(\boldsymbol{x})$ in (10.7) can be computed quickly even for a large number of regression models.

Rajagopal and del Castillo [42] use two examples from the literature to show that the Bayesian model averaging approach that maximizes $p(\boldsymbol{x})$ in (10.7) above appears to provide a more robust optimization in the sense that it produces an $\boldsymbol{x}$-point that provides good values of $p(\boldsymbol{x})$ for any of the individual models that have a relatively high posterior probability, i.e. relatively high $\Pr(M_i|\text{data})$.

Rajagopal et al. [43] present a modification to incorporate noise variables. They do this by averaging over the noise variables to compute

$$\Pr(Y \in A|\boldsymbol{x}_c, \text{data}, M_i) = E_{\boldsymbol{x}_n}(\Pr(Y \in A|\boldsymbol{x}_c, \boldsymbol{x}_n, \text{data}, M_i)).$$

Robust process optimization (within the Bayesian model averaging framework) can then be done by maximization of

$$p(\boldsymbol{x}_c) = \sum_i \Pr(Y \in A|\boldsymbol{x}_c, \text{data}, M_i) \Pr(M_i|\text{data}).$$

With or without noise variables, the extension to multivariate response models is more challenging. The multiple response framework may require the use of MCMC methods (such as Gibbs Sampling, Geman and Geman [16]). A nice review of Gibbs Sampling is given by Casella and George [8]. A related reference by Brown et al. [7] involves variable selection for the standard multivariate regression model using MCMC techniques. See also Chapter 2 in this volume for a discussion of Bayesian computational methods.

10.5 Seemingly Unrelated Regressions Model

An extension of the work of Peterson [36] to the seemingly unrelated regressions model was given by Peterson, et al. [35]. To compute $p(\boldsymbol{x}) = \Pr(\boldsymbol{Y} \in A|\boldsymbol{x})$ for multiresponse process optimization, we need to obtain the posterior predictive distribution for $\boldsymbol{Y}$ given $\boldsymbol{x}$. The regression model considered here is the one that allows the experimenter to use a different (parametrically) linear model for each response type. This will allow for more flexible and accurate modeling of $\boldsymbol{Y}$ than one would obtain with the standard multivariate regression (SMR) model.

Here, $\boldsymbol{Y} = (Y_1, \ldots, Y_r)'$ is a vector of r response-types and $\boldsymbol{x}$ is a $k \times 1$ vector of factors that influence $\boldsymbol{Y}$ by way of the functions

$$Y_i = z_i(\boldsymbol{x})'\boldsymbol{\beta}_i + e_i, \quad i = 1, \ldots, r \tag{10.8}$$

where $\boldsymbol{\beta}_i$ is a $q_i \times 1$ vector of regression model parameters and $z_i(\boldsymbol{x})$ is a $q_i \times 1$ vector of covariates which are arbitrary functions of $\boldsymbol{x}$. Furthermore, $\boldsymbol{e} = (e_1, \ldots, e_r)'$ is a random variable with a multivariate normal distribution having mean vector $\boldsymbol{0}$ and variance-covariance matrix, $\boldsymbol{\Sigma}$. The model in (10.8) has been referred to as the "seemingly unrelated regressions" (SUR) model (Zellner [52]). When $z_1(\boldsymbol{x}) = \cdots = z_r(\boldsymbol{x}) \equiv z(\boldsymbol{x})$, one obtains the SMR model.

In order to model all of the data and obtain a convenient form for estimating the regression parameters, consider the following vector-matrix form,

$$\tilde{\boldsymbol{Y}} = \boldsymbol{Z}\boldsymbol{\beta} + \tilde{\boldsymbol{e}}$$

where $\tilde{\boldsymbol{Y}} = [\tilde{\boldsymbol{Y}}_1', \ldots, \tilde{\boldsymbol{Y}}_r']$, $\boldsymbol{\beta} = [\boldsymbol{\beta}_1', \ldots, \boldsymbol{\beta}_r']'$, $\tilde{\boldsymbol{e}} = [\tilde{\boldsymbol{e}}_1', \ldots, \tilde{\boldsymbol{e}}_r']'$, and $\boldsymbol{Z}$ is a $nr \times q$ block diagonal matrix of the form diag$(\boldsymbol{Z}_1, \ldots, \boldsymbol{Z}_r)$, with $q = q_1 + \cdots + q_r$. Here, $\tilde{\boldsymbol{Y}}_i' = (Y_{i1}, \ldots, Y_{in})$, $\tilde{\boldsymbol{e}}'_i = (e_{i1}, \ldots, e_{in})$, and $\boldsymbol{Z}_i = [z_{i1}(\boldsymbol{x})', \ldots, z_{in}(\boldsymbol{x})']'$ for $i = 1, \ldots, r$. For a given $\boldsymbol{\Sigma}$, the maximum likelihood estimate (MLE) of $\boldsymbol{\beta}$ can be expressed as

$$\hat{\boldsymbol{\beta}} = [\boldsymbol{Z}'(\boldsymbol{\Sigma} \otimes \boldsymbol{I}_n)^{-1}\boldsymbol{Z}]^{-1}\boldsymbol{Z}'(\boldsymbol{\Sigma} \otimes \boldsymbol{I}_n)^{-1}\tilde{\boldsymbol{Y}}, \tag{10.9}$$

where $\boldsymbol{I}_n$ is the $n \times n$ identity matrix and $\otimes$ is the Kronecker direct product operator. The variance-covariance matrix of $\hat{\boldsymbol{\beta}}$ is

$$\mathrm{Var}(\hat{\boldsymbol{\beta}}) = [\mathbf{Z}'(\boldsymbol{\Sigma} \otimes \boldsymbol{I}_n)^{-1}\mathbf{Z}]^{-1}.$$

The variance-covariance matrix, $\boldsymbol{\Sigma}$, can be estimated by

$$\hat{\boldsymbol{\Sigma}}^* = \hat{\boldsymbol{\Sigma}}(\hat{\boldsymbol{\beta}}^{OLS}) = \frac{1}{n}\sum_{j=1}^{n} \hat{\boldsymbol{e}}_j(\hat{\boldsymbol{\beta}}^{OLS})\hat{\boldsymbol{e}}_j(\hat{\boldsymbol{\beta}}^{OLS})', \tag{10.10}$$

where $\hat{\boldsymbol{e}}'_j(\hat{\boldsymbol{\beta}}^{OLS}) = (\hat{e}_{1j}(\hat{\boldsymbol{\beta}}^{OLS}), \ldots, \hat{e}_{rj}(\hat{\boldsymbol{\beta}}^{OLS}))$ and $\hat{e}_{ij}(\hat{\boldsymbol{\beta}}^{OLS}) = y_{ij} - z_i(x_j)'\hat{\boldsymbol{\beta}}_i^{OLS}$, $i = 1, \ldots, r$. The estimator, $\hat{\boldsymbol{\beta}}_i^{OLS}$ is the ordinary least squares estimator of $\boldsymbol{\beta}_i$ for each response-type i independently of the other responses. The estimator of $\boldsymbol{\beta}$,

$$\hat{\boldsymbol{\beta}}^* = [\mathbf{Z}'(\hat{\boldsymbol{\Sigma}}^* \otimes \boldsymbol{I}_n)^{-1}\mathbf{Z}]^{-1}\mathbf{Z}'(\hat{\boldsymbol{\Sigma}}^* \otimes \boldsymbol{I}_n)^{-1}\tilde{\boldsymbol{Y}},$$

is called the two-stage Aiken estimator (Zellner [52]). For the SMR model, the MLE for $\boldsymbol{\beta}$ exists and becomes $\hat{\boldsymbol{\beta}} = [\boldsymbol{I}_n \otimes (\mathbf{Z}'\mathbf{Z})^{-1}\mathbf{Z}']\tilde{\boldsymbol{Y}}$.

For the SUR model no closed-form posterior density or sampling procedure exists. However, using Gibbs-sampling it is easy to generate random pairs of SUR model parameters from the posterior distribution of $(\boldsymbol{\beta}, \boldsymbol{\Sigma})$. See for example Griffiths [17] (pp 263–290). Using the SUR model in (10.8) it is then straightforward to simulate $\boldsymbol{Y}$ r.v.'s from the posterior predictive distribution of $\boldsymbol{Y}$ given $\boldsymbol{x}$.

In order to compute and maximize $p(\boldsymbol{x}) = \Pr(\boldsymbol{Y} \in A|\boldsymbol{x})$ over the experimental region, it is important to have a relatively efficient method for approximating $p(\boldsymbol{x})$ by Monte Carlo simulations. The approach taken in this paper is to simulate a large number of r.v.'s from the posterior distribution of $(\boldsymbol{\beta}, \boldsymbol{\Sigma})$, and use each $(\boldsymbol{\beta}, \boldsymbol{\Sigma})$ value to generate a $\boldsymbol{Y}$ r.v. for each $\boldsymbol{x}$. In this way, the sample of $(\boldsymbol{\beta}, \boldsymbol{\Sigma})$ values can be re-used for simulating $\boldsymbol{Y}$ values at each $\boldsymbol{x}$ point, instead having to do the Gibbs Sampling all over again for each $\boldsymbol{x}$-point.

Consider the noninformative prior for$(\boldsymbol{\beta}, \boldsymbol{\Sigma})$ which is proportional to $|\boldsymbol{\Sigma}|^{-(r+1)/2}$. Note that the posterior distribution of $\boldsymbol{\beta}$ given $\boldsymbol{\Sigma}$ is modeled by

$$\boldsymbol{\beta} \sim N(\hat{\boldsymbol{\beta}}(\boldsymbol{\Sigma}), (\mathbf{Z}'(\boldsymbol{\Sigma} \otimes \boldsymbol{I}_n)^{-1}\mathbf{Z})^{-1}),$$

where $\hat{\boldsymbol{\beta}}(\boldsymbol{\Sigma})$ has the form in (10.9). This follows from Srivastava and Giles [46] (pp 317–318). Note also that the posterior distribution of $\boldsymbol{\Sigma}^{-1}$ given $\boldsymbol{\beta}$ is described by

$$\boldsymbol{\Sigma}^{-1} \sim W(n,\ n^{-1}\hat{\boldsymbol{\Sigma}}^{-1}(\boldsymbol{\beta})),$$

where W is the Wishart distribution with n df and noncentrality parameter $n^{-1}\hat{\boldsymbol{\Sigma}}^{-1}(\boldsymbol{\beta})$. This follows from a slight modification of expressions in Percy [34]. Here, $\hat{\boldsymbol{\Sigma}}(\boldsymbol{\beta})$ has the form in (10.10) with $\hat{\boldsymbol{\beta}}^{OLS}$ replaced by $\boldsymbol{\beta}$. Sampling values from the posterior distribution of $(\boldsymbol{\beta}, \boldsymbol{\Sigma})$ can be done as follows using

Gibbs sampling. Generate a $\boldsymbol{\beta}$ value using

$$\boldsymbol{\beta} = \hat{\boldsymbol{\beta}}(\boldsymbol{\Sigma}) + \boldsymbol{R}'\boldsymbol{\varepsilon}_0,$$

where $\boldsymbol{R}'\boldsymbol{R} = [\boldsymbol{Z}'(\boldsymbol{\Sigma} \otimes \boldsymbol{I}_n)^{-1}\boldsymbol{Z}]^{-1}$ and $\boldsymbol{\varepsilon}_0$ is distributed as $N(\boldsymbol{0}, \boldsymbol{I}_r)$. Generate a $\boldsymbol{\Sigma}$ value using

$$\boldsymbol{\Sigma}^{-1} = \boldsymbol{\Gamma}'\boldsymbol{S}\boldsymbol{\Gamma},$$

where $\boldsymbol{\Gamma}'\boldsymbol{\Gamma} = n^{-1}\hat{\boldsymbol{\Sigma}}^{-1}(\boldsymbol{\beta})$, $\boldsymbol{S} = \sum_{i=1}^{n} \boldsymbol{\varepsilon}_i\boldsymbol{\varepsilon}_i'$, and $\boldsymbol{\varepsilon}_0, \boldsymbol{\varepsilon}_1, \ldots, \boldsymbol{\varepsilon}_n$ are iid $N(\boldsymbol{0}, \boldsymbol{I}_r)$. To compute $p(\boldsymbol{x})$, N $\boldsymbol{Y}$-vectors are generated for each $\boldsymbol{x}$. Each simulated $\boldsymbol{Y}$-vector, $\boldsymbol{Y}^{(s)}$, is generated using

$$\boldsymbol{Y}^{(s)} = \begin{bmatrix} z_1(\boldsymbol{x})' \\ \vdots \\ z_r(\boldsymbol{x})' \end{bmatrix} \boldsymbol{\beta}^{(s)} + \boldsymbol{e}^{(s)},$$

where $\boldsymbol{e}^{(s)}$ is sampled from $N(\boldsymbol{0}, \boldsymbol{\Sigma}^{(s)})$, $(\boldsymbol{\beta}^{(s)}, \boldsymbol{\Sigma}^{(s)})$ is sampled using the Gibbs sampler, and $s = 1, \ldots, N$. For each new $\boldsymbol{x}$-point, the same $N(\boldsymbol{\beta}^{(s)}, \boldsymbol{\Sigma}^{(s)})$ pairs are used. The Bayesian reliability, $p(\boldsymbol{x})$, can be approximated by

$$\frac{1}{N}\sum_{s=1}^{N} I(\boldsymbol{Y}^{(s)} \in A),$$

for large N.

Percy [34] provides a similar, but three-step, Gibbs sampling procedure that generates a $(\boldsymbol{Y}, \boldsymbol{\beta}, \boldsymbol{\Sigma})$ triplet for a given $\boldsymbol{x}$ value. However, this is not efficient for our purposes as this Gibbs sampling procedure might have to be re-done for many $\boldsymbol{x}$-points in order to optimize $p(\boldsymbol{x})$. Percy also proposes a multivariate normal approximation to the posterior predictive distribution of $\boldsymbol{Y}$ given $\boldsymbol{x}$. However, such an approximation may not be accurate for small sample sizes. This is because one would expect the true posterior predictive distribution of $\boldsymbol{Y}$ given $\boldsymbol{x}$ to have heavier tails than a normal distribution due to model parameter uncertainty; this is indeed the case with the SMR model.

As a comparison, Peterson, et al. [35] computed $p(\boldsymbol{x}) = \Pr(Y_1 \leq 234, Y_2 \leq 19|\boldsymbol{x})$ for the three-component mixture experiment discussed in Section 10.2, but this time using a SUR multivariate regression model instead of the standard multivariate model in (10.1). In this case Peterson, et al. [35] used a Becker-type model (Becker [3]) for the second response variable. The estimated model has the form

$$\begin{aligned}\hat{y}_2 = 18.8x_1 + 15.6x_2 + 35.4x_3 - 3.59\min(x_1, x_2) - 17.7\min(x_1, x_3) \\ + 10.0\min(x_2, x_3),\end{aligned}$$

which resulted in a mean squared error of 1.71, which is a 53% reduction over the mean squared error for the quadratic model for Y_2 in (10.5). For the standard multivariate regression model, $\Pr(Y_1 \leq 234, Y_2 \leq 19|\boldsymbol{x})$ maximized over the experimental region was determined to be 0.86, while for the SUR model it was determined to be only 0.54. One might expect the best posterior predictive probability to be better for the two SUR models when one of them has a much better fit than its counterpart using the standard multivariate regression model. However, the Becker model also had a generally higher mean for Y_2 in the region surrounding the best operating conditions.

10.6 The "Dual Response" Model

In the process optimization literature the term "dual response" (regression) model has come to mean a regression model where both the mean and variance of a (single) response is a function of the experimental factors. This model is nice in that it gives the experimenter the flexibility of assessing how both the mean and variance change over the experimental region. This is an important issue with regard to maximizing $p(\boldsymbol{x}) = \Pr(Y \in A|\boldsymbol{x}, \text{data})$as $p(\boldsymbol{x})$ is sensitive to both the mean and variance of the regression response model.

Chen and Ye [11] have proposed a Bayesian hierarchical model for the dual response situation. They show how to use Gibbs Sampling to simulate from the posterior of the regression model parameters. They do not make inferences of the form $p(\boldsymbol{x}) = \Pr(\boldsymbol{Y} \in A|\boldsymbol{x}, \text{data})$, nor do they compute the posterior predictive distribution of their regression model. However, such computations are straightforward if one can simulate values from the joint posterior of all the model parameters.

They do, however, solve the difficult aspects of the problem by using a hierarchical approach in their Bayesian formulation. The set-up for their regression model is as follows. It is assumed that there are m design points, and at each design point there are two or more responses measured. Let $\boldsymbol{x}$ represent a $k \times 1$ vector of control factors. For each $\boldsymbol{x}_i$ design point, $y_{i1}, \ldots, y_{in_i}$ responses are sampled. For each y_{ij} response we have

$$y_{ij}|\boldsymbol{\beta}, \sigma_i^2 \sim N\left(\boldsymbol{x}_i'\boldsymbol{\beta}, \sigma_i^2\right) \quad \text{for} \quad i = 1, \ldots, m; \quad j = 1, \ldots, n_i.$$

Also,

$$\ln \sigma_i^2 \left|\gamma, \delta^2 \sim N(z_i'\gamma, \delta^2) \quad \text{for} \quad i = 1, \ldots, m\right.$$

The following diffuse prior distributions are used for $\boldsymbol{\beta}$, γ, and δ^2:

$$\pi(\boldsymbol{\beta}) \propto 1, \quad \pi(\gamma) \propto 1, \quad \pi(\delta^2) \propto \frac{1}{\delta^2} \exp\left(-\frac{\lambda}{\delta^2}\right),$$

where λ is a small, positive number. (They also propose a proper prior distribution for $\boldsymbol{\beta}$ and for γ.) In addition to the usual assumption of independence of the $y_{ij}'s$ (given $\boldsymbol{\beta}$ and σ_i), it is further assumed that the $\sigma_i's$ are independent conditional on γ and δ^2. Furthermore, $\boldsymbol{\beta}$, γ, and δ^2 are assumed to be apriori independent.

Chen and Ye [11] sample from the joint posterior distribution of $(\beta, \gamma, \delta^2, \sigma_1^2, \ldots, \sigma_m^2)$ using Gibbs Sampling with a modification for the simulation of the σ_i^2's conditional on the data and other parameters, because this simulation is not direct. Chen and Ye [11] use a rejection algorithm to simulate the σ_i^2's conditional on the data and other parameters. The Gibbs Sampling posterior distribution forms are:

$$\pi(\beta|\text{others}) \sim N(\hat{\beta}, \hat{\Sigma}), \quad \text{where } \hat{\beta} = (\mathbf{X}'\mathbf{V}^{-1}\mathbf{X})^{-1}\mathbf{X}'\mathbf{V}^{-1}\bar{\mathbf{Y}}$$
$$\text{and } \hat{\Sigma} = (\mathbf{X}'\mathbf{V}^{-1}\mathbf{X})^{-1}$$
$$\pi(\gamma|\text{others}) \sim N(\hat{\gamma}, \hat{\Omega}), \quad \text{where } \hat{\gamma} = (\mathbf{Z}'\mathbf{Z})^{-1}\mathbf{Z}'\boldsymbol{d} \text{ and } \hat{\Omega} = (\mathbf{Z}'\mathbf{Z})^{-1}\delta^2$$
$$\pi(\delta^2|\text{others}) \sim \mathit{Inverse-gamma}\left(\frac{N}{2}, \frac{(\boldsymbol{d}-\mathbf{Z}\gamma)'(\boldsymbol{d}-\mathbf{Z}\gamma)+2\lambda}{2}\right)$$
$$\pi\left(\sigma_i^2|\text{others}\right) \propto \frac{1}{\left(\sigma_i^2\right)^{n_i/2+1}} \exp\left\{-\frac{1}{2\sigma_i^2}\left[(n_i-1)s_i^2 + n_i(\bar{y}-\boldsymbol{x}_i'\beta)^2\right] - \frac{1}{2\delta^2}\left(\ln\sigma_i^2 - z_i'\gamma\right)^2\right\},$$

for $i = 1, \ldots, m$.

The $\bar{\mathbf{Y}}$ above is the $m \times 1$ vector of sample means at each of the m design points, $\boldsymbol{V}$ is the variance-covariance matrix for $\bar{\mathbf{Y}}$, and $\boldsymbol{d} = (\ln\sigma_1^2, \ldots, \ln\sigma_m^2)$. Here, $\boldsymbol{X}$ is an $m \times p$ design matrix for the distinct $\boldsymbol{x}$'s and $\boldsymbol{Z}$ is an $m \times q$ design matrix for the distinct z's. While it is not possible to sample from $\pi(\sigma_i^2|\text{others})$directly, Chen and Ye [11] sample indirectly using a rejection algorithm. Using their improper prior distributions on β and γ, Chen and Ye [11] prove that the resulting joint posterior density of $(\beta, \gamma, \sigma_1^2, \ldots, \sigma_m^2, \delta^2)$ is proper for $n_i > 1$ for $i = 1, \ldots, m$.

Using the MCMC approach of Chen and Ye [11], one can get posterior samples of (β, γ) and use these to estimate $p(\boldsymbol{x}) = \Pr(\boldsymbol{Y} \in A|\boldsymbol{x}, \text{data})$ for any $\boldsymbol{x}$. Typically, $A = [L, U]$ for some specified lower and upper limits, respectively. The point $\boldsymbol{x}$ does not have to be a design point, although good statistical practice dictates that $\boldsymbol{x}$ should be within (or not far outside of) the experimental region. One can estimate $p(\boldsymbol{x}) = \Pr(\boldsymbol{Y} \in A|\boldsymbol{x}, \text{data})$ by simulating many response values of $y^{(s)}$, where

$$y^{(s)} = \boldsymbol{x}'\beta^{(s)} + \exp(z'\gamma^{(s)})\varepsilon^{(s)},$$

$(\beta^{(s)}, \gamma^{(s)})$ is a sample from the joint posterior distribution of β and γ, and $\varepsilon^{(s)} \sim N(0, 1)$ independently of β and γ.

10.7 Summary and Future Possibilities

The approaches discussed above for various statistical models all have a common theme in that one is trying to compute $p(\boldsymbol{x}) = \Pr(Y \in A|\boldsymbol{x}, \text{data})$ for various values of a set of controllable factors, $\boldsymbol{x}$, for the purpose of optimizing $p(x)$. Of course, it is not surprising that one can conceive of many more statistical models can be could be embedded into the $p(x) = \Pr(Y \in A|x,$ data) paradigm for optimization. In this final section, I discuss some further possibilities and leave it to the reader to try to think of others as well. (I would be very happy to hear about ideas for models not mentioned in this chapter!)

For many investigations, the experiment must be broken up into two or more parts or "experimental batches" due to time or physical constraints or the existence of certain hard-to-change factors. This modification of the design typically induces a type of split plot experiment which involves at least two sources of random error (aside from noise variables). Gibbs Sampling methods exist for univariate split-plot models (Hobert and Casella [19]), and hence Monte Carlo estimation of $p(x) = \Pr(Y \in A|x,$ data) is possible by sampling from the joint posterior distribution of all of the model parameters and using this distribution to sample from the posterior predictive distribution of Y(as shown in the immediately proceeding section). Gibbs Sampling can also be applied to multiple-response mixed-effects models (Schafer and Yucel [45]) for estimation of $p(x)$ as well. This notion of "batches" of data may also need to come into play for a preposterior analysis if the experimenter believes that simply replicating the experiment will induce a batch effect.

An important regression model for multivariate normally distributed responses has been proposed by Chaio and Hamada [12]. It is a generalization of the SUR model in that the variance-covariance matrix of the vector of errors, as well as the vector of response means, is a function of (controllable and noise) factors. A Bayesian solution to computing $p(x) = \Pr(Y \in A|x,$ data) for controllable factors, x, using this model would provide a very general model for process optimization for normally distributed data. Again, the Bayesian approach would adjust for the substantial number of unknown parameters that could occur with such a model.

Outside of the normally distributed response situation, one may prefer to use generalized linear models or survival data (time-to-response) models of either a univariate or multivariate form. Zeger and Karim [51] provide a Gibbs Sampling procedure for a (univariate) generalized linear mixed model. But one must take care with this approach with regard to having a proper joint posterior (Hobert and Casella [19]). Besag, Green, Higdon, and Mengersen [4] provide a review of MCMC that includes a discussion of MCMC applications to the generalized linear mixed model.

Chen, Shao, and Ibrahim [10] (Chapter 10) give a nice review of simulating values from the posterior of a semiparametric proportional hazards regression model (for univariate response-time data). Once such samples are available, simulation from the posterior predictive density can be done to compute $p(x) = \Pr(Y \in A|x)$. In many cases, investigators would be most interested in maximizing the survivorship function, $S(t_0; x) = \Pr(Y > t_0|x)$, for a specified time point, t_0, say. For the multivariate response-time case, Stefanescu and Turnbull [47] have proposed a Bayesian frailty model.

It is sometimes the case that when an experiment has several responses, some of these responses may naturally be of a mixed data type. For example, some responses could be continuous while others are ordinal or binary. Or, even if all responses were of the same basic type, some could have naturally different distributions (e.g. normal & gamma or Poisson & binary). Such difficulties, combined with the existence of random effects, pose a challenge to the Bayesian approach and, in some cases, even to constructing a tractable likelihood function. But since regression models are fundamental to so many experiments, it is certain that Bayesian statisticians will be continually adding to their repertoire of computational procedures for sampling from the posterior of the model parameters. Hence, computation of $p(x) = \Pr(Y \in A|\boldsymbol{x}, \text{data})$ should be a straightforward consequence of utilization of such new methods. However, it is up to applied statisticians to test such procedures on data from a wide variety of real experiments to fully assess the utility of such methods for response surface and process optimization.

References

1. Ames, A. E., Mattucci, N., and MacDonald, S., Szonyi, G., and Hawkins, D. M. (1997) "Quality Loss Functions for Optimization Across Multiple Response Surfaces", *Journal of Quality Technology*, 29, pp. 339–346.
2. Anderson, M. J. and Whitcomb, P.J. (1998), "Find the Most Favorable Formulations", *Chemical Engineering Progress*, April, pp. 63–67.
3. Becker, N. G. (1968), "Models for the response of a mixture", *Journal of the Royal Statistics Society – Series B*, 30, 349–358.
4. Besag, J., Green, P., Higdon, D., and Mengersen, K. (1995) Bayesian Computation and Stochastic Systems (with discussion), *Statistical Science*, 10, 3–66.
5. Borror, C. M., Montgomery, D. C., and Myers, R. (2002), "Evaluation of statistical designs for experiments involving noise variables", *Journal of Quality Technology*, 34, 54–70.
6. Box, G. E. P. and Jones, S. (1990a), "Designing Products that are Robust to the Environment", *Report Series in Quality and Productivity*, CPQI, University of Wisconsin, Number 56.
7. Brown, P. J. , Vannucci, M. , and Fearn, T. (1998), "Multivariate Bayesian variable selection and prediction", *Journal of the Royal Statistical Society, Series B, Methodological*, 60 , 627–641.
8. Casella, G. and George, E. I. (1992), "Explaining the Gibbs Sampler", *The American Statistician*, 46, 167–174.
9. Chatterjee, S. , Laudato, M. , and Lynch, L. A. (1996), "Genetic algorithms and their statistical applications: An introduction", *Computational Statistics and Data Analysis*, 22 , pp. 633–651
10. Chen, M., Shao, Q., and Ibrahim, J. G. (2000), *Monte Carlo Methods in Bayesian Computation*, Springer-Verlag, New York.
11. Chen, Y. and Ye, K. (2005), "A Bayesian Hierarchical Approach to Dual Response Surface Modeling", Technical Report 05-5, Department of Statistics, Virginia Polytechnic Institute and State University, Blacksburg, VA.

12. Chiao, C. and Hamada, M. (2001) "Analyzing Experiments with Correlated Multiple Responses", *Journal of Quality Technology*, 33, pp. 451–465.
13. Del Castillo, E. (1996) "Multiresponse Process Optimization via Constrained Confidence Regions", *Journal of Quality Technology*, 28, pp. 61–70.
14. Del Castillo, E., Montgomery, D. C., and McCarville, D. R. (1996), "Modified Desirability Functions for Multiple Response Optimization", *Journal of Quality Technology*, 28, pp. 337–345.
15. Derringer, G. and Suich, R. (1980), "Simultaneous Optimization of Several Response Variables", *Journal of Quality Technology*, 12, pp. 214–219.
16. Geman, A. and Geman, D. (1984), "Stochastic Relaxation, Gibbs Distributions, and the Bayesian Restoration of Images", *IEEE Transactions on Pattern Analysis and Machine Intelligence*, 6, 721–741.
17. Griffiths, W. E. (2003), "Bayesian Inference in the Seemingly Unrelated Regressions Model", in *Computer-Aided Econometrics*, Marcel-Dekker, New York.
18. Harrington, E. C. (1965), "The Desirability Function", *Industrial Quality Control*, 21 pp. 494–498.
19. Hobert, James P. , and Casella, George (1996), "The effect of improper priors on Gibbs sampling in hierarchical linear mixed models", *Journal of the American Statistical Association*, 91, 1461–1473.
20. Hoerl, R. W. (1985), "Ridge Analysis 25 Years Later", *The American Statistician*, 39, 186–192.
21. Hunter, J. S. (1999), Discussion of "Response Surface Methodology — Current Status and Future Directions", *Journal of Quality Technology*, 31, 54–57.
22. Khuri, A. I. and Conlon, M. (1987), "Simultaneous Optimization of Multiple Responses Represented by Polynomial Regression Function", *Technometrics*, 23, pp. 363–375.
23. Khuri, A. I. and Cornell, J. A. (1996), *Response Surfaces*, 2nd ed., Marcel-Dekker, New York.
24. Ko, Y., Kim, K., Jun, C. (2005), "A New Loss Function-Based Method for Multiresponse Optimization", *Journal of Quality Technology*, 37, 50–59.
25. Lind, E. E., Goldin, J., and Hickman, J. B. (1960), "Fitting Yield and Cost to Response Surfaces", *Chemical Engineering Progress*, 56, 62–68.
26. Miró-Quesada, G., del Castillo, E., and Peterson, J.J. (2004), "A Bayesian Approach for Multiple Response Surface Optimization in the Presence of Noise variables", *Journal of Applied Statistics*, 31, 251–270.
27. Miró-Quesada, G. and del Castillo, E. (2004), "A Dual-Response Approach to the Multivariate Robust Parameter Design Problem", *Technometrics*, 46, 176–187.
28. Miró-Quesada, G. and del Castillo, E. (2004), "Two Approaches for Improving the Dual Response Method in Robust Parameter Design", *Journal of Quality Technology*, 36, 154–168.
29. Montgomery, D.C. and Bettencourt, V.M. (1977), "Multiple Response Surface Methods in Computer Simulation", *Simulation*, 29, 113–121.
30. Myers, R. H., Khuri, A. I., and Vining, G. (1992), "Response Surface Alternatives to the Taguchi Robust Parameter Design Approach", *The American Statistician*, 46, 131–139.
31. Myers, R. H. and Montgomery, D.C. (1995), *Response Surface Methodology*, John Wiley and Sons Inc., New York.
32. Myers, R. H. and Montgomery, D.C. (2002), *Response Surface Methodology*, 2nd ed., John Wiley and Sons Inc., New York.

33. Nelder, J.A. and Mead, R. (1964), "A Simplex Method for Function Minimization", *The Computer Journal*, 7, pp. 308–313.
34. Percy, D. F. (1992), "Prediction for Seemingly Unrelated Regressions", *Journal of the Royal Statistical Society — Series B*, 54, pp. 243–252.
35. Peterson, J. J., Miró-Quesada, G., and del Castillo, E. (2003), "A Posterior Predictive Approach to Multiple Response Surface Optimization (with Seemingly Unrelated Regression Models)", presented at the *ASQ Fall Technical Conference*, October 2003, San Antonio, TX.
36. Peterson, J. J., (2004) "A Posterior Predictive Approach to Multiple Response Surface Optimization", *Journal of Quality Technology*, 36, 139–153.
37. Peterson, J. J. and Kuhn, A. M. (2005), "A Ridge Analysis with Noise Variables", *Technometrics*, 47, 274–283.
38. Pignatiello, Jr. J.J. (1993). "Strategies for Robust Multiresponse Quality Engineering", *IIE Transactions* 25, pp. 5–15.
39. Press, S. J. (2003), *Subjective and Objective Bayesian Statistics: Principles, Models, and Applications*, 2nd edition. John Wiley, New York.
40. Price, W. L. (1977), "A Controlled random Search Procedure for Global Optimization", *The Computer Journal*, 20, 367–370.
41. Raiffa, H. and Schlaiffer, R. (2000), *Applied Statistical Decision Theory*, John Wiley, New York
42. Rajagopal, R. and del Castillo, E, (2005), "Model-Robust Process Optimization using Bayesian Model Averaging", *Technometrics*, 47, pp. 152–163.
43. Rajagopal, R., del Castillo, E., Peterson, J. J. (2005), "Model and Distribution-Robust Process Optimization with Noise Factors", *Journal of Quality Technology* (to appear, July).
44. Romano, D., Varetto, M, and Vicario, G. (2004), "Multiresponse Robust Design: A General Framework Based on Combined Array", *Journal of Quality Technology*, 36, 27–37.
45. Schafer, J. L. and Yucel, R. M. (2002), "Computational Strategies for Multivariate Linear Mixed-Effects Models with Missing Values", *Journal of Computational and Graphical Statistics*, 11, 437–457.
46. Srivastava, V. K. and Giles, D. E. A. (1987), *Seemingly Unrelated Regression Equations Models: Estimation and Inference*, Marcel Dekker, New York.
47. Stefanescu, C. and Turnbull, B. W. (2005), "Multivariate Frailty Models for Exchangeable Survival Data with Covariates", unpublished paper (available at http://faculty.london.edu/cstefanescu/survival.pdf).
48. Steiner S. H. and Hamada, M. (1997), "Making Mixtures Robust to Noise and Mixing Measurement Errors", *Journal of Quality Technology*, 29, 441–450.
49. Vining, G. G. and Myers, R. H. (1990), "Combining Taguchi and Response Surface Philosophies: A Dual Response Approach", *Journal of Quality Technology*, 22, 38–45.
50. Vining, G. G. (1998), "A Compromise Approach to Multiresponse Optimization", *Journal of Quality Technology*, 30, pp. 309–313.
51. Zeger, Scott L. , and Karim, M. Rezaul (1991), "Generalized linear models with random effects: A Gibbs sampling approach", *Journal of the American Statistical Association*, 86, 79–86.
52. Zellner, A (1962), "An Efficient Method of estimating Seemingly Unrelated Regressions and Tests for Aggregation Bias", *Journal of the American Statistical Association*, 57, 348–368.

11

An Application of Bayesian Statistics to Sequential Empirical Optimization

Carlos W. Moreno
Ultramax Corporation, Cincinnati, Ohio, USA

CONTENTS

ABSTRACT One of the main motivations to use Bayesian statistical models in a sequential learning environment is to get useful knowledge sooner, and thus derive benefits sooner and/or achieve desired results with less work. A second important motivation is to avoid fitting noise and attempt to get a closer picture of the underlying input/output system operating characteristics — especially when there is limited data.

The methodology presented in this chapter was originally developed for adjusting production processes so as to achieve and maintain maximum value generation with existing equipment, which is one of management's perennial problems in industrial production. "Value" depends on the process, product and business conditions; and could include a combination of: better quality (means and variations); higher throughput (particularly valuable if capacity constrained); lower consumption costs (particularly valuable if sales constrained); lower losses and rework; and lower undesirable byproducts, such as pollution; all while maintaining safety, contractual, legal and capacity constraints.

An on-line methodology to achieve the above objectives is Sequential Empirical Optimization (SEO). SEO could be applied more generally to other input/output systems where operating performance is managed through adjustment of inputs, such as racing a sail boat, or the treatment of diabetes.

This chapter presents a commercial solution for generic sequential empirical optimization. It describes the problem of optimal I/O system adjustments; the generic SEO solution approach; and finally the relevant specifics of the Ultramax® SEO solution, where in addition to the benefits of a Bayesian approach, a computational method is described that is suitable for the quick processing of data in on-line applications. An example is illustrated.

11.1 Introduction to a Typical Situation in Industrial Production

The original problem that led to this application of Bayesian statistics was this perennial problem in industrial production:

Adjust a production process so as to achieve and maintain maximum value generation rate with existing equipment. "Value" depends on the process, product and business conditions; and could include a combination of: better quality (means and variations); higher throughput (particularly valuable if capacity constrained); lower consumption costs (particularly valuable if sales constrained); lower losses and rework; and lower undesirable byproducts, such as pollution; all while maintaining safety, contractual, legal and capacity constraints. The optimal adjustments usually depend on varying conditions, such as raw materials' characteristics, environmental and process conditions, and economic and other business conditions.

Our experience with the Ultramax®'s Sequential Empirical Optimization (SEO) on-line technology (www.ultramax.com) to achieve the above objectives illustrates that a fairly large percentage of processes underutilize their latent capabilities by over $1M/year, and a very small percentage are already relatively close to the optimum.

11.2 Process Operations and its Optimization

In general, imagine a generic Input/Output System represented in Figure 11.1, where a user or manager wishes to improve or optimize operating performance. The decisions available are the adjustments of adjustable inputs through which operating performance is controlled (the control inputs), whose effects most probably depend on the value of physical conditions (uncontrolled inputs).

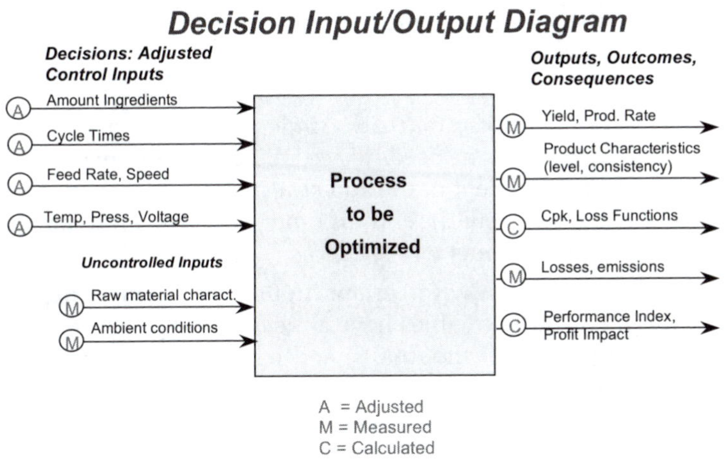

FIGURE 11.1
Decision Input/Output Diagram.

To be more specific, consider a production process as it looks to plant personnel. In particular, the "process" includes the built-in automatic logic to do process control and the sensors to collect measurements. The process control almost always includes regulatory or first-level feedback control (i.e., making controlled process parameters have desired setpoint values by manipulating some compensation for perturbations — e.g., add more steam to heat up to desired temperature). Process control may also contain some logic to adjust the setpoint values as a function of certain conditions such as uncontrolled inputs (this logic is very seldom optimal).

Better operating performance at the production floor is achieved by making better *adjustments of* the combination of:

- *manual settings*, if any, such as a manually set damper
- *setpoints* for first level feedback control (knobs or equivalent in control panels)
- *biases*, if any, for those setpoints set by the control logic — since the control logic is likely to represent only some basic engineering relationships, and is unlikely to have been refined and optimized in terms of contribution to current company objectives
- *control parameters*, possible (but done rarely) such as gains for first-level feedback control; and constants, factors, limits, etc. for control logic — if the parameter settings are accessible from the digital control system

An example of the above is shown by the adjusted variables (Role #1) in the Appendix. As explained later, the Appendix shows an Optimization Plan for operations of a boiler for power generation.

To formalize the above, imagine an Input/Output System characterized as follows:

1. The system has inputs x that can be independently adjusted; uncontrolled inputs u that one needs to accept (i.e., the current conditions) which are independent of the adjustments (but can be interdependent among themselves); and all other variables are their consequences, called outputs $y = \{y_0, y_1, y_2, \ldots\}$
2. A run of the system with certain inputs x, u produces an output vector y. The output values have a mean value which is a steady-state consequence of the inputs; and have stochastic components (noise) due to unknown uncontrolled inputs that change, errors in measurement, inherent system variability (e.g., bounded instabilities — chaos), leftover transient effects due to changes in x and continuing changes in u.

Operations with the I/O system is "equipment limited". That is, only one or a few operating runs can be made at a time, runs are sequential; e.g., an industrial production process, sailing a boat, or a personal treatment of diabetes. The opposite are "parallel runs" such as biological and agricultural experiments, where a large number of experiments run simultaneously.

System operations is managed by the decisions made on the adjustments. The objective is to maximize (or minimize) an objective function (a real-time production Performance Index in production) while variables operate within all defined constraints. The optimal adjustments depend on the values of the uncontrolled inputs (and the business conditions).

Having viable tools to define a Performance Index to represent multiple objectives is critical to doing practical optimization, and to avoid the syndrome of improving something while degrading something else at a higher cost.

The process is conceptually represented by the list of variables (some with one or two constraints), the Performance Index, and any known calculations. Putting all together this is called the Optimization Plan (OP) for operations. The Appendix displays an example for a coal-fired boiler producing steam for a power generating turbine.

The following mathematical formulation of an I/O Process and of Sequential Empirical Optimization is from Moreno [6] (pp. 138–142 in revision 5/27/05). Let:

$$y = f(x, u) + \varepsilon$$

where:

- $f(x, u)$ is the steady-state mean (i.e., after transients due to readjustments in continuous processes) process output vector, or the "response". The output vector **f** also has the indices $0, 1, 2, 3, \ldots$ where $f_0(x, u)$ is the value of the Objective Function (Performance Index).

As we shall see, the "empirical" part of the SEO solution is because the form of f and the coefficients for any approximate fit are unknown, as well as the values of ε below.

f, as usual in production processes, is relatively "smooth" in the area of interest. f could possibly be changing slowly with time (or equivalently, changing because of slow changes in unknown uncontrolled inputs not included in u). Fast changes are represented by noise.

- ε is the noise vector produced by a bell-shaped distribution with mean zero and covariance matrix Σ. The square root of the diagonal elements are the standard deviations, sigma or **noise** vector n; most likely affected by x,u.

  ε is a property of the process and of the Optimization Plan, in particular, of the inputs included and how y is measured (e.g., averages of more raw data may have lower noise, especially if u does not change too quickly).

  This concept of noise is smaller than in most quality control analyses because major changes in the known uncontrolled inputs u do not contribute to noise, while they do in regular methods.

  Almost no industrial processes produces data following a normal distribution (except averages of data obtained similarly — Central Limit Theorem). There are almost no physical principles that indicate that process outputs should be normally distributed — one exception being the energy emitted by a black object vs. the log of the frequency. On the contrary, for instance, a process often places limits — resulting in trimming tails — which destroys a normal distribution. This has been confirmed by experience, where with sufficient data one almost always can prove that the process output distributions are not normal.

Note that in this model "errors" are assigned only to the outputs. The inputs are presumed to be perfectly correct.

We define Optimal Operations as to Maximize the Objective Function (the Performance Index) — while all constraint requirements are satisfied — by making adjustments **x** for given the values of **u**. (It could be Minimize if the Objective Function is a cost or loss function.)

The constraint requirements are:

a) *The mean process values almost never violate any upper* (UC) *and lower* (LC) *constraints.*

   Note: In this Chapter "mean" is used to express the average for the same values of the inputs for one run; "average" is used for the average of several runs across time (with possible changes in inputs).

b) *The actual data for each run is almost never beyond a constraint by more than the amount "Minimum Important Difference" (MID) (defined by the user). An MID>0 allows for a "gray" area for the actual noisy data constraint violation.*

Mathematically, given the value of $\boldsymbol{u}$:

$\text{Max}_{x/u}\{f_0(\boldsymbol{x}, \boldsymbol{u})\}$	f_0 is the objective function
s.t.	(subject to; that is, while satisfying these constraint requirements)

The subscript "i" below applies to all variables with respective upper or lower constraints

Requirement (a)

$x_i \leq UC_i$	(as a decision, it has to obey constraints)
$x_i \geq LC_i$	
$f_i(\boldsymbol{x}, \boldsymbol{u}) \leq UC_i$	most of the time
$f_i(\boldsymbol{x}, \boldsymbol{u}) \geq LC_i$	most of the time

Requirement (b)

$$y_i = f_i(\boldsymbol{x}, \boldsymbol{u}) + \varepsilon_i \leq \text{UC}_i + \text{MID}_i \quad \text{most of the time}$$
$$y_i = f_i(\boldsymbol{x}, \boldsymbol{u}) + \varepsilon_i \geq \text{LC}_i - \text{MID}_i \quad \text{most of the time}$$

The "most of the time" requirement is translated into requiring that the condition be satisfied for $\varepsilon_i \leq 3\text{n}_i$ (3sigmas), called the *3sigma protection* (we are not using the criterion of the probability of violating constraints in order to avoid having to make assumptions about the distribution of the data).

Similarly, *practical optimization* is when the achieved f_0 is within MID_0 of the optimal one defined above.

So, bringing both constraint requirements (a) and (b) into one composite set of equations, the optimum we are searching is defined as:

$$\text{Max}_{x/u}\{f_0(\boldsymbol{x}, \boldsymbol{u})\}$$
$$s.t. x_i \leq \text{UC}_i$$
$$x_i \geq \text{LC}_i$$
$$f_i(\boldsymbol{x}, \boldsymbol{u}) \leq \text{UC}_i - \max\{3n_i - \text{MID}_i, 0\}$$
$$f_i(\boldsymbol{x}, \boldsymbol{u}) \geq \text{LC}_i + \max\{3n_i - \text{MID}_i, 0\}$$

which defines the *optimum adjustment* $\boldsymbol{x}^*$ and the *mean optimum outputs* $\boldsymbol{y}^* = f(\boldsymbol{x}^*, \boldsymbol{u})$; or $[\boldsymbol{x}^*, \boldsymbol{y}^*]$ for each value of $\boldsymbol{u}$.

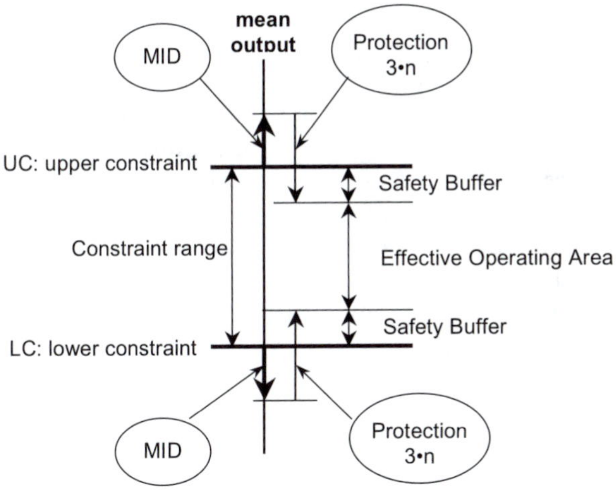

FIGURE 11.2
The "Effective Operating Area" to satisfy (b) requirement; the area of acceptable output means.

The last two inequalities define the Effective Operating Area for the mean outputs $f_i(x, u)$, where the Safety Buffer = max{3n$_i$ – MID$_i$, 0}, are illustrated in Figure 11.2. If the Safety Buffer is larger than zero this constitutes a loss in Operating Area which results in a loss of opportunities for higher f_0 if there are outputs with active constraints.

The practical objective is not just to find the optimal x, but to define x's that satisfy all constraints and produces mean outputs no further away from the optimal $f_0(x^*,u)$ than MID$_0$. The set of such x's defines the *Window of Operations*.

If there are outputs with upper and lower constraints, a sufficiently large Safety Buffer could result in *no* Operating Range, and then the process is totally incapable. In this case optimization makes little sense — the problem should be fixed by engineering or other such teams. The following procedures would apply also to reduce the losses in Operating Area due to high noise:

- find the cause of noise and reduce or eliminate its effects (reduce noise)
- find the cause of noise and include it as an uncontrolled input (recognize the source in the Optimization Plan, that is, eliminate it as a cause of noise, and enable SEO to compensate for its known value)
- relax the Constraints and/or enlarge the MIDs (relax requirements)

If process runs with Adjustments at Baseline do not satisfy some output constraints, doing the above will not necessarily assure that all constraints can be obeyed; this can be determined only after optimizing the adjustments — and there are many success stories in this respect.

11.3 A Solution: Sequential Empirical Optimization (SEO)

SEO starts with a customer-defined optimization plan that includes the variables, their roles, any constraints, the maximum reasonable adjustment changes (under MID in the Appendix), the MID for the output constraints, and any calculations for the calculated outputs, as illustrated in the Appendix.

Operations with SEO go through the following cycles to improve performance as it builds up knowledge: (1) adjust inputs; (2) produce and collect data (which will be stored); (3) collect predicted values of new uncontrolled inputs (usually the current values); (4) update prediction models; and (5) calculate new adjustments for the next cycle. The first cycle adjustments are at Baseline or current practices, from which improvements will be measured.

In the case of production processes we see that SEO is a way to automate the decision processes — the adjustments — made by plant personnel to manage performance. This is usually part of "supervisory control", including what is called "tweaking". Supervisory control practices tend to be distinctive for each crew — an undesirable source of variation.

The main characteristics of an SEO approach to optimize an I/O system are:

- It is *Empirical*: that is, series of run data $\{x, \boldsymbol{u}, y\}_t$ are known, but $f(x, \boldsymbol{u})$ is not (some may be known), and neither is ε. In particular the form of f is also not known except to assume that it is reasonably "smooth".
- It is *Sequential*: that is, it continually stores operating (run) data and continually accesses all the stored run data to extract information and knowledge from it to create updated sequential advice to adjust the process for the next cycle. Then a process run produces a new $\{x, u, y\}$ which in turn is stored in the database to repeat the cycle. Thus SEO creates the series of $\{x, u, y\}_t$ stored in a data base. The first x_t is the adjustments at Baseline, defined by the user.

  Note how the sequential analysis above emulates the process of a mind gaining experience through repeated action; and using the remembered actions, conditions and outcomes to make better action decisions in the future, an aspect of learning and skill development. Sequential analysis has a *very valuable advantage* (Wald [8]): it uses the information in newly collected run data *immediately* in order to refine knowledge and to decide where to run next, thus increasing value generated right away. By comparison, in traditional more parallel empirical studies (DOE, Neural Networks) the information is not obtained until the data is analyzed at the end, and turned into increased value even later. The simple advantage of early use of information is the most important reason why a properly implemented SEO technology such as Ultramax is in our view the fastest empirical optimizer available today.

The other synergistic reason ("bootstrapping") is that since SEO converges the optimum faster, the database has relatively more data around the optimum, which is the most valuable data to understand the location of the optimum (recall, a generic SEO does not require including the form of the response or process transfer function).

To be suitable for *on-line production optimization* SEO is meant to aim at a *local optimum* at the end-path of continuous improvements from the starting adjustments, and ideally jumping over minor local optima. On the other hand, making adjustments for finding "separate other mountains" to climb — if they exist at all — is taken as a responsibility of engineering or R&D, not of daily production.

The *performance of a SEO solution* should be evaluated, basically, by:

- How *quickly* the series $\{x, u, y\}_t$ converges to the optimum $\{x^*, u, y^*\}$.
- How *closely* the series $\{x, u, y\}_t$ converges the optimum $\{x^*, u, y^*\}$.
- In production, which is a case of continual value generation, the evaluation is the *cumulative value generated* since starting the optimization implementation project. Equivalently, one would take into account the poor performance and time that it takes, e.g., to collect experimental data for model-based alternative approaches (such as DOE and Neural Networks); and to design and implement the first-principle models and collect data to validate them.

The "quick" and "cumulative" requirements are the ones that motivate the use of Bayesian statistics as will be described below. The "closely" is aided by the "locally accurate models" briefly discussed in Section 11.5.

Depending on the SEO technology used (or on how it is adjusted), achievements in terms of the objective function f_0 may frequently be increased at the sacrifice of increasing the incidence of violating some constraints, and thus the SEO needs to proceed carefully. The sequence of x_t needs to be "intelligent": low risk, effective, and responsive to changes in conditions.

11.4 The Ultramax Approach for SEO

So far, we have defined the problem that needs to be solved, and the generic characteristics of a SEO solution. The "quick" and "cumulative" requirements described above are the ones that motivate the use of Bayesian statistics as will be described below.

Now we will develop the aspects of the Ultramax approach which takes advantage of Bayesian Statistics, as it suits the topic of this book. Some introductory aspects, and other aspects of the approach not described here are in the Blue Book [6].

11.4.1 General Aspects and Introduction to Bayesian Models

Every time a new adjustment is required, the Ultramax SEO solution is based on the following.

- Creating prediction model(s) $m(x, u)$ which is a suitable approximation to $f(x, u)$ around the best known and predicted running conditions
- Generating adjustment decisions (advice) by optimizing the prediction models

The SEO optimization, like most others, is based on prediction models, but it is unique in that it depends on sequential empirical learning to quickly extrapolate towards the optimum. A technique that is sequential but is not based on prediction models is the Simplex search method, Walters et al [9].

At the same time the technology creates $m(x, u) \cong f(x, u)$ based on the whole set of accumulated historical run data $\{x, u, y\}_t$, it also creates:

- An Area of Confidence (AOC), which is the region in x, u where $m(x, u)$ is most accurate — within the region covered by $\{x, u\}_t$. It is calculated by an elaborate pattern application of the Mahalanobis [2] distance which allows the AOC to be concave or even composed of disjointed areas.

  Optimum Estimates and Adjustment Advices are given only within the AOC — where an optimum estimate at the edge of the AOC is an indicator that optimal operating conditions are seen outside the AOC.
- An estimate of the covariance matrix Σ of ε, and therefore of the noise vector n

The following two properties about the model m for SEO yield significant simplification and effectiveness in calculations:

1. It is not required to understand the effects of each input separately; it is just necessary to be able to predict results in the AOC region. Thus, with the protection afforded by the AOC, confounding of input effects is of little consequence.
2. Trying to make m accurate away from the optimum (e.g., by making its generic form too involved or by fitting data away from the optimum) is not only unnecessary, but it distorts the fitted models resulting in lesser accuracy around the optimum.

These two properties are less demanding on $m(x, u)$ than the usual requirements for DOE, Neural Networks and First Principle models (each for different reasons).

When not using user-supplied calculations, the generic mathematical models of m are a linear or quadratic approximation of the response surface $f(x, u)$ for each output as a function of all inputs.

Models are created with Bayesian statistics when there is little data. Bayesian approaches enable us to utilize the information available in as little as two runs (with different input values) to increase the probability of moving sequentially in a direction of improvements.

What is desirable is illustrated by this thought problem. Let us suppose that:

- A process (I/O system) with two adjustments and one result is to be maximized
- There are two process runs (with different adjustments and different output values), which correspond to two points in the three dimensional space.
- Now, we ask the question: in which direction would we change the adjustments next with the highest likelihood of improving results? Obviously we would move the adjustments towards and beyond the high performing run away from the low performing run. Now, what prediction model would yield such result?

The simplest prediction model form would be a linear model with three parameters (a constant and a linear coefficient for each input).

Classical statistics cannot create a prediction model with the above data (there are two data points to estimate three parameters).

Bayesian statistics can create such a model. When using non-informative priors the result matches what our intuition above indicates. This holds true for any dimensionality of the input space! Thus Ultramax's sequential Bayesian models will start increasing the probability to obtain performance improvements starting from the *third run*. The example in Section 11.5 below with 40 inputs illustrates this behavior in Figure 11.3, where in this application the TPL is the Total Performance Loss and the symbols indicate the degree of constraint satisfaction by all variables. Note that there are improvements in the first 40 runs while classical statistics cannot even venture a guess until the 42nd run. Neural Networks, having many more coefficients, would be much worse in this respect.

11.4.2 A Short-Cut to Calculate Bayesian Prediction Models

The "pure" Bayesian, generic approach would assume a non-informative prior for the model coefficients of $\boldsymbol{m}(x, \boldsymbol{u})$, add the actual operating data, to end up with a posterior distribution of the coefficients. The prediction model used would have coefficients at the mean of the posterior distribution (there are new developments taking advantage of the distribution of the coefficients — see Rajagopal and del Castillo [7]).

Now, such true Bayesian analysis, involving multidimensional integrals, is very CPU time consuming — for instance as compared to classical regression analysis. Thus, to be practical for an on-line application, a *quasi-Bayesian* analysis was developed around 1983 and published by Hurwitz [1], 1993. In summary:

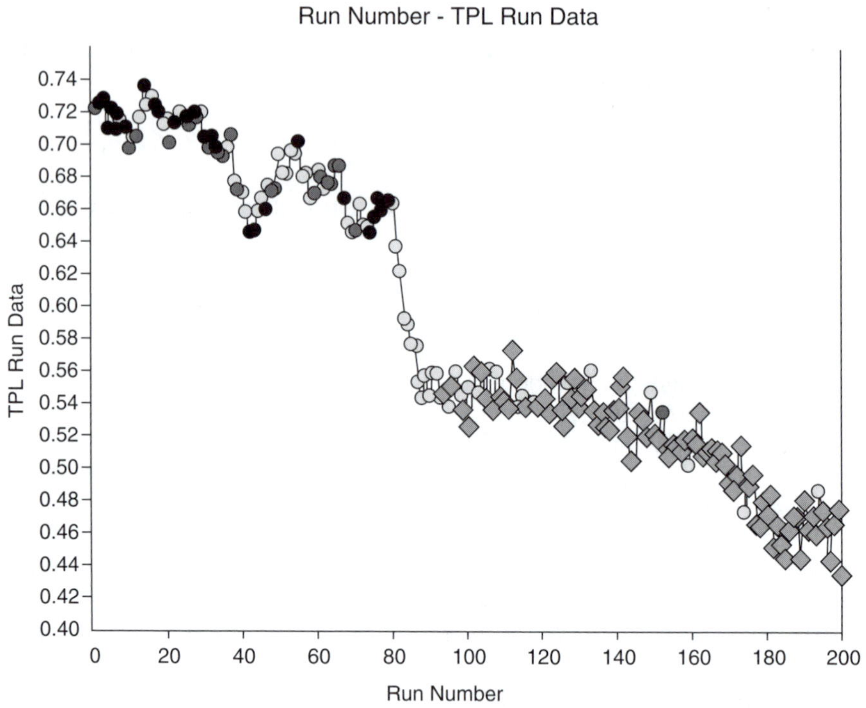

FIGURE 11.3
Plot of Sequential Performance Index values for the Example.

Models are created applying classical regression analysis using (1) dummy non-informative "prior data" and (2) the actual operating data.

This alternate approach was inspired by the solution of optimal sequential sampling plans in binomial sampling. The approach used Markov chains to represent each sequential sample, and the transition probabilities calculated with Bayesian statistics and optimized with Dynamic Optimization (see Moreno [4] and [5]). It turns out that the closed form of the Bayesian transition probabilities are a conceptual set of prior data plus the outcome from observations from the last sampling step.

It has been proven that under certain conditions Bayesian regression has a form of ridge regression (see Hurwitz [1] and Melluish et al. [3]). Since the solution here does not follow the assumptions above (it does not assume a normal distribution) we call it quasi-Bayesian.

Using prior non-informative data also ends up in a form of ridge regression, as follows:

For one run, the inputs $\boldsymbol{x}, \boldsymbol{u}$ (of total size n, the number of inputs) are expanded into the single vector $\boldsymbol{p}$ of size $(1 + n)$ for only linear terms or

$\frac{(n+1)(n+2)}{2}$ for linear plus quadratic terms. For the "linear" model coefficient matrix $\boldsymbol{B}$ the prediction equation is $\boldsymbol{m} = \boldsymbol{pB}$. To be clear, this is linear regression in form, but the model can really be quadratic in terms of the process inputs in Figure 11.1.

Let us expand the conceptual vectors $\boldsymbol{p}$ and $\boldsymbol{y}$ into matrices $\boldsymbol{P}$ and $\boldsymbol{Y}$ of data, one row per run or operating experience, but with standardized data.

There will be two sets of "data": the *prior data* and the *actual data*. Let $\boldsymbol{P}_p$ be the partition with the *prior* data and $\boldsymbol{P}_a$ be the partition with the *actual* run data. Similarly for the outputs $\boldsymbol{Y}_p$ and $\boldsymbol{Y}_a$.

$\boldsymbol{P}_a$ and $\boldsymbol{Y}_a$ will be created with standardized data for each variable (column) with mean $= 0$ and standard deviation $= 1$.

Note that if $\boldsymbol{P}$ is partitioned into separate sets of run data, $\boldsymbol{P}'\boldsymbol{P} = \boldsymbol{P}_p{}'\boldsymbol{P}_p + \boldsymbol{P}_a{}'\boldsymbol{P}_a$.

The least-square solution for the matrix $\boldsymbol{B}$ for all outputs is the traditional "normal equation":

$$\boldsymbol{B} = (\boldsymbol{P}'\boldsymbol{P})^{-1}\boldsymbol{P}'\boldsymbol{Y}$$

where eventually $\boldsymbol{B}$ will be transformed so as to use it with the original scales.

The non-informative **prior data** used has:

- An orthogonal distribution, thus $\boldsymbol{P}_p{}'\boldsymbol{P}_p = \boldsymbol{D}$ is a diagonal matrix of positive values.
- Zero for all output $\boldsymbol{Y}_p$ values (that is, representing the average actual output).

Thus,

$$\boldsymbol{P}'\boldsymbol{P} = \boldsymbol{P}_p{}'\boldsymbol{P}_p + \boldsymbol{P}_a{}'\boldsymbol{P}_a = \boldsymbol{D} + \boldsymbol{P}_a{}'\boldsymbol{P}_a$$
$$\boldsymbol{P}'\boldsymbol{Y} = \boldsymbol{P}_a{}'\boldsymbol{Y}_a$$

This is the identical form of $\boldsymbol{P}'\boldsymbol{P}$ as for ridge regression, except that $\boldsymbol{D}$ has a different interpretation: it is from prior input data rather than a penalty for estimating high values of B_i. The effect of $\boldsymbol{D}$ is practically the same: to have reasonable values of $\boldsymbol{B}$ ...

1. when otherwise it could not be calculated because $\boldsymbol{P}_a{}'\boldsymbol{P}_a$ is singular (rank-deficient), such as fewer runs than number of inputs
2. for ill-conditioned distribution of actual inputs, as it will happen for a sequence of SEO runs that approach the optimum quickly
3. to reduce fitting the noise rather than the underlying actual response surface when there is too little data, i.e., few excess data (degrees of freedom) to calculate the noise

The models created in this manner are good for predicting, but not necessarily for understanding the effects of each input separately (the partial derivatives); but as already explained the latter is not necessary for optimization.

Note that if $\boldsymbol{D} = \mathbf{0}$ then we have a totally uninformative prior, that is, only the actual data determines the estimation of coefficients, and it is regular regression. This is true separately for each component of D_i as well. The larger the component in $\boldsymbol{D}$ the more the coefficient is tempered towards zero.

Note that $\boldsymbol{D}$ is all that matters for this Bayesian regression approach. The $\boldsymbol{D}$ used here is internal to the software, the user is not aware of it.

$\boldsymbol{D}$ is the product of: (1) a diagonal matrix which is the *distribution of one prior data*, times (2) a factor which is the *amount of prior data* which is lower the larger the amount of actual data. In particular:

- The *distribution of one prior data* is a constant diagonal array that has:
  - zero for the constant term (totally non-informative, relies exclusively on actual data)
  - a uniform distribution with range ± 0.10 for linear terms
  - a uniform distribution with range ± 1.00 for quadratic terms

  This approach "tempers" quadratic terms more than the linear terms, which is useful to keep even "soft" information about direction to improvements.
- A factor that is the *amount of prior data required*, defined as follows: The *driving equation* is the "number of degrees of freedom" NDF

  $$\text{NDF} = \#\text{prior data} + \#\text{actual data} - \#\text{coefficients}$$

  (Note that the meaning of NDF is relative because for the same $\boldsymbol{D}$ the # prior data could be different depending on the meaning of the "distribution of one prior data".)

  The key heuristic applied is to force having a "Minimum NDF" (MNDF), which in turn determines the "# prior data" required. MNDF is a function of n (the number of process inputs). MNDF is n times a sigmoid transition between 1.0 with few data points to 2.0 with a lot of data. (These and all such arbitrary numbers undergo periodic refinement in the course of more empirical studies and experience.)

  Thus, the *amount of prior data* factor used is such that NDF $\geq$ MNDF; or

  $$\#\text{prior data} = \max\{\text{MNDF} + \#\text{coefficients} - \#\text{actual data}, 0\}$$

  Finally, the "# coefficients" or size of $\boldsymbol{p}$ depends on the model being fit:
  - 1 for (the model is the average) for only one actual run data
  - $n+1$ for linear model for the first as-many runs plus MNDF
  - $\frac{(n+1)(n+2)}{2}$ for full quadratic model thereafter (all second order terms including interactions between inputs)

Note that the transition from linear to quadratic models is also determined by MNDF: as soon as there is sufficient data to require no prior data for linear models, then continue with Bayesian models but with full quadratics.

Until version 6.0, Ultramax built up the quadratic models one component at a time, as explained by Hurwitz [1]. We have found that often enough the most useful quadratic components appeared later, and that is why now we use them all from the very beginning of quadratic models. To do this involves hundreds of prior data. So, we had to solve another problem: the NDF as defined above when there is prior data is not the best divider to the sum of squares of residuals. The resolution was a fitted curve nearly optimized with empirical studies.

So, after we turn to full quadratics, the large D is gradually reduced as actual run data is accumulated, until by the time that there is enough data to calculate all quadratic coefficients plus the desired MNDF. From then on there is zero prior data and thus we gradually revert to normal least-square linear regression.

The factors that define the size of one prior data and the MNDF were determined heuristically through common sense and through optimizing rounds of SEO with simulated processes. The factor optimization takes place using another copy of Ultramax acting as a higher lever optimizing supervisory control controlling those factors in the working Ultramax.

11.5 An Example

Let us now illustrate the notion of achieving performance improvements with SEO; in particular at the beginning with few run data, the reason for applying Bayesian statistics.

This I/O system has:

- 40 adjusted (decision) inputs (e.g., several Crude Feeds, Ingredients, Temperatures)
- No uncontrolled inputs (if it had changing uncontrolled inputs it would be much more difficult to illustrate the gains just from better adjustments)
- 55 measured outputs (e.g., several % Solids, Viscosities, Pressures) plus 4 calculated outputs (sums of loss function for: Throughput, all Solids, Viscosities, and Pressures, plus the addition of these sums for a Total Loss Function (TPL) to be minimized.)

- 31 explicit constraints, of which 10 are active. 16 are constraints on Crude Feeds (decision inputs), and 15 are constraints on Pressures (consequent outputs).
- 66 (upper and lower) quality specifications which are represented as losses from not being at ideal target in the Performance Index. This is done because outstanding production quality is *not* just to satisfy specification constraints, but to be consistently as close as possible to nominal design values — a significant awareness in world-class quality considerations. (It would also work *in addition* to include the quality specifications limits as constraints.)

While ideally the user would construct an economic impact calculation, in this particular application that was impossible. Here the Performance Index (Objective Function) is a quasi-economic "Total Performance Loss" (TPL) of a "cost" metric to be minimized, representing 11 Crude Feeds to be maximized and 33 outputs to be as close to nominal (target) values as possible. The latter is actually a way to use optimization to apply regulatory process control to outputs which cannot be controlled directly by process control systems.

The adjustments at baseline satisfy the explicit constraints but frequently violate one of the quality specification limits — one of the reasons SEO was applied.

In Figure 11.3 we see a plot of the TPL Performance Index values for the first 200 sequential operating runs with SEO optimization. At 45 min. per optimization adjustment cycle the plot represents about 150 hours or 20 shifts of "normal" operations — no: startup, shutdown, changeover, cleaning, or while doing maintenance.

Note in particular that the gains obtained during the first 40 runs are possible because of the Bayesian models, while the simplest standard regression models cannot even be created (more linear coefficients than actual data).

Until about run # 81 the model form was linear, and from then on a full quadratic. Then, until about 900 runs (not shown) the driving equation calls for Bayesian models (with decreasing priors), and thereafter it will be regular regression. Note, as well, that the gains due to quadratic models up to the 900 runs are possible because of the Bayesian models.

At the level of using regular regression, another novel approach is used aimed at the objective of getting closer to the optimum: weighted regression to create locally accurate models around the optimum — and reducing or avoiding altogether the distortion created by fitting data away from the optimum (Property #2 in Section 4a above).

The rate of SEO improvement is determined by several factors, including:

- the maximum change in adjustments that the user considers prudent from a production operating run to the next, the input MIDs

- the noise level n
- the gradient of the response surface (vis-à-vis the noise level and the maximum allowed changes in adjustments)

The lighter diamonds in Figure 11.3 are an indication of pushing against at least one explicit output constraint. Had the quality requirements **also** be included as constraints, then the first few runs would have been darker diamonds (none in the plot), to indicate definite constraint violation.

Eventually, constraint sensitivity analysis will lead to understanding their impact on business, and to constraint relaxation or elimination through re-engineering.

11.6 Summary

Designing a Sequential Empirical Optimization tool that delivers likely improvements even in very early sequential cycles is solved by creating prediction models which are reasonable when:

- there is not sufficient data to generate classical least-square models
- the matrix of input values is ill- conditioned
- there is too little data to discriminate between underlying process behavior and noise

The method described here is to create Bayesian regression models with appropriate priors.

In addition, this chapter also offers a quasi-Bayesian solution which is computationally much faster (and simpler) and thus suitable for on-line applications. The solution is simply to expand the normal equations by adding a Bayesian diagonal matrix to the standardized $P_a'P_a$ of actual data to represent non-informative prior data according to rules such as discussed above. This approach turns out to be similar to ridge regression.

The benefit of enabling the user to derive improvements sooner than otherwise is a great complement to the "sequential" aspect of SEO that enables faster improvements and optimization than all alternative approaches.

Since all this logic is built in and applied automatically, this makes starting and restarting SEO very easy for the customer, such as restarting after making such changes to the process when the old data is no longer representative. This benefit also applies when restarting due to adding new input variables to the Optimization Plan.

All this adds up to a practical, effective and easy-to-use solution suitable for generic on-line analysis for optimizing production operations through better adjustments.

Appendix

Optimization Plan for combustion optimization of a power plant boiler.

PROCESS OPTIMIZATION PLAN

15:41 10 APR 2005

APPLICATION: xxxxxx.umax #7 Boiler Optimisation
ULTRAMAX® 7.0.02 ©1982-2004, Ultramax Corporation. All rights reserved.

VAR #	NAME	UNITS	DEPENDENCY	ROLE	MODE	MID	CONSTRAINTS LO	CONSTRAINTS HI
1	MILL_A_run	on/off		2	A	1.	--	--
2	MILL_B_run	on/off		2	A	1.	--	--
3	MILL_C_run	on/off		2	A	1.	--	--
4	MILL_D_run	on/off		2	A	1.	--	--
5	MILL_E_run	on/off		2	A	1.	--	--
6	MILL_F_run	on/off		2	A	1.	--	--
7	MILL_A_man	Yes-No	(MILL_A_run	2	H	1.	--	--
8	MILL_B_man	on/off	MILL_B_run	2	H	1.	--	--
9	MILL_C_man	on/off	MILL_C_run	2	H	1.	--	--
10	MILL_D_man	on/off	MILL_D_run	2	H	1.	--	--
11	MILL_E_man	on/off	MILL_E_run	2	H	1.	--	--
12	MILL_F_man	on/off	MILL_F_run	2	H	1.	--	--
13	MILL_A_bias	%	MILL_A_run	1	H	4.	80.	100.
14	MILL_B_bias	%	MILL_B_run	1	H	4.	80.	100.
15	MILL_C_bias	%	MILL_C_run	1	H	4.	80.	100.
16	MILL_D_bias	%	MILL_D_run	1	H	4.	80.	100.
17	MILL_E_bias	%	MILL_E_run	1	H	4.	80.	100.
18	MILL_F_bias	%	MILL_F_run	1	H	4.	80.	100.
19	MILL_A_feed	%	MILL_A_man	1	A	5.	30.	90.
20	MILL_B_feed	%	MILL_B_man	1	A	5.	30.	90.
21	MILL_C_feed	%	MILL_C_man	1	A	5.	30.	90.
22	MILL_D_feed	%	MILL_D_man	1	A	5.	30.	90.
23	MILL_E_feed	%	MILL_E_man	1	A	5.	30.	90.
24	MILL_F_feed	%	MILL_F_man	1	A	5.	30.	90.
25	O2_sp	% dry vo		1	H	0.6	2.8	4.
26	RHTR_A_sp	deg c		1	H	3.	530.	545.
27	RHTR_B_sp	deg c		1	H	3.	530.	545.
28	RHTR_SPRAY_s	Deg C		1	H	8.	450.	580.
29	SAH_A_CET_sp	deg c		1	H	3.	55.	65.
30	SAH_B_CET_sp	deg c		1	H	3.	55.	65.
31	PAH_A_TP_sp	deg C		1	H	20.	--	--
32	PAH_B_TP_sp	deg c		1	H	20.	--	--
39	MILL_TP_sp	deg c		1	H	3.	--	90.
40	ID_FAN_bias	% bias		1	H	4.	-10.	10.
41	FD_FAN_bias	% bias		1	H	4.	-10.	10.
42	PA_BUS_PR_sp	kPa		1	H	1.5	7.5	12.

```
 43 MAIN_A_sp      deg c                     1   H    3.   500.       545.
 44 MAIN_B_sp      deg c                     1   H    3.   500.       545.
 45 MW_LOAD        MW                        2   H    60.    --         --
 46 SOOTBLOW_HR    hours                     2   H    100.   --         --
 49 FD_IL_TEMP     deg c                     2   H    12.    --         --
 50 COAL_SE        MJ/kg                     2   H    1.2    --         --
 51 COAL_MOIST     %                         2   H    1.5    --         --
 52 COAL_ASH       %                         2   H    1.2    --         --
 53 FEED_W_FLOW    kg/s                      4   H    1.     --         --
 54 FEED_W_TEMP    deg c                     5   H    1.     --         --
 55 MAIN_A_TMP     deg c                     5   H    0.5  528.       545.
 56 MAIN_B_TMP     deg c                     5   H    0.5  528.       545.
 57 MAIN_TEMP_av   deg c                     5   C    0.5  528.       545.
 58 RHT_A_TMP      deg c                     5   H    0.5  528.       545.
 59 RHT_B_TMP      deg c                     5   H    0.5  528.       545.
 60 RHT_TMP_avg    deg c                     5   C    0.5  528.       545.
 61 RHT_SPRAY      kg/s                      5   H    0.5    --        20.
 62 MAIN_PRESS     MPa                       5   H    0.2     8.       17.
 63 GAS_EX_TMP     deg c                     5   H    0.5    80.      150.
 64 GAS_DIF_TMP    deg c                     5   C    0.5    70.      130.
 65 FLUE_A_O2      %                         5   H    0.1  2.5          5.
 66 FLUE_B_O2      %                         5   H    0.1  2.5          5.
 67 FLUE_O2_avg    %                         5   C    0.1  2.5          5.
 68 FLUE_B_CO      % CO                      5   H    10.    --       150.
 69 DUST_A         %                         5   H    3.8    --        30.
 70 DUST_B         %                         5   H    3.8    --        30.
 71 CinD_B         % carbon                  5   H    1.     --        20.
 72 CinD_C         % carbon                  5   H    1.     --        20.
 73 CinD_avg       % carbon                  5   C    1.     --        20.
 74 UNIT_FUEL      kg/s                      4   H    --     --         --
 75 BOILER_HEAT    MW                        4   H    --     --         --
 76 BN_TILT_A      deg                       4   H    --     --         --
 77 BN_TILT_B      deg                       4   H    --     --         --
 78 B_EFF_LOSS     %                         5   C    0.1    --         --
 79 TURB_EFF       % corr                    5   C    0.01   --         --
 80 TOTAL_EFF      % blr ef                  5   C    0.1    --         --

---| ----------| -------| ------------| --| --| -----| -----| ------|
  MAXimizing objective function, variable # 80, "TOTAL_EFF" (Type 6)

 Calculations:
MAIN_TEMP_av = ( MAIN_A_TMP + MAIN_B_TMP )/2
RHT_TMP_avg  = ( RHT_A_TMP + RHT_B_TMP )/2
GAS_DIF_TMP  = ( GAS_EX_TMP - FD_IL_TEMP )
FLUE_O2_avg  = ( FLUE_A_O2 + FLUE_B_O2 )/2
CinD_avg     = ( CinD_B + CinD_C ) /2
B_EFF_LOSS   = (88.99
               + (( COAL_SE * 0.004876 - 0.12032)
               + ( CinD_avg *(-0.00258) + 0.015482)
               + ( GAS_DIF_TMP*(-0.000612) + 0.057354)
               + ( COAL_ASH * 0.000273 - 0.00568)
               + (FLUE_O2_avg *(-0.002215) + 0.006165)
               )*100)
TURB_EFF     = (1
               +(((0.0006043* MAIN_PRESS * 1000)-9.64479)/100)
```

```
                +(((0.022703* MAIN_TEMP_av )-12.2109)/100)
                +(((0.0272571* RHT_TMP_avg )-14.655)/100)
                )
TOTAL_EFF     = B_EFF_LOSS * TURB_EFF

       Definitions:
Role
    Role = 1 Input - Control
    Role = 2 Input - Uncontrolled
    Role = 4 Output - monitored
    Role = 5 Output -- important (has constraints or is used in
               calculations)
    Role = 6 Objective Function -- Performance Index
    Role = 8 Input -- ruled
Entry Mode
    Mode = H Manual Entry (by Hand)
    Mode = C Calculated (within ULTRAMAX with equations given by the
               user)
    Mode = A Automatic, entered through integration
```

References

1. Hurwitz, Arnon M. (1993) "Sequential Process Optimization with a Commercial Package", Arizona State University; 6th National Symposium: Statistics and Design in Automated Manufacturing, Phoenix, AZ, February 1993.
2. Mahalanobis P.C. (1936). "On the generalized distance in statistics", Proceedings of Natural Sciences, India 2:49–55.
3. Melluish, Thomas; Saunders, Craig; Nouretdinov, Ilia, and Vovk, Volodya (>2000) "Comparing the Bayes and Typicalness Frameworks". Computer Science, Royal Holloway, University of London, Egham, Surrey, TW20 0EX; website http://www.ecs.soton.ac.uk/~cjs/publications/TypBayes_ECML01.pdf
4. Moreno, C.W. (1979). "A Performance Approach to Attribute Sampling and Multiple Action Decisions," AIIE Transactions, September, pp. 183-97.
5. Moreno, C.W. (1981) "Scientific Common-sense in Acceptance Sampling", unpublished, but the base for an quality control sampling system applied to production at Procter & Gamble.
6. Moreno, C.W. (1995–2005) "Maximizing Profits through Production. *The ULTRAMAX Method: Concepts and Procedures*" (The Blue Book) Ultramax Corporation, Cincinnati, Ohio, USA., Ed. Nov 2004.
7. Rajagopal, Ramkumar and del Castillo, Enrique (2005): Model-Robust Process Optimization Using Bayesian Model Averaging. Technometrics May 2005, Vol 46, No. 2.
8. Wald, A. (1947). *Sequential Analysis*, John Wiley & Sons, New York, New York.
9. Walters, F.H. et al. (1991). *Sequential SIMPLEX Optimization*, CRC Press, Inc., Boca Raton, Florida.

12

Bayesian Estimation from Saturated Factorial Designs

Marta Y. Baba and Steven G. Gilmour
School of Mathematical Sciences, Queen Mary, University of London

CONTENTS

ABSTRACT Saturated designs are very useful for screening factors and in experiments where the observations are very difficult or expensive to obtain. Due to lack of degrees of freedom available to make inferences about the parameters in the model, the frequentist approach is extremely limited, so the Bayesian approach is a good way to analyse data from saturated designs. However, because of the reliance on prior information, Bayesian methods must be used carefully. Conjugate priors are shown to be somewhat inflexible, whereas priors using finite mixtures of densities yield more natural posterior densities. Also shown is that non-conjugate priors, with independence between the effect parameters and the variance, can be useful.

12.1 Introduction

Factorial designs of various types are among the most commonly used statistical methods in industrial research and development. They can be used whenever a product or process is to be improved or optimized experimentally.

In the early stages of experiments, often a large number of candidate factors exist to test and the initial objective is to identify the factors that have the largest influences on the outcome of this experiment, i.e., those in which the factors are active. Saturated designs allow the screening of these factors by testing the largest number of factors with the least number of observations. They are popular as economical designs when a large number of factors must be tested on an expensive process, e.g., when testing is destructive.

A *saturated design* is a fractional factorial design in which the number of parameters in the main effects model is equal to the number of runs. Here we concentrate on two-level saturated designs in which $k = n - 1$ main effects are considered in n experimental units without replicates. In such designs, all information is used to estimate the effect parameters, leaving no degrees of freedom to estimate the error variance. The classical analysis of data from saturated designs allows only the estimation of main effects under the assumption that interactions are negligible.

The frequentist approach is extremely limited, being confined to the construction of the normal probability plot of main effect estimates, and there being no degrees of freedom left to estimate the experimental error variance. Consequently, standard tests cannot be used to identify the significant factors in the experiment without making very strong prior assumptions. On the other hand, in a Bayesian framework it is possible to estimate all these parameters is possible, where the marginal posterior distribution of each effect can show if that effect is active or not. Also, graphical displays of the posterior distributions of effects can be developed, making their interpretation easy. For all these reasons, a Bayesian approach is a good way to analyse data from saturated designs.

Most work on two-level saturated designs is based on the groundbreaking work of Plackett and Burman [12], who constructed designs from Hadamard matrices of order n, where n is a multiple of 4, that is, $n \equiv 0 \bmod 4$. Such Plackett-Burman designs use a set of orthogonal columns, thus giving optimal estimation of all main effects.

Our goal in this chapter is to show practical applications of Bayesian methods applied to data from saturated designs. It is shown that the conjugate prior gives some undesirable results and the prior using a finite mixture of densities improves the results of the posterior densities, although it produces some unexpected results. This chapter is structured as follows: Section 12.2 describes the statistical model for the data from a saturated design used in this work. In Section 12.3, a brief review of the frequentist approach and some comments about the subjectivity of the methods used are presented. Section 12.4 describes the Bayesian approach, where estimation is discussed using conjugate priors, priors that are finite mixtures of conjugate densities and non-conjugate priors. Finally, the last section is dedicated to final comments about Bayesian methods for saturated designs.

TABLE 12.1

Saturated Design with 8 Runs for 7 Factors

Factor						
A	B	C	D	E	F	G
−	−	−	−	−	−	−
−	−	+	−	+	+	+
−	+	−	+	−	+	+
−	+	+	+	+	−	−
+	−	−	+	+	−	+
+	−	+	+	−	+	−
+	+	−	−	+	+	−
+	+	+	−	−	−	+

12.2 The Model for Saturated Designs

The model for the data from a two-level saturated design is defined as

$$E(Y) = \beta_0 + \beta_1 x_1 + \beta_2 x_2 + \cdots + \beta_k x_k, \tag{12.1}$$

where x_i represents the ith factor's main effect. The model can be expressed in matrix notation as

$$E(\mathbf{Y}) = \mathbf{X}\beta, \tag{12.2}$$

with $\mathbf{X}$ an $n \times p$ matrix showing the levels at which the factors are fixed, β a $p \times 1$ vector of parameters (intercept included), and $\mathbf{Y}$ an $n \times 1$ vector of observations. A simple example of a design matrix of a saturated design with 8 runs for 7 factors is presented in Table 12.1.

12.3 Classical Analysis

The difficulty that arises in saturated factorial designs is that the usual analysis of variance (ANOVA) can no longer be used because, although main effect estimates can be obtained by least squares, no degrees of freedom are left to estimate error. Hence, frequentist techniques can allow estimation but cannot be used to make inferences about parameters. Assuming that the matrix $\mathbf{X}'\mathbf{X}$ is nonsingular, as for a Plackett-Burman design, the least squares estimators of the main effects are obtained from

$$\widehat{\beta} = (\mathbf{X}'\mathbf{X})^{-1}\mathbf{X}'\mathbf{y}. \tag{12.3}$$

A graphical method has been proposed [6] to identify the important effects on the response variable, that is, to show up the active factors that produce a change in the response. The estimated effects are arranged on a half-normal probability plot. A similar method using the full normal plot to show the important effects [2, 7] is now more popular.

The idea behind these methods is to give a simple exploratory analysis to make judgements about the estimated effects. If the distribution of the response variable $\mathbf{Y}$ is assumed Normal with mean $\mathbf{X}\beta$ and dispersion matrix $\sigma^2\mathbf{I}$, then the sampling distribution of $\widehat{\beta}$ is Normal with mean β and dispersion matrix $\sigma^2(\mathbf{X}'\mathbf{X})^{-1}$. Therefore, under the null hypothesis that all the effects are zero, the means of all the estimated effects are zero and the estimated effects will be close to a straight line. When some of the effects are nonzero, the corresponding estimated effects will tend to be larger and fall off the straight line. For positive effects, the estimated effects fall above the line whereas those for negative effects fall below the line.

The problem with this method is the subjectivity inherent in it because the visual judgement used to analyse the active factors is particular to each individual. From Figure 12.1, clearly the factors J and L are active but the effect B is not so easy to judge. Furthermore, no frequentist inference of any kind can be determined and this analysis can only ever be regarded as exploratory.

A detailed review of the analysis of unreplicated factorial experiments is given in [9], where many attempts to carry out formal inference for saturated designs are described. Among the methods presented in that paper, that of Lenth [10] is by far the most often used due to its computational simplicity. The performance of this method has been shown to be reasonable [9]. Lenth's method can be presented as a formal test of effect significance for unreplicated experiments [15] where sample variance is absent. Lenth's method uses robust estimation of the standard deviation of estimated factorial effects to define a measure called the pseudo-standard error, that is, a trimmed median that attempts to remove contrasts corresponding to active effects. It can be shown to have reasonable frequentist properties in many cases but note that it cannot be considered to be a strictly correct frequentist procedure. It is possible to construct examples where it fails completely, so that by the formal definition, the size of the test is always 100%.

The first Bayesian method proposed for analyzing data from saturated designs was due to Box and Meyer [3]. They suggested using priors for each effect that are a mixture of two Normal distributions, both having mean zero, one with a small variance and one with a large variance. They concentrated on the mixing parameter, which they interpreted as the prior probability of an effect being active. The priors were then updated using the estimated effects

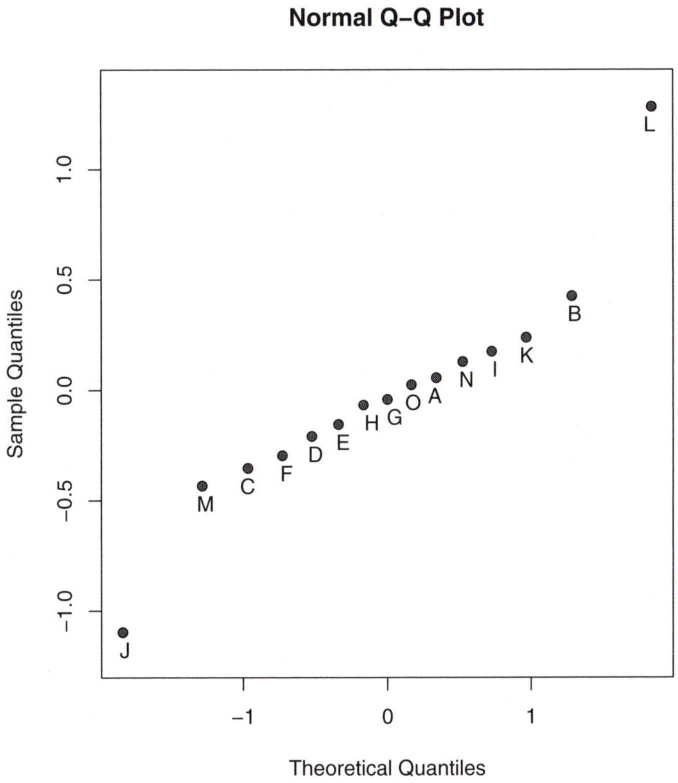

FIGURE 12.1
Normal plot of main effects using hypothetical data.

as data to obtain the posterior probability of each effect being active. Here we take this approach further by using the full prior and posterior distributions and concentrating on estimating the effects rather than just identifying them as being active or inactive.

12.4 Bayesian Estimation

Bayesian inference differs from classical procedures by formalizing the prior knowledge in the form of subjective probability distributions, incorporating the information from the data, and then providing updated posterior probability distributions based on these two sources. To carry out this approach, it is necessary to specify the prior distribution for the parameters that represent the prior beliefs about the possible values of parameters before obtaining the data. Bayesian inference has been termed "subjective" inference because it allows a certain subjectivity in the selection of the prior distribution. In situations where the data do not provide much information, the prior distribution can strongly affect the posterior distribution, as we will see.

If care is taken over the choice of prior, Bayesian inference methods can give very satisfactory answers when a saturated design is being used. Therefore, in the next sections some approaches to the use of Bayesian methods are presented, such as conjugate priors, priors as mixtures of conjugate densities, and non-conjugate priors.

12.4.1 Conjugate Priors

The model for the data from a saturated design is a particular case of the general linear model, and several works have proposed its analysis using Bayesian methods [1, 4, 11]. However, a particular detail of a saturated design appears in the likelihood function that shows what is different from the usual general linear model as follows.

The likelihood function of $\boldsymbol{\beta}$ and σ^2 is

$$f(\mathbf{y}|\boldsymbol{\beta}, \sigma^2) = (2\pi\sigma^2)^{-n/2} \exp\left\{-\frac{1}{2\sigma^2}(\mathbf{y} - \mathbf{X}\boldsymbol{\beta})'(\mathbf{y} - \mathbf{X}\boldsymbol{\beta})\right\}. \tag{12.4}$$

Another way to write this expression is by expanding the quadratic form $(\mathbf{y} - \mathbf{X}\boldsymbol{\beta})'(\mathbf{y} - \mathbf{X}\boldsymbol{\beta})$ to justify the choice of the conjugate prior density, i.e.,

$$f(\mathbf{y}|\boldsymbol{\beta}, \sigma^2) = (2\pi\sigma^2)^{-n/2} \exp\left\{-\frac{1}{2\sigma^2}[(\boldsymbol{\beta} - \widehat{\boldsymbol{\beta}})'\mathbf{X}'\mathbf{X}(\boldsymbol{\beta} - \widehat{\boldsymbol{\beta}}) + Q]\right\}, \tag{12.5}$$

where $\widehat{\boldsymbol{\beta}} = (\mathbf{X}'\mathbf{X})^{-1}\mathbf{X}'\mathbf{y}$ is the well-known classical maximum likelihood estimator of $\boldsymbol{\beta}$, which is the same as the least squares estimator, and $Q = (\mathbf{y} - \mathbf{X}\widehat{\boldsymbol{\beta}})'(\mathbf{y} - \mathbf{X}\widehat{\boldsymbol{\beta}})$ is the residual sum of squares.

In a saturated design, estimating the vector of parameters $\boldsymbol{\beta}$ is feasible but the amount Q is always zero, which means that the error variance cannot be estimated from the likelihood and so the analysis used in the general linear model is not proper for saturated designs.

From Equation (12.5) and following the same structure of the likelihood function expression, it is natural to propose a joint prior density of $\boldsymbol{\beta}$ and σ^2 as

$$f(\boldsymbol{\beta}, \sigma^2) = \frac{\left(\frac{a}{2}\right)^{\frac{d}{2}} (\sigma^2)^{-\frac{(d+p+2)}{2}}}{(2\pi)^{p/2} |\mathbf{V}|^{1/2} \Gamma\left(\frac{d}{2}\right)} \exp\left\{-\frac{1}{2\sigma^2}\left[(\boldsymbol{\beta} - \mathbf{m})'\mathbf{V}^{-1}(\boldsymbol{\beta} - \mathbf{m}) + a\right]\right\}, \tag{12.6}$$

with hyperparameters $a > 0$, $d > 0$, $\mathbf{m} \in \Re^p$ and $\mathbf{V}$ a $p \times p$ positive definite matrix.

The prior density in Equation (12.6) is known as the Normal-Inverse Gamma, denoted by $N\text{-}IG(a, d, \mathbf{m}, \mathbf{V})$, and it is useful to explore some features of this density to see the dependence among the parameters $\boldsymbol{\beta}$ and σ^2. For instance, the conditional distribution of $\boldsymbol{\beta}$ given σ^2 is Normal with mean $\mathbf{m}$

and dispersion matrix $\sigma^2\mathbf{V}$. It is not complicated to see that the marginal density of σ^2 is that of the Inverse-Gamma distribution with shape parameter $d/2$ and scale parameter $a/2$. Also, the marginal distribution of $\boldsymbol{\beta}$ is a multivariate t distribution with d degrees of freedom, location vector $\mathbf{m}$, and scale matrix $\mathbf{V}$. A summary of means and dispersions related to the parameters is as follows:

$$\begin{aligned}
&\text{(i)}\ E(\boldsymbol{\beta}) = \mathbf{m}\\
&\text{(ii)}\ Var(\boldsymbol{\beta}) = \tfrac{a\mathbf{V}}{d-2}, \quad d > 2\\
&\text{(iii)}\ E(\sigma^2) = \tfrac{a}{d-2}, \quad d > 2\\
&\text{(iv)}\ Var(\sigma^2) = \tfrac{2a^2}{(d-2)^2(d-4)}, \quad d > 4\\
&\text{(v)}\ E(\boldsymbol{\beta}|\sigma^2) = \mathbf{m}\\
&\text{(vi)}\ Var(\boldsymbol{\beta}|\sigma^2) = \sigma^2\mathbf{V}
\end{aligned} \tag{12.7}$$

Note that $Var(\boldsymbol{\beta}) = E(\sigma^2)\mathbf{V}$ and $Var(\boldsymbol{\beta}|\sigma^2) = \sigma^2\mathbf{V}$, showing the complete dependence of $\boldsymbol{\beta}$ on σ^2.

Therefore, the posterior density can be written as

$$f(\boldsymbol{\beta}, \sigma^2|\mathbf{y}) = \frac{\left(\frac{a^*}{2}\right)^{\frac{d^*}{2}} (\sigma^2)^{-\frac{(d^*+p+2)}{2}}}{(2\pi)^{p/2}\, |\mathbf{V}^*|^{1/2}\, \Gamma\left(\frac{d^*}{2}\right)} \exp\left\{-\frac{1}{2\sigma^2}[\,(\boldsymbol{\beta} - \mathbf{m}^*)'(\mathbf{V}^*)^{-1}(\boldsymbol{\beta} - \mathbf{m}^*) + a^*]\right\}, \tag{12.8}$$

where $\mathbf{V}^* = (\mathbf{V}^{-1} + \mathbf{X}'\mathbf{X})^{-1}$, $d^* = d + n$, $\mathbf{m}^* = \mathbf{V}^*(\mathbf{V}^{-1}\mathbf{m} + \mathbf{X}'\mathbf{y})$, and $a^* = a + \mathbf{m}'\mathbf{V}^{-1}\mathbf{m} + \mathbf{y}'\mathbf{y} - \mathbf{m}^{*\prime}(\mathbf{V}^*)^{-1}\mathbf{m}^*$.

To understand the effect of the dependence, it is interesting to draw some pictures of posterior densities and see the influence of the choice of hyperparameters in the prior distribution. Example 1 shows a numerical application where a set of data is simulated to illustrate the prior and posterior densities.

Example 1:
Consider a 12-run Plackett-Burman design and also consider simulated data from the true model $Y = 2x_1 + x_2 + 1.5x_3 + \varepsilon$ with $\varepsilon \sim N(0, 0.25^2)$, where

$$x_1 = \begin{cases} -1 & \text{when factor A uses "low" level } (-) \\ 1 & \text{when factor A uses "high" level } (+). \end{cases}$$

A similar coding is used for factors B to K. This means that the factors A, B, and C are active in the experiment. The simulated data are shown in Table 12.2.

From Equation (12.3), the least squares estimates of the main effects and the intercept can be easily calculated and are as follows:

TABLE 12.2

Twelve-Run Plackett-Burman Design

Factor											
A	B	C	D	E	F	G	H	I	J	K	response
+	+	−	+	+	+	−	−	−	+	−	1.7502
+	−	+	+	+	−	−	−	+	−	+	2.7130
−	+	+	+	−	−	−	+	−	+	+	0.4377
+	+	+	−	−	−	+	−	+	+	−	4.4825
+	+	−	−	−	+	−	+	+	−	+	1.4302
+	−	−	−	+	−	+	+	−	+	+	−0.5429
−	−	−	+	−	+	+	−	+	+	+	−4.4129
−	−	+	−	+	+	−	+	+	+	−	−1.4384
−	+	−	+	+	−	+	+	+	−	−	−2.4942
+	−	+	+	−	+	+	+	−	−	−	2.5350
−	+	+	−	+	+	+	−	−	−	+	0.8998
−	−	−	−	−	−	−	−	−	−	−	−4.6797

$$\widehat{\beta}_0 = 0.057,\ \widehat{\beta}_1 = 2.005,\ \widehat{\beta}_2 = 1.028,\ \widehat{\beta}_3 = 1.548,$$
$$\widehat{\beta}_4 = 0.031,\ \widehat{\beta}_5 = 0.091,\ \widehat{\beta}_6 = 0.071,\ \widehat{\beta}_7 = 0.021,$$
$$\widehat{\beta}_8 = -0.069,\ \widehat{\beta}_9 = -0.010,\ \widehat{\beta}_{10} = -0.011,\ \widehat{\beta}_{11} = 0.031.$$

Figure 12.2 shows the normal plot of the least squares estimates and clearly the three points at the top right-hand side are distant from the other nine, which means the effects associated with these estimates are active. These three points are the estimates of β_2, β_3, and β_1 in a sequence from left to right.

In this example, it will be assumed that $\mathbf{V} = c\mathbf{I}_p$, which can be interpreted as the prior beliefs about the β_i's being mutually independent. Figure 12.3 shows the marginal posterior density of β_1 and the posterior density of σ^2, fixing the hyperparameters $c, d, \mathbf{m} = \mathbf{0}$, and a taking values 0.1 (top), 1 (middle), and 10 (bottom). The prior and posterior variance of β_1, given by the t distribution, increase as the hyperparameter a increases for c, d, and $\mathbf{m}$ fixed, confirming the linear relationship shown in the properties of the marginal distributions given above. Also, the prior and posterior distributions of σ^2 have the same linear relationship, that is, when the value of a increases, the mean and posterior variance of σ^2 increase as well. A similar interpretation can be given to the hyperparameter d, that is, the larger the value of d is, the smaller are the prior and posterior variance of β_1.

Figure 12.4 is constructed in a similar way and shows the marginal posterior density of β_1 and the posterior density of σ^2, fixing the hyperparameters $a, d, \mathbf{m} = \mathbf{0}$, and c taking values 0.1 (top), 1 (middle), and 10 (bottom). In this case, the interpretation the effect of changing the value of c on the posterior distribution of β_1 and σ^2 is not obvious because the posterior hyperparameters a^* and $\mathbf{V}^*$ are functions of c and the relationship is not easy to determine. In this case, the posterior mode of β_1 tends to the true value 2 when c increases and in general, the variability of β_1 also decreases as c increases.

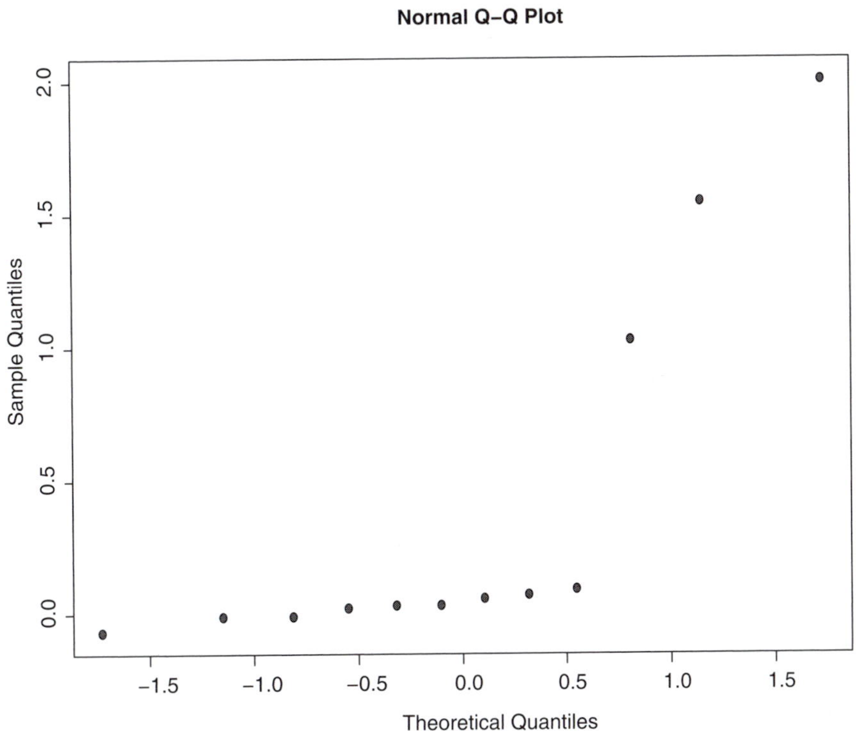

FIGURE 12.2
Normal plot of main effects of hypothetical data from Example 1.

The general message from these and other examples is that conjugate priors are a very inflexible way of describing prior opinion. Very informative priors can lead to the posterior being very vague and centered in the wrong place, whereas very vague priors make the dependence between the β_i's and σ^2 persistent. We now seek more flexible, but still simple, families of priors that might be more useful.

12.4.2 Prior as a Finite Mixture of Densities

Due to the dependence among the parameters, the use of conjugate prior distributions is often criticized because it is too restrictive. However, we can increase the flexibility of these families by considering mixtures of conjugate prior distributions.

Some general theory for finite mixtures of distributions [5] shows that the class of conjugate priors can be enlarged using this method. To describe the theory, start by assuming that ϕ is a discrete random variable taking values $\phi_1, \phi_2, \ldots, \phi_m$ such that $P(\phi = \phi_i) = p_i$. A mixture of distributions $f_i(\theta)$ with

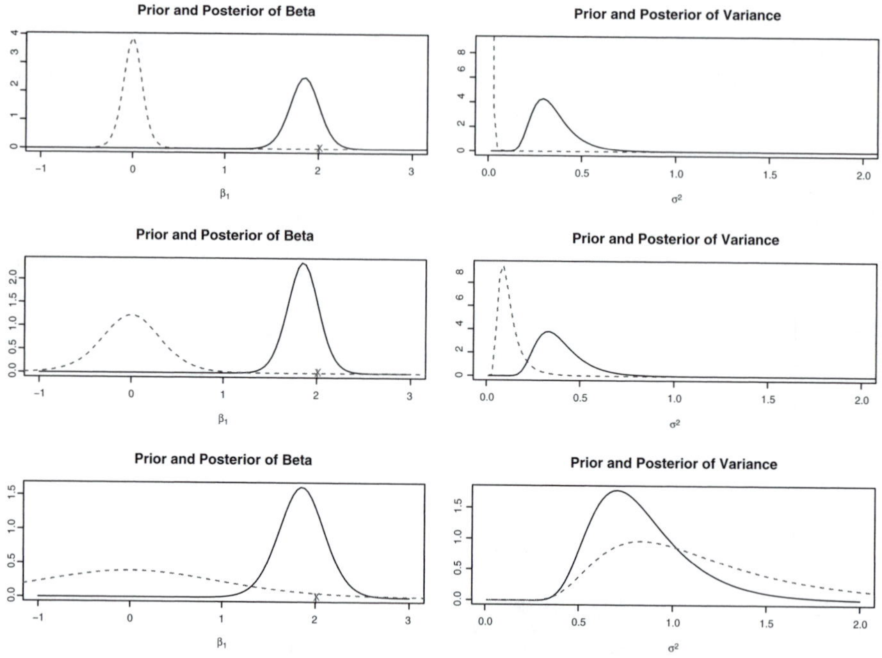

FIGURE 12.3
Marginal prior (dashed) and posterior (solid) of β_1 and σ^2 using hyperparameters $a = 0.1, 1, 10$ (from top to bottom), $d = 10$, and $c = 1$.

weights p_i is given by

$$f(\theta) = \sum_{i=1}^{m} p_i f_i(\theta), \tag{12.9}$$

where each component of the mixture $f_i(\theta) = f(\theta|\phi = \phi_i)$ is the conditional distribution of θ given $\phi = \phi_i$.

If $f(\theta)$ represents the prior distribution of θ, then the posterior distribution of θ given data x is

$$f(\theta|x) = \sum_{i=1}^{m} p_i^* f_i(\theta|x), \tag{12.10}$$

where

$$f_i(\theta|x) = \frac{f_i(\theta)\ f(x|\theta)}{\int f_i(\theta) f(x|\theta)\, d\theta} \tag{12.11}$$

and

$$p_i^* = \frac{p_i \int f_i(\theta) f(x|\theta)\, d\theta}{\sum_{j=1}^{m} p_j \int f_j(\theta) f(x|\theta)\, d\theta}, \tag{12.12}$$

which means that p_i^* is proportional to $p_i \int f_i(\theta) f(x|\theta)\, d\theta$.

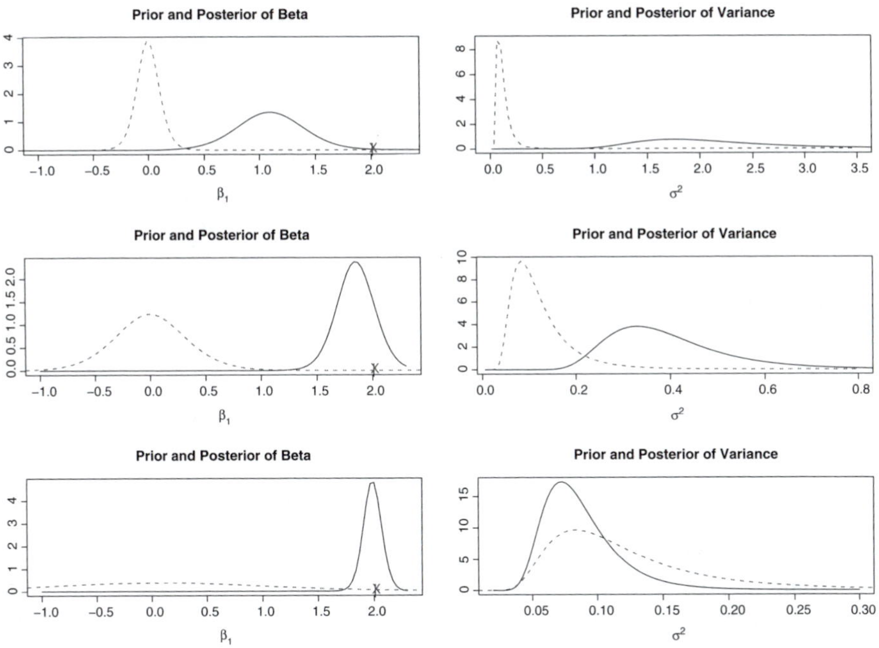

FIGURE 12.4
Marginal prior(dashed) and posterior (solid) of β_1 and σ^2 using hyperparameters $a = 1, d = 10$, and $c = 0.1, 1, 10$ (from top to bottom).

The posterior distribution expressed in Equation (12.10) is also a mixture of distributions $f_i(\theta|x)$ where the weights p_i^* are updated after incorporating the information from the data. Applying Equation (12.9), the prior of $(\boldsymbol{\beta}, \sigma^2)$ can be expressed as a mixture of N-IG densities, i.e.,

$$f(\boldsymbol{\beta}, \sigma^2) = \sum_{i=1}^{m} p_i f_i(\boldsymbol{\beta}, \sigma^2),$$

where

$$f_i(\boldsymbol{\beta}, \sigma^2) = \frac{\left(\frac{a_i}{2}\right)^{\frac{d_i}{2}} (\sigma^2)^{-\frac{(d_i+p+2)}{2}}}{(2\pi)^{p/2} |\mathbf{V}_i|^{1/2} \Gamma\left(\frac{d_i}{2}\right)} \exp\left\{-\frac{1}{2\sigma^2}[\,(\boldsymbol{\beta} - \mathbf{m_i})'\mathbf{V}_i^{-1}(\boldsymbol{\beta} - \mathbf{m}_i) + a_i]\right\}. \tag{12.13}$$

This prior density has N-IG densities as the components of the mixture, which have different sets of hyperparameters $a_i, d_i, \mathbf{m}_i$, and $\mathbf{V}_i$ to define the distribution suggested by the experimenter.

Using the likelihood function of Equation (12.4), the posterior density from Equation (12.10) can be formulated as

$$f(\boldsymbol{\beta}, \sigma^2|\mathbf{y}) = \sum_{i=1}^{m} p_i^* f_i(\boldsymbol{\beta}, \sigma^2|\mathbf{y}),$$

where the component of the posterior mixture is given as a posterior density of the N-IG density, i.e.,

$$f_i(\beta, \sigma^2|\mathbf{y}) = \frac{\left(\frac{a_i^*}{2}\right)^{\frac{d_i^*}{2}} (\sigma^2)^{-\frac{(d_i^*+p+2)}{2}}}{(2\pi)^{p/2} \left|\mathbf{V}_i^*\right|^{1/2} \Gamma\left(\frac{d_i^*}{2}\right)} \exp\left\{-\frac{1}{2\sigma^2}[\,(\beta - \mathbf{m}_i^*)'(\mathbf{V}_i^*)^{-1}(\beta - \mathbf{m}_i^*) + a_i^*]\right\} \quad (12.14)$$

assuming that a_i^*, d_i^*, $\mathbf{m}_i^*$, and $\mathbf{V}_i^*$ are the same as in Equation (12.8).

The updated weights are calculated using

$$\iint f_i(\beta, \sigma^2) f(\mathbf{y}|\beta, \sigma^2) \mathrm{d}\beta\, \mathrm{d}\sigma^2 = \frac{\left(\frac{a_i}{2}\right)^{\frac{d_i}{2}} \Gamma\left(\frac{d_i^*}{2}\right) \left|\mathbf{V}_i^*\right|^{1/2}}{\left(\frac{a_i^*}{2}\right)^{\frac{d_i^*}{2}} \Gamma\left(\frac{d_i}{2}\right) |\mathbf{V}_i|^{1/2} (2\pi)^{\frac{n}{2}}}. \quad (12.15)$$

Observe that the quotient above uses the prior, likelihood, and posterior information. Actually, this quotient is composed of the normalizing constant from the prior density, the likelihood function, and the posterior density. Consequently,

$$p_i^* \;\propto\; p_i \,\frac{(a_i)^{\frac{d_i}{2}} \Gamma\left(\frac{d_i^*}{2}\right) \left|\mathbf{V}_i^*\right|^{1/2}}{(a_i^*)^{\frac{d_i^*}{2}} \Gamma\left(\frac{d_i}{2}\right) |\mathbf{V}_i|^{1/2} \pi^{\frac{n}{2}}}. \quad (12.16)$$

In order to compare the impact of the hyperparameters on the prior and posterior, the marginal density of β_i will be rewritten below. Note that from Equation (12.13), the marginal density of the vector of parameters β is

$$f_i(\beta) = \frac{\Gamma\left(\frac{d_i+p}{2}\right)}{(d_i\pi)^{p/2}\Gamma\left(\frac{d_i}{2}\right)\left|\left(\frac{a_i\mathbf{V}_i}{d_i}\right)\right|^{1/2}} \left[1 + \frac{1}{d_i}(\beta - \mathbf{m}_i)'\left(\frac{a_i\mathbf{V}_i}{d_i}\right)^{-1}(\beta - \mathbf{m}_i)\right]^{-\frac{d_i+p}{2}}$$

and then the marginal density of each β_j is given by

$$f_i(\beta_j) = \frac{\Gamma\left(\frac{d_i+1}{2}\right)}{\Gamma\left(\frac{d_i}{2}\right)(\pi a_i\, v_{jj,i})^{1/2}} \left[1 + \frac{1}{d_i}\left(\frac{d_i}{a_i v_{jj,i}}\right)(\beta_j - m_{j,i})^2\right]^{-\frac{d_i+1}{2}}, \quad (12.17)$$

where $v_{jj,i}$ is the $(j+1)$th diagonal element of the matrix $\mathbf{V}_i$ and $m_{j,i}$ is the $(j+1)$th element of the mean vector $\mathbf{m}_i$. Therefore, the marginal posterior density of β is

$$f_i(\beta|\mathbf{y}) = \frac{\Gamma\left(\frac{d_i^*+p}{2}\right)}{\Gamma(\frac{d_i^*}{2})(d_i^*\pi)^{p/2}\left|\frac{a_i^*}{d_i^*}\mathbf{V}_i^*\right|^{1/2}} \left[1 + \frac{1}{d_i^*}(\beta - \mathbf{m}_i^*)'\left(\frac{a_i^*\mathbf{V}_i^*}{d_i^*}\right)^{-1}(\beta - \mathbf{m}_i^*)\right]^{-\frac{d_i^*+p}{2}}$$

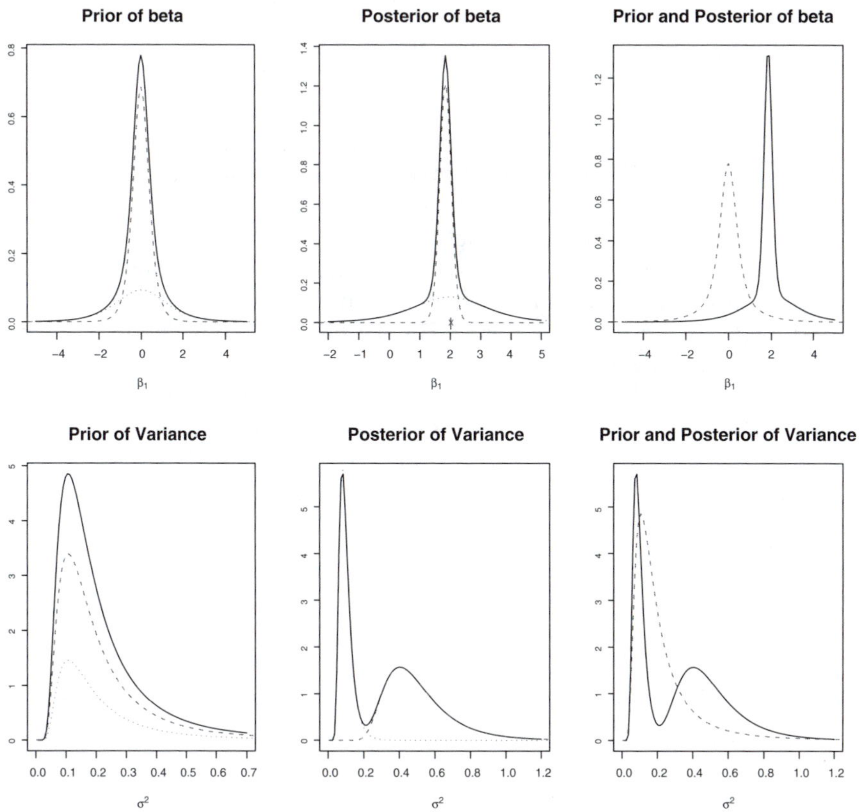

FIGURE 12.5
Prior of β_1 and σ^2 using weights 0.7 and 0.3 for the mixture, $a = 0.75$, $d = 5$, and c is 1 and 10 to compose the mixture.

and consequently, the marginal posterior density of each β_j is

$$f_i(\beta_j|\mathbf{y}) = \frac{\Gamma\left(\frac{d_i^*+1}{2}\right)}{\Gamma\left(\frac{d_i^*}{2}\right)(\pi\ a_i^*\ v_{jj,i}^*)^{1/2}}\left[1 + \frac{1}{d_i^*}\left(\frac{d_i^*}{a_i^*\ v_{jj,i}^*}\right)(\beta_j - m_{j,i}^*)^2\right]^{-\frac{d_i^*+1}{2}}, \tag{12.18}$$

where $v_{jj,i}^*$ is the $(j+1)$th diagonal element of the matrix $\mathbf{V}_i^*$.

Example 1: (continued)

The aim here is to show the marginal prior and posterior densities of β_i under different values of the hyperparameters. First, let the prior density be a mixture of two densities with weights 0.7 and 0.3. Figure 12.5 shows the prior and posterior density of β_1 in the first line and the prior and posterior density of the variance in the second line. The graph on the top left is composed of the prior density (solid) and the two components of the mixture, where one component is given by N-IG(0.75, 5, 0, 1) with weight 0.7 (dashed) and the second one is N-IG(0.75, 5, 0, 10) (dotted). For the graph in the middle, the first line is the posterior density (solid), which is the mixture of N-IG(7.669, 17, 1.850, 0.077) and N-IG(1.493, 17, 1.988, 0.083) with updated weights 0.581 and 0.419. The graph in the first line on the right side shows the prior (dashed) and posterior (solid) densities. The hyperparameters used in this case are the smallest for a and d such that $E(\sigma^2) = 0.25$, and the values of c are used for the dispersion matrix $\mathbf{V}$. In the second line, from left to right, is the prior density (solid) of the variance σ^2 with its two mixture components, the posterior density (solid) of the variance with its two mixture components, and the prior (dashed) and posterior (solid) densities of variance. The pattern found here in the posterior is something of a surprise.

Figure 12.6 uses the same values of c but the values of a and d are larger so that $E(\sigma^2)$ is preserved. The posterior density of β_1 in Figure 12.6 has smaller variance than in Figure 12.5, and this difference is explained by the choice of higher values of the hyperparameters a and d. The prior expectation of σ^2 is

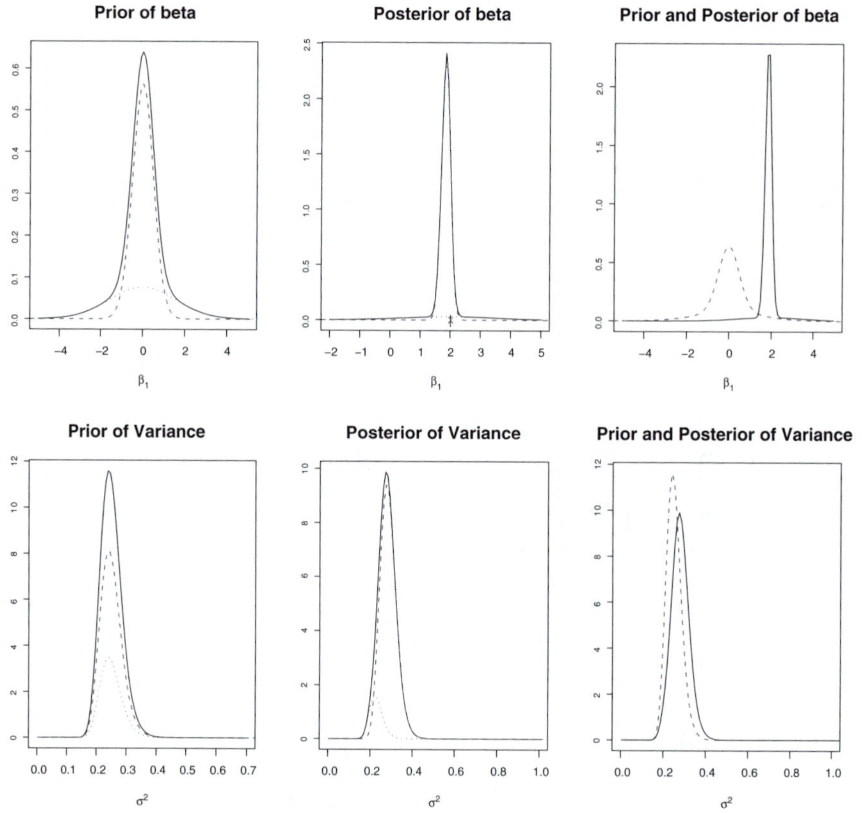

FIGURE 12.6
Prior of β_1 and σ^2 using weights 0.7 and 0.3 for the mixture, $a = 24.5$, $d = 100$, and c is 1 and 10 to compose the mixture.

the same but the dispersion on the t distribution of β_i depends on the quotient a/d instead of $a/(d-2)$.

The change in the posterior induced by changing the values of a and d is not as drastic as when the value of c is changed. The posterior distribution is much more sensitive to changes in c than in other hyperparameters. What happens to the marginal density of β when the values for c are smaller? Figure 12.7 and Figure 12.8 show the change in the posterior density with hyperparameter c. Clearly, these figures show the change in the variance for the marginal prior density of β_1 when the values of c decrease and the prior and posterior densities have the same behavior, i.e., the posterior density does not change much and the prior density is thinner when the hyperparameters for a and d are smaller.

Again the conclusion is that great care is needed in choosing the prior distributions. However, the extra flexibility gained by using mixtures of conjugate distributions seems to be worthwhile. Using more than two components in the mixture of distributions is perfectly possible, but little advantage seems evident in doing so. Each effect having a moderate to large probability of being close to zero, with the rest of the probably being spread widely around zero, seems to describe the assumptions of effect sparsity, and with little being known about the specific effects, that is usually the case when saturated designs are being used.

12.4.3 Non-Conjugate Priors

The advantage of using conjugate priors is that an analytic expression for the posterior density can be obtained that does not need any method of approximating or simulation. Alternately, the prior density of $\boldsymbol{\beta}$ depends on σ^2 completely, which forces the results of posterior estimation of $\boldsymbol{\beta}$ to be dependent on the prior distribution of this parameter.

The prior information on $\boldsymbol{\beta}$ and σ^2 might be obtained independently, for example, that for σ^2 might come from the routine running of the process, whereas that for $\boldsymbol{\beta}$ might come from experiments on a pilot plant. In such cases, one might assume that the prior distribution of $\boldsymbol{\beta}$ is multivariate Normal with mean $\mathbf{b}$ and dispersion matrix $\mathbf{W}$, and that the prior distribution of σ^2 is Inverse-Gamma with shape parameter α and scale parameter δ. Therefore, the joint prior density of $(\boldsymbol{\beta}, \sigma^2)$ is given by

$$f(\boldsymbol{\beta}, \sigma^2) \propto (\sigma^2)^{-(\alpha+1)} \exp\left\{-\frac{1}{2}(\boldsymbol{\beta}-\mathbf{b})'\mathbf{W}^{-1}(\boldsymbol{\beta}-\mathbf{b}) - \frac{\delta}{\sigma^2}\right\}. \tag{12.19}$$

The posterior density of $(\boldsymbol{\beta}, \sigma^2)$ will be determined by incorporating the prior knowledge of Equation (12.19) with data information, given by the likelihood function of Equation (12.5), and is

$$f(\boldsymbol{\beta}, \sigma^2|\mathbf{y}) \propto (\sigma^2)^{-(\frac{n}{2}+\alpha+1)}$$

$$\exp\left\{-\frac{1}{2}(\boldsymbol{\beta}-\mathbf{b})'\mathbf{W}^{-1}(\boldsymbol{\beta}-\mathbf{b}) - \frac{\delta}{\sigma^2} - \frac{1}{2\sigma^2}(\boldsymbol{\beta}-\widehat{\boldsymbol{\beta}})'\mathbf{X}'\mathbf{X}(\boldsymbol{\beta}-\widehat{\boldsymbol{\beta}})\right\}. \tag{12.20}$$

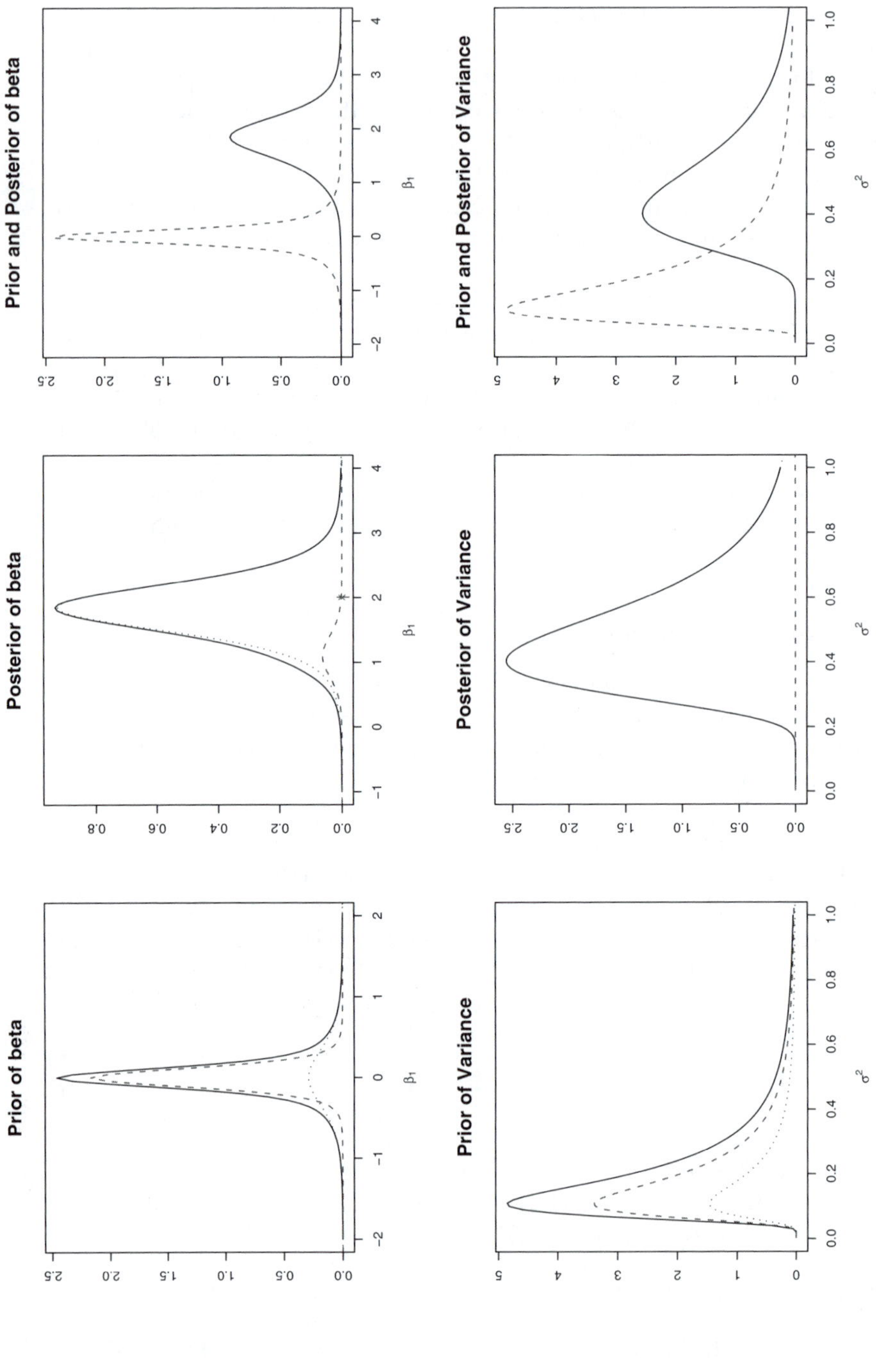

FIGURE 12.7
Prior of β_1 and σ^2 using weights 0.7 and 0.3 for the mixture, $a = 0.75$, $d = 5$, and c is 0.1 and 1 to compose the mixture.

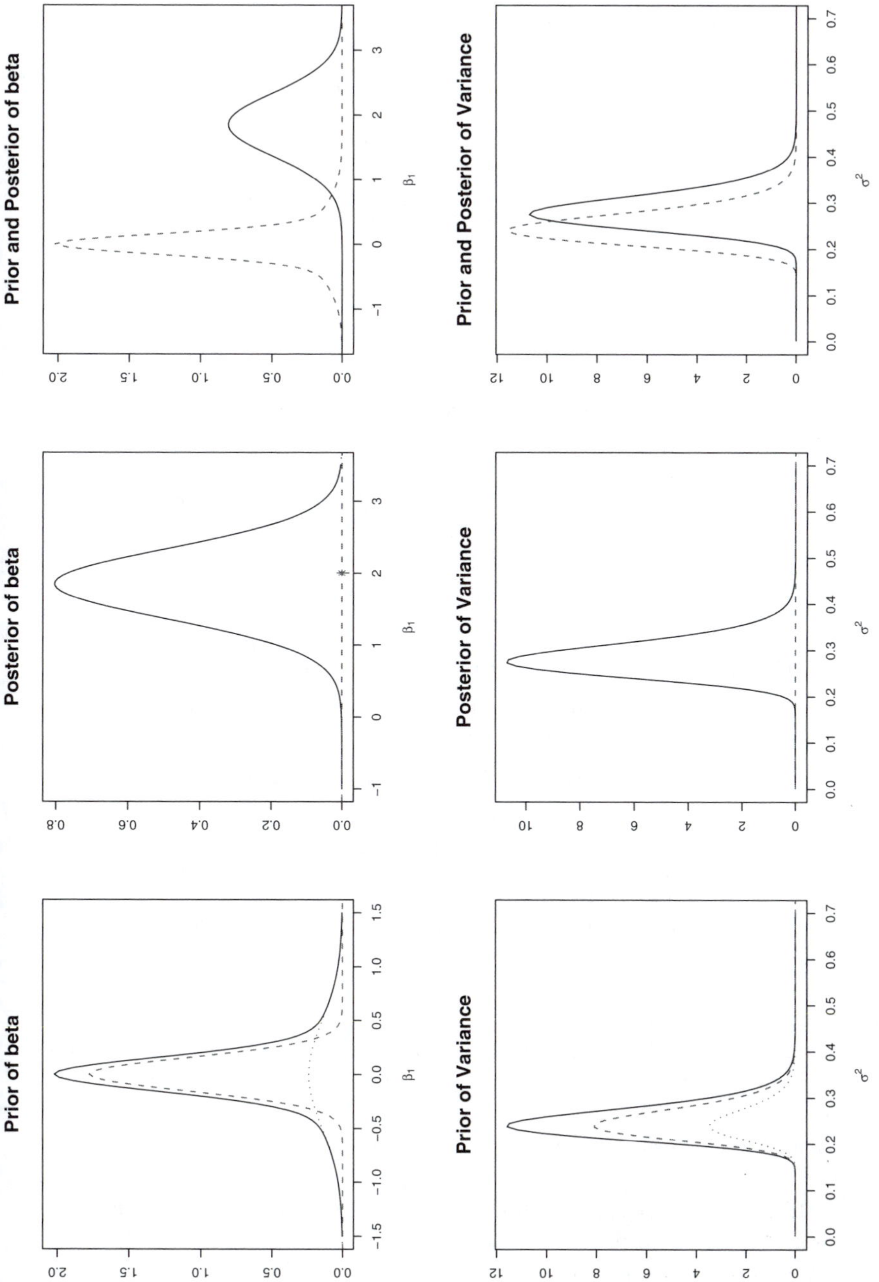

FIGURE 12.8
Prior of β_1 and σ^2 using weights 0.7 and 0.3 for the mixture, $a = 24.5$, $d = 100$, and c is 0.1 and 1 to compose the mixture.

TABLE 12.3

Statistics from Simulated Data Using Non-Conjugate Prior for (β, σ^2) with $h = 0.125$, $\alpha = 5$, and $\delta = 0.5$

Parameter	Mean	SD	MC Error	2.5%	Median	97.5%
β_0	0.040	0.1949	0.000883	−0.3721	0.045	0.439
β_1	1.391	0.4487	0.002432	0.4145	1.483	1.989
β_2	0.713	0.2841	0.001482	0.0484	0.776	1.109
β_3	1.074	0.3697	0.001921	0.2434	1.152	1.574
β_4	0.022	0.1971	0.000856	−0.3961	0.023	0.431
σ^2	0.930	1.0000	0.005164	0.0740	0.556	3.589

TABLE 12.4

Statistics from Simulated Data Using Non-Conjugate Prior for (β, σ^2) with $h = 0.125$, $\alpha = 10$, and $\delta = 10$

Parameter	Mean	SD	MC Error	2.5%	Median	97.5%
β_0	0.026	0.2596	0.001210	−0.4837	0.028	0.532
β_1	0.923	0.3094	0.002185	0.2953	0.934	1.504
β_2	0.471	0.2716	0.001442	−0.0829	0.477	0.990
β_3	0.712	0.2916	0.001847	0.1219	0.722	1.255
β_4	0.012	0.2590	0.001134	−0.4969	0.013	0.519
σ^2	1.885	0.6995	0.005586	0.8988	1.758	3.601

TABLE 12.5

Statistics from Simulated Data Using Non-Conjugate Prior for (β, σ^2) with $h = 1$, $\alpha = 10$, and $\delta = 10$

Parameter	Mean	SD	MC Error	2.5%	Median	97.5%
β_0	0.053	0.2862	0.001338	−0.515	0.054	0.612
β_1	1.842	0.2881	0.001434	1.258	1.848	2.394
β_2	0.941	0.2874	0.001268	0.357	0.945	1.494
β_3	1.422	0.2890	0.001377	0.834	1.428	1.976
β_4	0.026	0.2851	0.001257	−0.536	0.027	0.587
σ^2	1.081	0.3668	0.002457	0.579	1.013	2.000

Example 1 (continued):
The following results show the output of the posterior density simulated using WinBUGS [14], where $\mathbf{b} = \mathbf{0}$ and $\mathbf{W} = h\mathbf{I}_p$ were assumed. Thus, a different set of values for h, α, and δ is used to produce the summary of 50,000 simulated points from the posterior and the graphs of the marginal posteriors for the parameters.

Table 12.3 summarizes the statistics of the parameters β_0, β_1, β_2, β_3, β_4, and σ^2 using $h = 0.125$, $\alpha = 5$, and $\delta = 0.5$. Table 12.4 summarizes the statistics of parameters β_0, β_1, β_2, β_3, β_4, and σ^2 using $h = 0.125$, $\alpha = 10$, and $\delta = 10$, and Table 12.5 summarizes the statistics of parameters β_0, β_1, β_2, β_3, β_4, and σ^2 using $h = 1$, $\alpha = 10$, and $\delta = 10$. The corresponding simulated marginal posterior distributions are shown in Figure 12.9, Figure 12.10 and Figure 12.11.

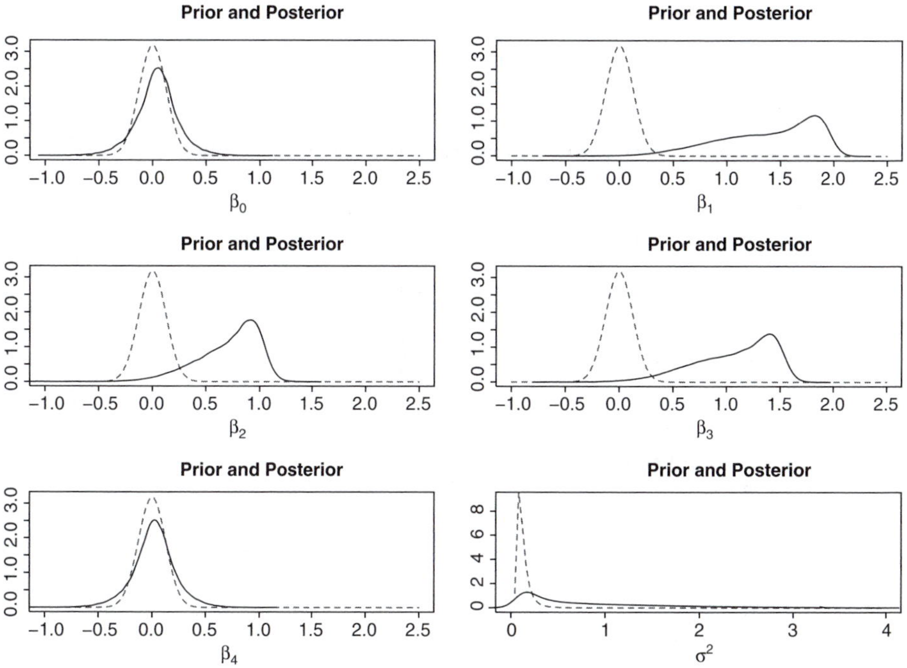

FIGURE 12.9
Posterior from Non-Conjugate Prior for β_0, β_1, β_2, β_3, β_4, and σ^2 Using $h = 0.125$, $\alpha = 5$, and $\delta = 0.5$.

These results seem to be less sensitive to changes in the prior distributions than those from the conjugate priors. In particular, if the prior for σ^2 is badly misspecified, the effect on the posterior of β_i seems to be less damaging. In particular, Figure 12.9 shows that the main impact of underestimating σ^2 with a highly informative prior does not stop the posterior for β_1 covering the true value. However, it does mean that β is widely spread.

12.5 Conclusions

The main message from this chapter is that great care is needed when using Bayesian estimation with saturated designs. This requirement is only to be expected. It is well known that the data from saturated designs provide only limited information, so naturally the interpretation of these data depends heavily on the prior assumptions made. This is also true of the classical analysis, where the interpretation of the normal plots depends on how strongly the experimenter believes in factor sparsity.

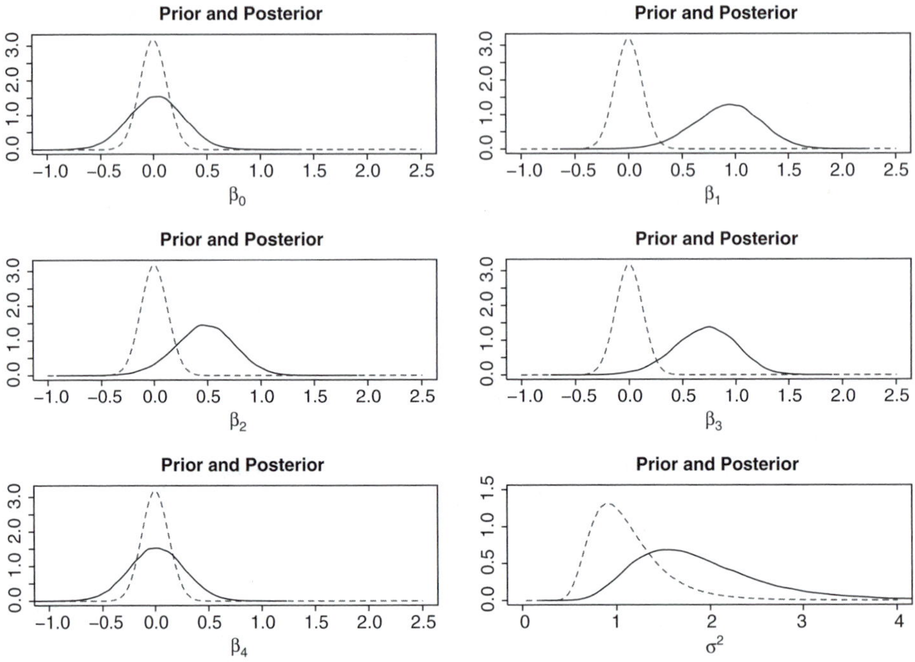

FIGURE 12.10
Posterior from Non-Conjugate Prior for β_0, β_1, β_2, β_3, β_4, and σ^2 Using $h = 0.125$, $\alpha = 10$, and $\delta = 10$.

We have presented results in which the priors for β are noninformative as well as those in which they are moderately or highly informative. However, the priors for σ^2 have been either moderately or highly informative. Using noninformative priors for all parameters simultaneously does not seem appropriate for several reasons. First, noninformative priors do not realistically describe the true state of experimenters' prior knowledge, as something is always known about the order of magnitude of run-to-run variation. Second, finding a prior that is noninformative for σ^2 is notoriously difficult [13]. Most importantly, however, is the fact that in a saturated design, we are using n observations to estimate $n + 1$ parameters. In the absence of any prior knowledge, the data do not provide any information. This condition is also true for classical analyses in which a point prior is required for σ^2 to obtain finite standard errors for the other parameters. The Bayesian analysis has the advantage that useful conclusions can still be drawn even if the prior knowledge on σ^2 is only moderately informative.

The conjugate priors seem to be too inflexible for most applications but mixtures of conjugate priors greatly increase the flexibility. However, advantages also exist in having priors for the factors' effects that are independent of those for the variance. The choice must depend on the experimenters' prior beliefs about their process. If the run-to-run variation is bigger than expected,

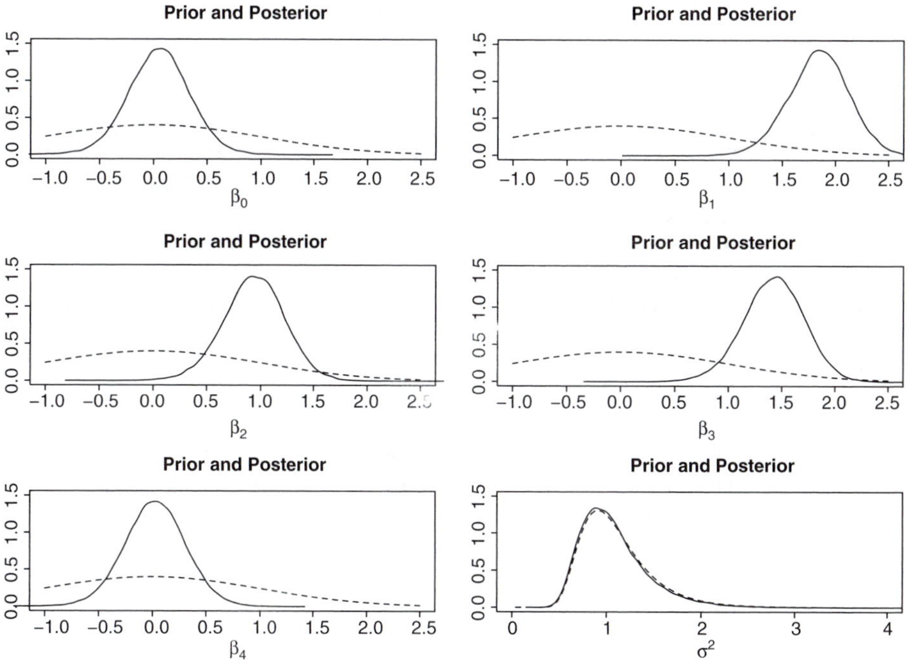

FIGURE 12.11
Posterior from Non-Conjugate Prior for β_0, β_1, β_2, β_3, β_4, and σ^2 using $h = 1, \alpha = 10$, and $\delta = 10$.

are the effects of factors likely to be also bigger than expected? In summary, the Bayesian analysis provides the opportunity to get more information from saturated designs at the cost of having to be more careful about the specification of the prior assumptions. When more than a vague indication of which factors are active is needed, this method seems like a good deal.

References

1. Bernardo, J.M. and Smith, A.F.M., *Bayesian Theory*, New York: Wiley, 1994.
2. Box, G.E.P., Hunter, W.G., and Hunter, J.S., *Statistics for Experimenters: An Introduction to Design, Analysis and Model Building*, New York: John Wiley & Sons, 1978.
3. Box, G.E.P. and Meyer, R.D., "An analysis for unreplicated factorials," *Technometrics*, 28, 11–18, 1986.
4. Box, G.E.P. and Tiao, G.C., *Bayesian Inference in Statistical Analysis*, Reading, MA: Addison-Wesley, 1973.
5. Dalal, S. and Hall, W.J., "Approximating priors by mixtures of natural conjugate priors," *Journal of the Royal Statistical Society*, Series B, 45, 278–286, 1983.

6. Daniel, C., "Use of half-normal plot in interpreting factorial two-level experiments," *Technometrics*, 1, 149, 1959.
7. Daniel, C., *Applications of Statistics to Industrial Experimentation*, New York: John Wiley & Sons, 1976.
8. Gelman, A., "Prior distributions for variance parameters in hierarchical models," *Bayesian Analysis*, 2, 1–19, 2005.
9. Hamada, M. and Balakrishnan, N., "Analyzing unreplicated factorial experiments: A review with some new proposals," *Statistica Sinica*, 8, 1–41, 1998.
10. Lenth, R.V., "Quick and easy analysis of unreplicated fractional factorials," *Technometrics*, 31, 469–473, 1989.
11. O'Hagan, A. and Forster, J.J., *Kendallıs Advanced Theory of Statistics*, 2B: Bayesian Inference, 2nd ed., London: Arnold Publishers, 2004.
12. Plackett, R.L. and Burman, J.P., "The design of optimum multifactorial experiments," *Biometrika*, 33, 305–325, 1946.
13. Spiegelhalter, D.J., Abrams, K.R., and Myles, J.P., *Bayesian Approaches to Clinical Trials and Health-Care Evaluation*, New York: John Wiley & Sons, 2004.
14. Spiegelhalter, D., Thomas, A., Best, N., and Lunn, D., *WinBUGS User Manual*, Version 1.4., `http://www.mrc-bsu.cam.ac.uk/bugs`, 2004.
15. Wu, C.F.J. and Hamada, M., *Experiments: Planning, Analysis and Parameter Design Optimization*, New York: John Wiley & Sons, 2000.

Index